Drinking-Water Distribution, Sewage, and Rainfall Collection

François G. Brière

POLYTECHNIC INTERNATIONAL PRESS

Drinking-Water Distribution, Sewage, and Rainfall Collection
François G. Brière

For information on distribution and points of sale, see our Web site:
http://www.polymtl.ca/pub
E-mail of Presses internationales Polytechnique:
pip@mail.polymtl.ca

Legal deposit: 1st quarter 2000
Bibliothèque nationale du Québec
National Library of Canada

ISBN 2-553-796-5
Printed in Canada
1 2 3 4 5 03 02 01 00 99

PREFACE

Many authors have written textbooks on treatment of drinking water and wastewater. Many of these books deal with the design and construction of water-treatment units. Very few, however, address the subjects of urban hydraulics, drinking-water-distribution networks, and wastewater and runoff collection systems.

This book is the result of many years of teaching civil engineering students at École Polytechnique de Montréal and of a close collaboration with Pierre Éthier, consulting engineer from BMST-Richelieu Experts Conseil inc. It was written for undergraduate students at École Polytechnique de Montréal, but it is also aimed at engineers and technicians responsible for the design and maintenance of urban hydraulics networks, college and technical institute instructors, civil engineering professors, and city administrators.

Chapter 1 presents the water-management process. It deals with the collection of natural water, its transformation into drinking water, and its distribution to consumers. It also discusses rainwater and wastewater collection, treatment, and discharge.

Chapter 2 reviews hydraulics concepts needed to understand water flows in various networks, water pressure and water level in equipment, and flowrates.

Chapters 3, 4, and 5 present essential information for the design and choice of equipment needed to supply water in adequate quantity and pressure. Procedures and criteria are presented for the design of water intakes, intake pipes, raw-water reservoirs, clear-water reservoirs, and distribution networks. Frost prevention is also discussed.

In chapters 6, 7, 8, and 9, information concerning the evaluation and variation in flows of rainwater and domestic, sanitary, and combined wastewater is presented in order to determine global maximum and minimum flows. Procedures and criteria required for the design of different types of sewage systems are also discussed. For example, the design of different types of pipe networks is discussed in terms of flowrates, minimum depth, and position in the ground.

Chapter 10 discusses the risks encountered by water-management workers in general, and by sewer workers more specifically. Chapter 11 describes how to evaluate dead and live loads on buried pipes.

Each chapter has examples and problem sets; answers are appended at the end of the book. With the exception of chapter 2, which discusses the basic hydraulics concepts necessary for understanding this book, all chapters may be read independently from one another.

Finally, although this book was written with a North American approach, it can easily be used in other parts of the world as long as essential data are obtained, such as precipitation curves, wastewater flows, air temperature, and materials available.

François G. Brière
November, 1999

ACKNOWLEDGMENTS

This book would not have been possible without the input and support of many of my colleagues.

It first required the trust of Lucien Foisy, director of Presses internationales Polytechnique, and the efficient work of Diane Ratel who ensured good co-ordination of the original french version of this book (*Distribution et collecte des eaux*) as well as this english translation.

I would also like to stress the collaboration of Frederic Monette who proofread each chapter to eliminate errors, vagueness and incoherence, and Khalil Foukay, whom I frequently relied upon when hunting for references.

This book was also made possible by the hard work of Martine Aubry who did the editing and made numerous (and often late) corrections.

I will never forget the excellent collaboration of Louise Daoust who wrote the first version of this translation: the quality of her text was so good that Käthe Roth, editor-translator, and Stephen Schettini did not have much to do to bring it to its final form.

Finally, I want to acknowledge the special contribution of Pierre Éthier, P.Eng., consulting engineer, without whom this book would not have reached the practical and professional style I was aiming for.

TABLE OF CONTENTS

Chapter 4 Collection and Distribution of Drinking Water

Chapter 5 Design of a Drinking-Water Distribution System

Chapter 6 Volumes and Flows of Sanitary Sewage

Chapter 7 Runoff in Urban Areas

Chapter 8 Design of Sewer Systems

Chapter 9 Hydraulic Design of Sanitary, Storm and Combined Sewer Systems

Chapter 10 Disease and Other Risks Related to Water

Chapter 11 Loads on Pipes and Selection of Pipes Able to Sustain such Loads

LIST OF SYMBOLS

Abbreviations

ILLUDAS	Illinois Urban Drainage Area Simulation
SIRDU	(simulation of an urban draining network) Simulation d'un réseau de drainage urbain
STORM	Storage, Treatment and Overflow Runoff Model
SWMM	Storm Water Management Model
WATCAD	Water Computer Assisted Design

Alphabet

		Si units	*Units of measurement*
A	area; cross-sectional area	m^2	ha
A_{imp}	fraction of the area of a sub-basin made of surfaces supposedly impermeable	–	
C_c, C_d, C_n, C_s	load coefficient	–	
C_C	Chezy's coefficient	–	
C_{fire}	fire flow coeffcient	–	
C_{HW}	Hazen-Williams' coefficient	–	
d	diametre	m	mm, cm, inch
d	average thickness of the ice layer	m	cm
d_{equ}	equivalent pipe diametre	m	mm
d_{out}	outside diametre	m	mm
f	Darcy-Weisbach's friction factor	–	
f	friction factor	–	
F	force	N	kN
F_c	force due to the backfill and other dead loads on the pipe	N/m	
FP	peak factor	–	
FP_{max}	maximum peak factor	–	
FP_{min}	minimum peak factor	–	
F_{res}	net (resulting) force	N	kN
FS	safety coefficient (factor)	–	
F_{tot}	total force	N	
F_v	force due to a live load	N/m	
g	acceleration due to gravity or gravitational constant	m/s^2 or 9,81 m/s^2	
G	freezing index		°F-d, °C-d

		Si units	Units of measurement
h	height of water	m	
h	freezing depth	m	inch, ft
h_e	water level at the inlet (entrance)	m	
h_o	water level at the outlet	m	
H	head losses		m of water/100 m, kPa/100 m, m of water
H_L	head losses on a pipe of length L		m of water, kPa
$(H_L)_{equ}$	head losses in the equivalent pipe		m of water, kPa
H_m	minor head losses	m	
H_{tot}	total head losses		m of water, kPa
I	rain intensity		mm/h, inch/h
I	impact factor	–	
K	decreasing growth constant		a^{-1}
K_a	arithmetic growth constant		a^{-1}
K_g	geometric growth constant		a^{-1}
K_L	coefficient of the structure	–	
ln	base e logarithm	–	
log	base 10 logarithm	–	
L	length	m	mm, km
L_{equ}	length of the equivalent pipe	m	mm
L_f	bedding factor	–	
m	uniformly distributed horizontal load	kg	kg/m^2
m_b	mass of the backfill	kg	kg/m
m_p	mass of the pipe	kg	kg/m
m_w	mass of the water	kg	kg/m
n	Manning's coefficient, Kutter's coefficient	–	
n	number of time periods	–	
N_R	Reynolds number	–	
p	pressure	Pa	kPa
P	live load	kg	
P	population	–	
Q	flowrate	m^3/s	m^3/d, L/s, L/d
Q_d	daily flowrate	m^3/s	m^3/d, L/s, L/d
$(Q_d)_{avg}$	average daily flowrate for a year	m^3/s	m^3/d, L/s, L/d
$(Q_d)_{max}$	maximum daily flowrate for a year	m^3/s	m^3/d, L/s, L/d
$(Q_d)_{min}$	minimum daily flowrate for a year	m^3/s	m^3/d, L/s, L/d
Q_{dom}	domestic flowrate	m^3/s	m^3/d
$(Q_{dom})_{avg}$	average annual domestic flowrate	m^3/s	m^3/d
$(Q_{dom})_{max}$	maximum annual domestic flowrate	m^3/s	m^3/d

		Si units	*Units of measurement*
$(Q_{dom})_{min}$	minimum annual domestic flowrate	m^3/s	m^3/d
Q_f	flow in pipe flowing full	m^3/s	L/s
Q_h	hourly flowrate	m^3/s	m^3/h, L/s
$(Q_h)_{avg}$	average hourly flowrate for a year	m^3/s	m^3/h, L/s
$(Q_h)_{max}$	maximum hourly flowrate for a year	m^3/s	m^3/h, L/s
$(Q_h)_{min}$	minimum hourly flowrate for a year	m^3/s	m^3/h, L/s
Q_i	inflow	m^3/s	m^3/d
Q_{inf}	infiltration flowrate	m^3/s	m^3/d
Q_{max}	maximum flowrate	m^3/s	L/s
Q_{min}	minimum flowrate	m^3/s	L/s
Q_{par}	parasitic flowrate	m^3/s	L/s
$(Q_{san})_{avg}$	average annual sanitary flowrate	m^3/s	m^3/d
$(Q_{san})_{max}$	maximum annual sanitary flowrate	m^3/s	m^3/d
$(Q_{san})_{min}$	minimum annual sanitary flowrate	m^3/s	m^3/d
r	growth rate of the population by time periods		hundredth/y
r_{sd}	settlement ratio	–	
R	runoff coefficient	–	
R_{emer}	emergency storage	m^3	
R_{equ}	equilibrium reserve (service storage)	m^3	
R_{fire}	fire-fighting storage	m^3	
R_h	hydraulic radius	m	
R_{max}	maximum storage	m^3	
R_{min}	minimum storage	m^3	
R_{prod}	production storage	m^3	
R_{recom}	recommended storage	m^3	
R_L	sewer line runoff coefficient	–	
s	slope	m/m	
S	saturation population	–	
t	time	s	min, h, d, y
t_c	time of concentration	s	min
t_e	inlet time	s	min
t_f	time of flow	s	min
v	velocity	m/s	
v_f	velocity in pipe flowing full	m/s	
v_{min}	minimum velocity	m/s	
w	width	m	
w_{tr}	width of the trench	m	mm
$(w_{tr})_{trans}$	width of the transitional trench	m	mm
WP	wetted perimeter	m	

Greek alphabet

		Si units	*Units of measurement*
η	dynamic viscosity of water	N·s/m^2 or Pa·s	
θ	angle	rad	°
ρ_e	density of water	kg/m^3	
ϕ	friction angle	rad	°

Chapter 1

BACKGROUND KNOWLEDGE REQUIRED FOR THE MANAGEMENT OF DRINKING WATER AND WASTEWATER

INTRODUCTION

This book focuses on water and its transportation through collection systems, distribution networks, and sewage systems. This chapter gives a general overview of the use of water and some of the treatments it undergoes. It also discusses population-growth projection methods used to determine the water flows and volumes necessary to design networks and installations. First, we deal with water-transportation cycles, then we discuss sources of drinking water; lastly, we present population-growth projection models.

1.1 WATER-TRANSPORTATION CYCLE

Water used for consumption is basically:

- surface fresh water, found in lakes and waterways;
- groundwater, often highly mineralized.

1.1.1 Surface-Water Utilization Cycle

Figure 1.1 shows the utilization cycle of surface water briefly described in this section.

First, fresh water is collected from a lake or a river using various techniques, ranging from a simple pipe submerged in a river to a collection tower, equipped with selective intakes at different depths, installed in a lake or deep reservoir. In some cases, a dam must be built to retain water and regulate river flows, especially those with minimal flows.

Collection systems (channels or pipes) transport raw water toward the targeted communities. If the water is not potable, it goes through a treatment plant before being distributed to consumers. However, even if the water source is of good quality, disinfection must be performed (in Quebec, chlorination is mandatory.)

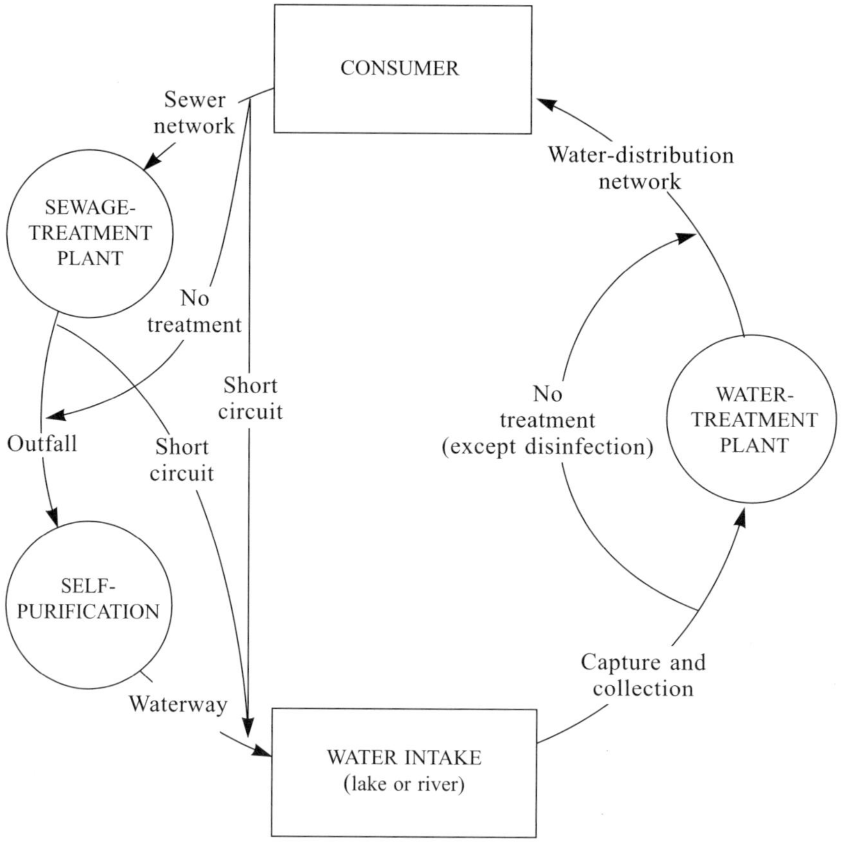

Figure 1.1 Surface-water utilization cycle

When water is suitable for consumption, a distribution network, consisting of pipes under pressure, brings the water to the community, all the way to consumers' taps. After water has been used, it loses its quality and becomes wastewater. The sewage system is then used to evacuate the wastewater from the community.

Wastewater is sent to a treatment plant to avoid pollution of the environment. This allows cities located downstream from points of discharge to use water for different purposes, including drinking. Treated water is sent back to the environment through a pipe called the outfall, whose role is to ensure the best possible dilution into waterways.

Finally, in a river, the water is subjected to a natural self-purification process in which residual organic and microbial pollution is gradually and partially reduced.

When treated wastewater is discharged into a river or a lake, however, it may reach a water intake before the self-purification process is completed. Reasons for this short circuit may be that the treated wastewater does not mix well with the surrounding waters because of either different densities or the hydraulic characteristics of the body of water. As a result, the water intake may be abnormally contaminated.

1.1.2 Groundwater Utilization Cycle

Figure 1.2 shows the groundwater utilization cycle. Groundwater is usually collected using pumps submerged at the bottom of a bored well; often, it is sheltered from any pollution and can usually be distributed and used after a simple disinfection. In some cases, however, the concentration of chemical substances must be reduced through treatments such as water softening (Ca and Mg) and the removal of iron (Fe), manganese (Mn) and sulphur (S).

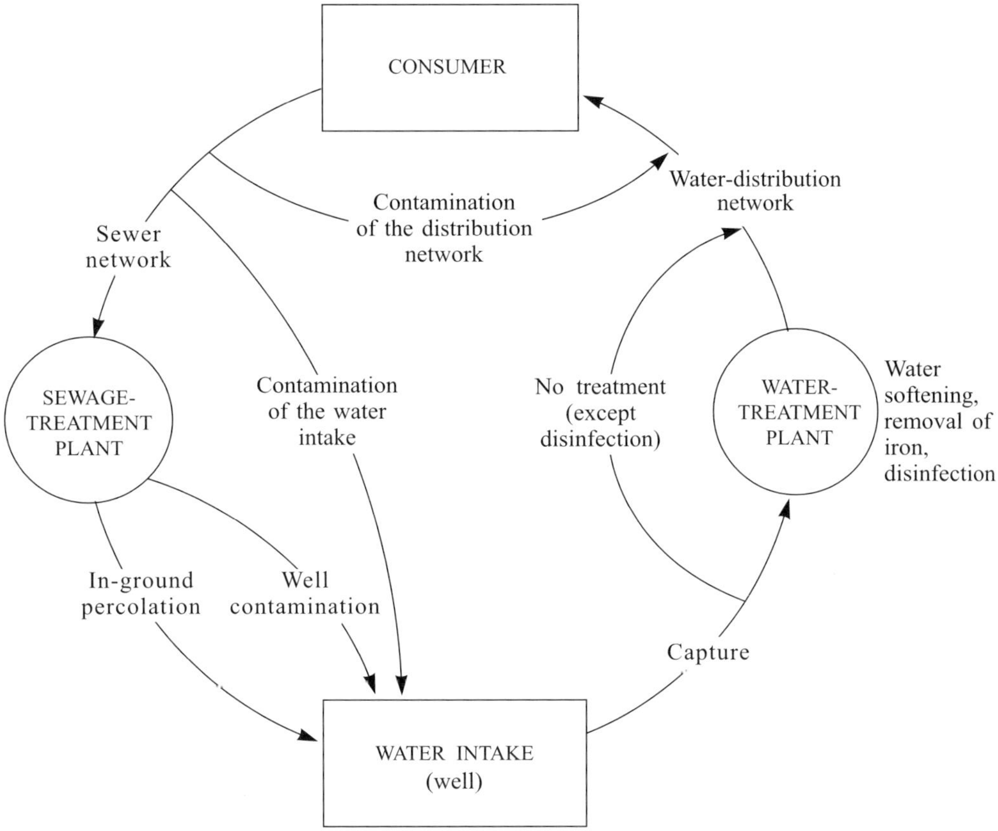

Figure 1.2 Groundwater utilization cycle

The groundwater utilization cycle is very similar to the surface-water utilization cycle. However, there is more chance of contamination of the water intake if the well or water table is not protected against exfiltration of polluted water, which may be untreated or partially treated wastewater. Contamination may be due to exfiltration from sewage systems or runoff from solid-waste landfills, septic tanks, or even water-treatment plants. The sources of pollution may be located too close to the wells, or polluted water may have flowed into streams or faults and contaminated distant wells through short-circuiting.

During its underground journey and before reaching the water table, water undergoes a natural treatment (percolation) that provides groundwater with its quality. In fact, in comparison with the few hours it spends in the sewage system, wastewater may, in cases where the ground dictates a very slow flowrate, take many years to reach the water table; this water will probably be of very good quality.

1.2 SOURCES OF DRINKING WATER

Depending on the circumstances, the engineer may consider using the following sources of drinking water:

- surface water;
- groundwater;
- rainwater;
- sea water and brackish water.

In most cases, surface water and groundwater is used. When such water is not available, the engineer turns to rainwater or sea water.

1.2.1 Surface Water

In areas that have abundant lakes and rivers (including many regions of Canada), most cities will use surface water as their main source of drinking water.

River water. The upstream of a river (mountain areas) is often less populated and surrounded by forests and agricultural lands. The flow of the river is fast and turbulent and the water is often turbid (lacks clarity) as it transports suspended matter from the surrounding forests. The water is often cold, as it originates in springs, melting snow, or melting glaciers. Finally, its colour index is low as it does not usually carry much organic matter in solution.

The downstream of a river has a higher population base; surrounding lands are farmed and industries are present. The water is therefore contaminated by pathogenic (disease-transmitting) bacteria and viruses, toxic matter, biodegradable organic matter, and suspended solids responsible for turbidity and colour.

Depending on the season and the rainfall in a given region, variations in stream flow will cause variations in the concentration of pollutants, which is why the operations of a treatment plant must constantly be reviewed.

Lake water. Lakes are natural water reservoirs. They have the following effects:

- Because of their low turbulence, water turbidity of a lake is reduced as suspended solids tend to settle at the bottom of the lake.
- Concentration of pathogenic bacteria and viruses is also reduced because of the combined effects of sedimentation and the longer time that the water remains in the lake, which is not a favourable place for these organisms to thrive. For example, many of them are of intestinal origin and cannot survive in cold temperatures.
- Concentration of mineral nutrients, such as phosphorus and nitrogen, is increased, causing eutrophication, or proliferation of algae and aquatic plants.

In deep lakes in temperate zones, stratification of the water temperature can be observed. In summer, water temperature is higher at the surface than at the bottom, and the reverse is true in winter. In the fall, when the surface temperature becomes lower than the bottom temperature, surface water, now denser, moves toward the bottom and bottom water rises to the surface, carrying sediment deposits, causing the surface water to become turbid. This phenomenon is called overturn of lake waters. In cold climates, lakes undergo winter stratification and spring overturn, as well.

1.2.2 Groundwater

Groundwater originates from springs or wells. Usually, it has the following characteristics:

- relatively constant temperature, in all seasons, because the ground acts as a buffer;
- low in colour, as it contains little organic or colloidal matter in solution;
- low turbidity, as the water is filtered through the ground;
- often, low bacterial or viral pollution if the water is filtered through the ground or has remained in the ground for a long time;
- presence of iron and manganese in solution (the main disadvantage of these elements is that they stain clothes and plumbing);
- presence of calcium and magnesium, which react with soap to form a precipitate;
- presence of sulfhydric acid, H_2S, a nauseating gas that emanates when water is drawn and slightly stirred, and generates a taste that many find offensive.

1.2.3 Rain Water

Rain water is usually of good quality for human consumption and other domestic purposes. However, many consumers find the absence of taste displeasing.

This water is also very soft. In other words, it does not contain any calcium, magnesium, iron, strontium, or manganese salts in solution. Because it contains few minerals, it is corrosive and can wear away copper, cast iron, and concrete pipes. In the past, people who collected rain water for laundry knew that soft water uses little soap and produces a proliferation of suds.

Rain water is saturated with oxygen, O_2, nitrogen, N_2, and carbon dioxide, CO_2. These three gases are found in the atmosphere. In industrialized regions, however, rain water can be contaminated by atmospheric particles and foreign gases such as carbon monoxide, CO, sulphur dioxide, SO_2, nitrate oxides, NO and NO_2, and sulfhydric acid, H_2S. Gases containing nitrogen and sulphur are responsible for the creation of acid rain, with a pH as low as 4.0.

The city of Gibraltar uses rain water as its main source of drinking water, through a series of collection areas installed on the hillside around the famous Rock of Gibraltar.

1.2.4 Sea Water and Brackish Water

When surface water, groundwater, and rain water are not available, brackish water (about 15,000 mg/L of dissolved salts) or even sea water (25,000 mg/L or more of dissolved salts) can be used. There are two main desalinization techniques which use membranes: electrodialysis and reverse osmosis. These techniques are expensive, as they require a large amount of energy. Other techniques used are solar distillation and freezing, where sun and weather conditions are appropriate.

1.3 POPULATION-GROWTH PROJECTION

Civil-engineering facilities used for water distribution and collection in urban areas must meet the needs of the population over a certain period of time, called the *design period*. The design engineer's first task is to estimate the size of the population so that the planned structure will be adequate throughout the design period. This is not an easy task. Depending on the projection needs, there are two types of population projections: short-term projection, from 5 to 10 years, and long-term projection, from 10 to 50 years.

Making long-term projections is a challenge because of the great uncertainties involved in predicting the distant future. In this case, it is appropriate to rely on past growth curves to predict trends, even though economic and social factors (war, recession, installation or relocation of large companies, etc.) affect the growth of urban areas.

Most projection methods can predict the behaviour of populations only over a relatively short period. They are, however, useful models and can be instrumental to the designer when forming an opinion, which will then be clarified by interviews with responsible authorities, examination of the cities' technical service files, and information gathered from professionals in other fields (urban planning, economics, demography, sociology, etc.).

In Quebec, rapid population growth is now over, and its demographic growth is now lower than that of Canada as a whole. Urban centres are no longer attracting rural populations, and there is major outmigration to other provinces. In fact, Quebec has reached almost zero demographic growth. The engineer must therefore be cautious in terms of projections, as there may be population decreases in some cities and regions.

In many parts of the world, such as Africa and South America, very high population growth is still observed. In the wealthiest Western countries, however, demographic growth has dropped.

1.3.1 Sources of Information

To better estimate population growth, engineers can gather information from a wide range of sources:

- *national censuses;*
- *immigration and emigration data*, on the national and regional levels;
- *birth and death statistics*;
- *urban planning documents*, from which we know the actual and planned use of the land and can estimate the probable density of the population.

1.3.2 Models for Population Projection

In a finite environment, and considering the social and economic factors and available food supply, the behaviour of the human population is similar to the behaviour of a bacterial population living in a confined environment where food resources are limited (fig. 1.3). After the initial period of adjustment to the environment, the number of bacteria grows quickly as food is readily available; then, the growth rate decreases as resources are used up. Finally, overcrowding and depletion of food resources cause the population to decline. For the human population living in an urban area, however, the evolution is not as systematic. Resources may be injected from outside, or part of the population may leave. Moreover, human adaptations to the environment, such as birth control, have significant effect on the population-growth curve.

Finally, and most importantly, the recycling of resources allows for an *equilibrium* to be reached without going through a decline phase due to the depletion of resources. The use of organic wastes and natural fertilizers allows for a constant food supply and makes it possible for the population of an urban area to tend toward its maximum. In this case, the shape of the growth curve looks like an inclined S, as shown in figure 1.9 (most long-term projection methods attempt to evaluate the growth of this type of curve.) In any case, although most methods are appropriate for short-term projections, the choice must be carefully made and be based on the available information. Historical data must be carefully analyzed in order to detect extraordinary events that may have influenced population growth but have little chance of reoccurring; these events may induce projection errors.

Graphic extrapolation. Using this method, extrapolations are made from the growth curve of the population plotted on graph paper (fig. 1.4). Past events that have influenced population variations are used for the population prediction for the years to come.

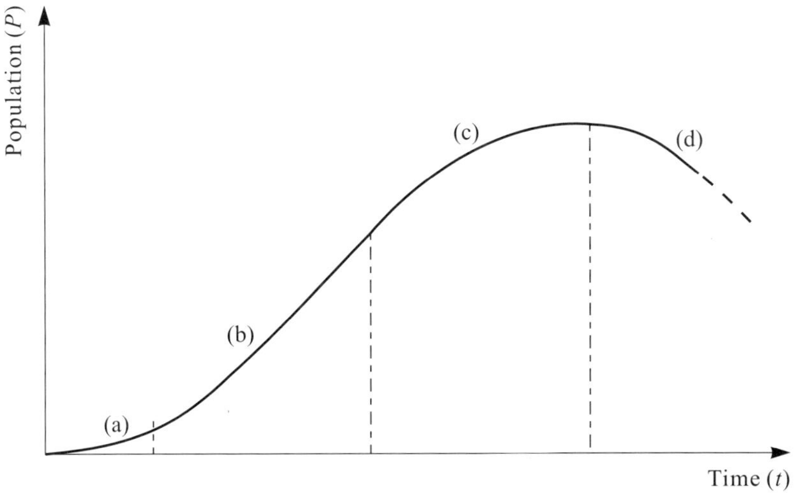

Figure 1.3 Evolution of a bacterial population in a confined environment: a) initial adjustment period; b) rapid growth; c) slowed growth; d) decline

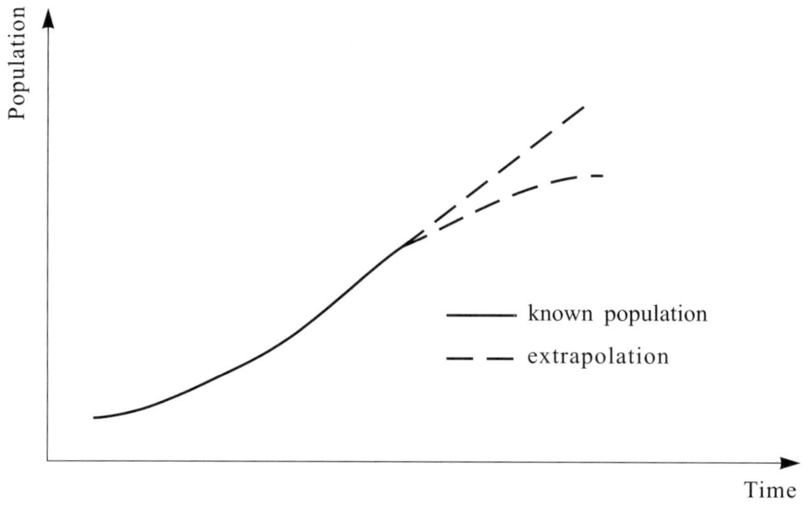

Figure 1.4 Graphic extrapolation of population growth.

Comparison of population growth in different cities. Using this method, the growth curve of the city where the population is to be estimated is compared with those of cities that showed similarities in terms of population in the past. If cities *X*, *Y*, and *Z* evolve in a similar fashion but at different times, the demographic growth of city *X*, for example, can be projected by assuming that it corresponds to the average population growth of cities *Y* and *Z* (fig. 1.5.) However, it is important to make sure that all compared cities have similar social and economic characteristics for the periods during which the populations were a match for the population of the city being studied; needless to say, very few reference cities meet all of the requirements for this type of analysis.

Projection based on arithmetic growth. The growth of population *P* is said to be arithmetic when the growth rate, d*P*/d*t*, of this population is constant:

$$\frac{dP}{dt} = K_a \tag{1.1}$$

therefore

$$\int_{P_1}^{P_2} dP = K_a \int_{t_1}^{t_2} dt$$

thus

$$P_2 - P_1 = K_a(t_2 - t_1) \tag{1.2}$$

where P = population
t = time
K_a = arithmetic growth rate constant

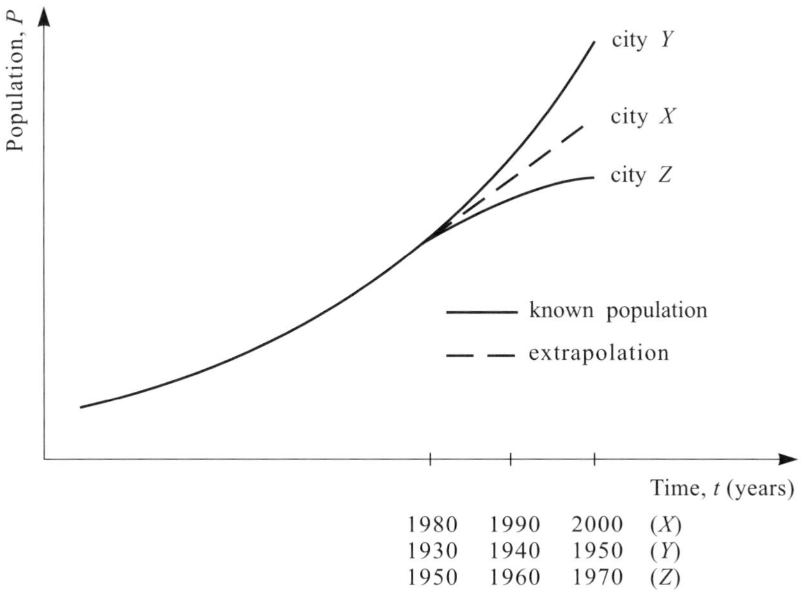

Figure 1.5 Population evolution projection of city X by graphic comparison with cities Y and Z

Equation 1.2 is represented by a straight line on the arithmetic scale graph (fig. 1.6). K_a is calculated from known populations, then the projected population, P_n, is evaluated at time t_n using equation 1.2, where K_a is replaced by the calculated value. It yields:

$$K_a = \frac{P_2 - P_1}{t_2 - t_1} \tag{1.3}$$

and

$$P_n = P_2 + K_a(t_n - t_2) \tag{1.4}$$

where t_n = year for which the population is projected
P_n = population for the year t_n
P_2 = known population for the year t_2

Population projection using the arithmetical growth method is well suited to long-established cities and to regions where development has stabilized, as is often the case in an agricultural setting.

Projection based on geometric growth. The growth of population P is said to be geometrical when the growth rate, $\mathrm{d}P/\mathrm{d}t$ is proportional to the population itself:

$$\frac{\mathrm{d}P}{\mathrm{d}t} = K_g P \tag{1.5}$$

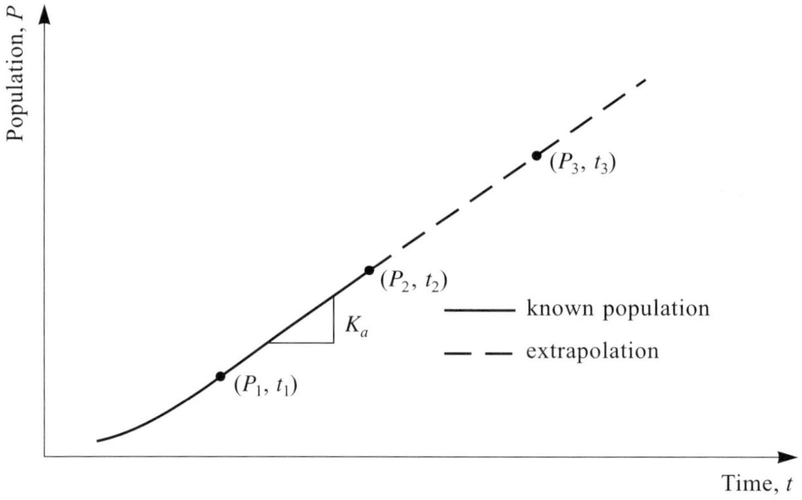

Figure 1.6 Population-growth projection when population growth is arithmetic

therefore

$$\int_{P_1}^{P_2} \mathrm{d}P/P = K_g \int_{t_1}^{t_2} \mathrm{d}t$$

thus

$$\ln P_2 - \ln P_1 = K_g(t_2 - t_1) \tag{1.6}$$

where K_g is the geometric growth rate constant.

The graphic representation of the logarithm variation for population P, as a function of time, is a straight line with a slope of K_g (fig. 1.7.) The projection of a future population is obtained using the following equation:

$$\ln P_n = \ln P_2 + K_g(t_n - t_2) \tag{1.7}$$

If a base 10 logarithm is used instead of a natural logarithm, the value of the constant is multiplied by 2,3 to obtain the value of K_g.

Geometric growth can also be written using the compound-interest equation, if the annual growth percentage of the population is known. It then yields:

$$P_n = P_1(1 + r)^n \tag{1.8}$$

where n = number of periods (generally years) during which geometric growth is observed $(t_n - t_1)$

r = growth rate for each period (expressed as decimals, e.g., 0,05 instead of 5%)

P_1 = population at time t_1

As the unit of time used is generally the year, when a different time unit is used, equation 1.8 must be adjusted so that n is the number of time units and r is the growth rate during the specified unit of time.

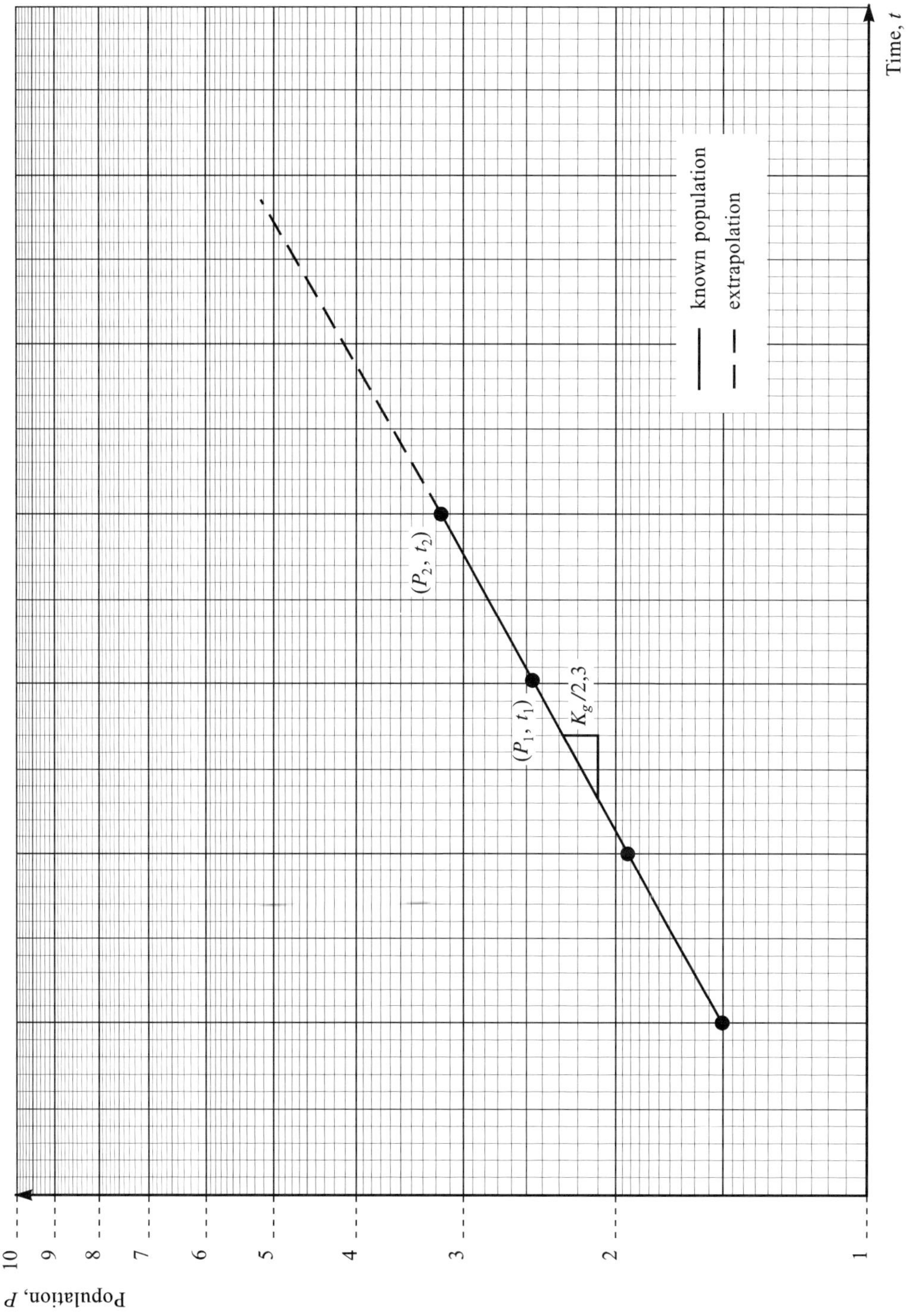

Figure 1.7 Projection of the population evolution when the growth is geometric

EXAMPLE 1.1

The annual growth rate of a population of 25 000 inhabitants is 5%.

a) In how many years will the population reach 50 000 inhabitants?
b) Compare the values of r and K_g.

Solution

a) As only the present population is known (25 000 inhabitants), the compound-interest equation is used (equation 1.8). It yields:

$$P_n = P_1(1+r)^n$$

$$50\,000 = 25\,000\,(1+0{,}05)^n$$

so that

$$n = 14{,}2 \text{ a}$$

b) Here, equations 1.7 and 1.8 must be compared knowing that n is equal to $(t_n - t_1)$; it then yields:

$$\frac{P_n}{P_1} = (1+r)^n$$

and

$$\ln\left(\frac{P_n}{P_1}\right) = K_g(t_n - t_1)$$

thus

$$\frac{P_n}{P_1} = e^{K_g(t_n - t_1)}$$

As a result

$$(1+r)^n = e^{K_g(t_n - t_1)} = e^{K_g n}$$

and

$$n\ln(1+r) = K_g n$$

$$\ln(1+r) = K_g$$

Finally, since $r = 0{,}05$, it yields:

$$K_g = 0{,}049$$

Moreover, the growth rate can be calculated using the following equation, which is a variation of equation 1.8:

$$r = \left(\frac{P_2}{P_1}\right)^{1/n} - 1 \tag{1.9}$$

In some cases, we may want to estimate the time required for a population with a geometrical growth rate to double. For example, in a developing country where there is a population spurt and the annual growth rate is 3,4%, 21 years are required to double the population, assuming that the trend remains unchanged over the period.

To calculate the time required for the population to double, equation 1.8 is modified in the following manner:

$$P_2 = P_1(1+r)^n$$

where $P_2 = 2P_1$.

Similarly, since $r = 0{,}034$, it yields:

$$2 = (1 + 0{,}034)^n$$

thus

$$n = 20{,}7 \text{ a}$$

Geometrical growth of a given population is observed only when an economic factor has a great influence, as when large companies set up shop close to a community. Population growth is caused by the arrival of new employees and their families (in Quebec, for example, in Boisbriand with the arrival of GM or in Bromont of IBM). It is important to note that geometrical growth occurs only over short periods of time; as a result, the use of this method for long-term projections may lead to unrealistic overestimates.

Projection based on a decreasing growth rate. In the third and last phase of population growth (fig. 1.8), growth tends toward a maximum called the saturation population, *S*. This growth phase can be modelled by assuming that, at all times, the growth rate is a function of the difference between the saturation population, *S*, and the current population, *P*. In other words:

$$\frac{\mathrm{d}P}{\mathrm{d}t} = K(S - P) \tag{1.10}$$

where S = saturation population
K = decreasing growth rate constant

The equation is rewritten:

$$\int_{P_1}^{P_2} \mathrm{d}P/(S-P) = K\int_{t_1}^{t_2} \mathrm{d}t$$

therefore

$$-\ln\left(\frac{S-P_2}{S-P_1}\right) = K(t_2 - t_1) \tag{1.11}$$

thus

$$(S-P_2) = (S-P_1)\left[e^{-K(t_2-t_1)}\right] \tag{1.12}$$

Each member of the equation is subtracted from $(S - P_1)$; it then yields:

$$(S-P_1)-(S-P_2) = (S-P_1)-(S-P_1)\left[e^{-K(t_2-t_1)}\right]$$

thus

$$P_2 - P_1 = (S - P_1)\left[1 - e^{-K(t_2 - t_1)}\right]$$

and, in general:

$$P_n = P_2 + (S - P_2)\left[1 - e^{-K(t_n - t_2)}\right] \tag{1.13}$$

This projection method is appropriate for communities approaching saturation population.

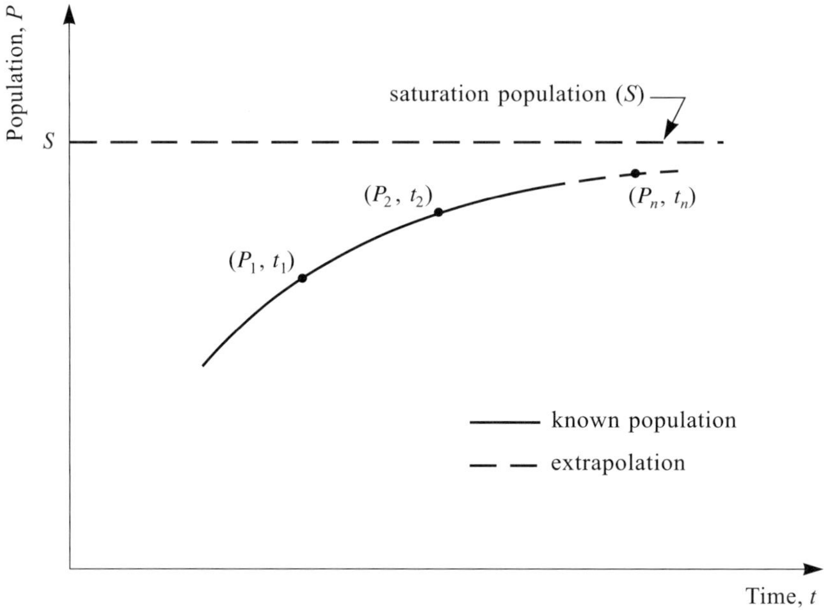

Figure 1.8 Projection of population growth when the growth rate is decreasing

EXAMPLE 1.2

The growth rate of the population of a city is decreasing. Ten years ago, the population was 65 145 inhabitants; currently, it is 70 000 inhabitants, and it should ultimately reach 100 000 inhabitants.

a) Calculate the annual growth rate (decreasing) of this population.
b) Estimate what the population will be in 12 years.

Solution

a) Calculation of the annual growth rate, K

Equation 1.13 can be used to solve this problem, as the value of all parameters and variables, except for K, are known.

$$P_n = P_2 + (S - P_2)\left[1 - e^{-K(t_n - t_2)}\right]$$

$$70\ 000 = 65\ 145 + (100\ 000 - 65\ 145)\left[1 - e^{-K(10)}\right]$$

$$K = 0{,}015$$

b) Calculation of the population in 12 years

Again, this population is calculated using equation 1.13.

$$P_n = P_2 + (S - P_2)\left[1 - e^{-K(t_n - t_2)}\right]$$

$$= 70\ 000 + (100\ 000 - 70\ 000)\left[1 - e^{-0{,}015(12)}\right]$$

$$= 74\ 942 \text{ inhabitants}$$

Logistic projection. This long-term projection method relies on the previously discussed hypothesis (fig. 1.3) that in a given community, population growth goes through three distinct periods before reaching its peak; a slow start, followed by rapid growth, then gradual slowing as the population tends toward saturation. This phenomenon is graphically represented by an S curve when the variation of the population is plotted as a function of time on graph paper (fig. 1.9).

This method is appropriate when the population of the given community is known over a certain number of years. One chooses three populations, P_1, P_2, and P_3, evaluated at regularly spaced intervals ($t_2 - t_1 = t_3 - t_2 = n$). Population P_1 corresponds to the slow growth period; P_2, to the rapid growth period; and P_3, to the decreasing growth-rate period.

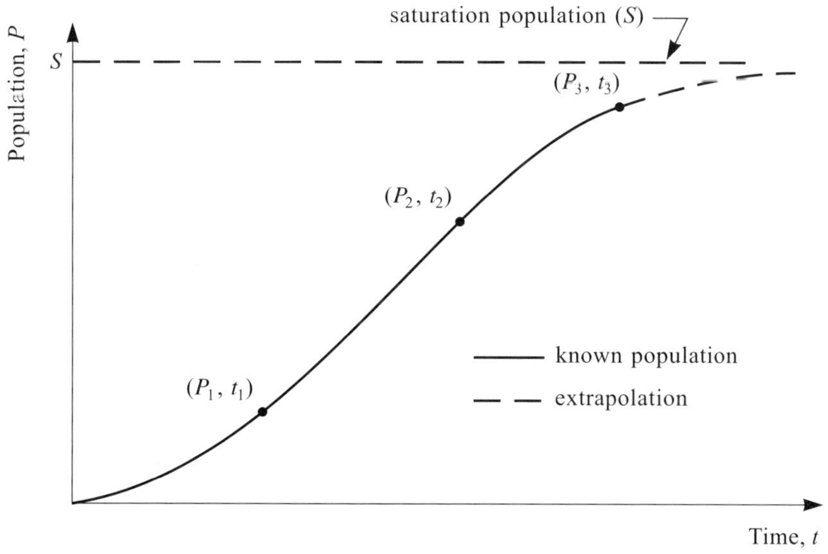

Figure 1.9 Projection of population growth using the logistic method

Population P_n is calculated at time t_n using the following equation:

$$P_n = \frac{S}{1 + ae^{bt_n}} \tag{1.14}$$

where

$$S = \frac{2P_1P_2P_3 - P_2^2(P_1 + P_3)}{P_1P_3 - P_2^2} = \text{saturation population} \tag{1.15}$$

$$a = \log\frac{(S - P_1)}{P_1} \tag{1.16}$$

$$b = \frac{1}{n}\left[\log\frac{P_1(S - P_2)}{P_2(S - P_1)}\right] \tag{1.17}$$

t_n = time elapsed from the beginning of logistical growth

EXAMPLE 1.3

The population of a city was 10 000 in 1920 (P_1), 50 000 in 1940 (P_2), and 80 000 in 1960 (P_3). In 1900, the population was zero. Calculate the population in 1980 using the logistic method (fig. 1.10).

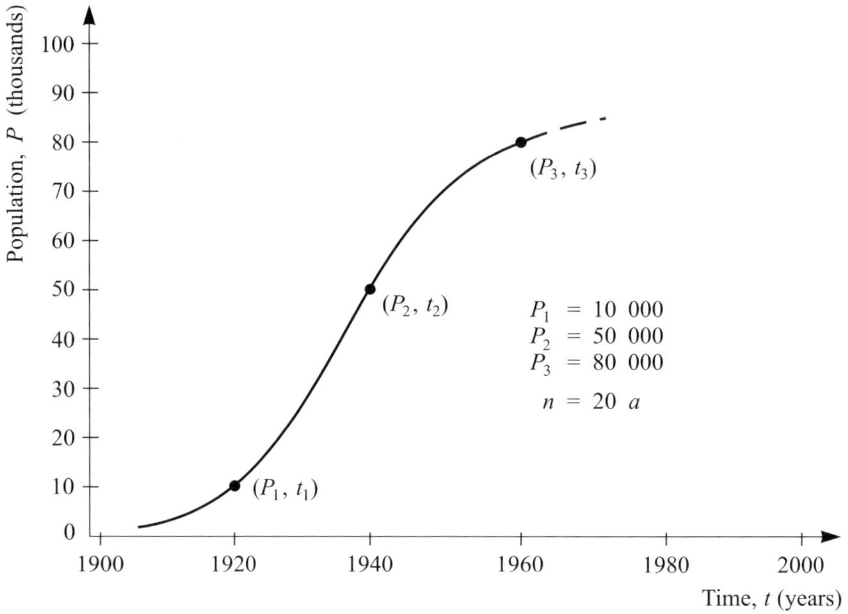

Figure 1.10 (ex. 1.3) Projection of population growth when the growth rate follows a logistic curve

Solution

- Calculation of the saturation population, S

$$S = \frac{2P_1P_2P_3 - P_2^2(P_1 + P_3)}{P_1P_3 - P_2^2}$$

$$= \frac{2(10\,000)(50\,000)(80\,000) - (50\,000)^2(10\,000 + 80\,000)}{(10\,000)(80\,000) - (50\,000)^2}$$

$$= 85\,294 \text{ inhabitants}$$

- Calculation of the population in 1980

$$a = \log\left(\frac{S - P_1}{P_1}\right) = \log\left(\frac{85294 - 10000}{10000}\right)$$

$$= 0{,}88$$

$$b = \frac{1}{n}\left[\log\frac{P_1(S - P_2)}{P_2(S - P_1)}\right] = \frac{1}{20}\log\left[\frac{10000(S - 50000)}{50000(S - 10000)}\right]$$

$$= -0{,}05$$

$$n = 1960 - 1940 = 1940 - 1920 = 20 \text{ a}$$

$$\text{Population in 1980} = \frac{S}{1 + ae^{b(1980-1900)}}$$

$$= \frac{85\ 284}{1 + 0{,}88e^{-0{,}05(80)}}$$

$$= 83\ 931 \text{ inhabitants}$$

Comparison between populations of different communities and regions. This method is similar to the graphic extrapolation method. Here, the extrapolation is done on the ratio P_A/P_B, between the population of the community studied (or local population, P_A) and the population of other communities or regions, or a regional or national population, P_B (fig. 1.11). We then assume that there must be a relation between the local population and the regional or national population. Extrapolations are made from this relationship, then the future population of the studied community is calculated using the future population of the reference community. Practically, this method is used as follows:

- The ratios P_A/P_B are calculated for the years when the populations are known.
- From these ratios, extrapolations are made to a later date using one of the methods presented earlier (arithmetical, geometrical, etc.)

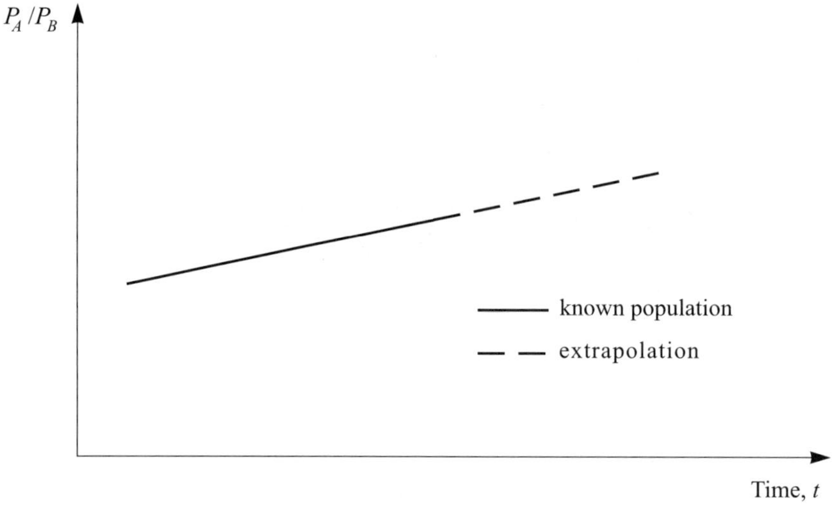

Figure 1.11 Projection of population growth using growth comparison of different regions

– The future population(s) is/are calculated using these ratios and the population projections of the reference communities.

Example 1.4 illustrates this comparative method.

EXAMPLE 1.4

Between 1980 and 1990, the populations of a city, (P_c), and its region, (P_r), have increased as shown in the following table.

Year	City	Region	P_r/P_c
1990	10 000	100 000	10
1995	12 000	132 000	11
2000	14 000	168 000	12

What will the population of this city be in 2005 and in 2010?

Solution

Extrapolation of the P_r/P_c (population region/population city) ratio yields 13 and 14, respectively, for 2005 and 2010; as a result, if the regional population is projected to be 208 000 in 2005 and 252 000 in 2010, the city population can be projected as follows:

Population of the city in 2005

Population of the region in 2005/13 = 208 000/13
= 16 000

Population of the city in 2010

Population of the region in 2010/14 = 252 000/14
= 18 000

Table 1.1 shows a summary of the different projection methods of the targeted populations and their main characteristics.

Table 1.1 Projection methods for targeted populations

Method	Equation	Description of the parameters	Constant value
Arithmetic growth	$\frac{dP}{dt} = K_a$	P = population t = time* K_a = arithmetic-growth-rate constant	$K_a = \frac{P_2 - P_1}{t_2 - t_1}$
Geometric growth	$\frac{dP}{dt} = K_g P$	K_g = geometric-growth-rate constant	$K_g = \frac{\ln P_2 - \ln P_1}{t_2 - t_1}$
Growth with a decreasing rate	$\frac{dP}{dt} = K(S - P)$	S = saturation population K = decreasing-growth-rate constant	$K = \frac{-\ln\left[(S - P_2)/(S - P_1)\right]}{t_2 - t_1}$
Logistic projection	$P_n = \frac{S}{1 + ae^{bt_n}}$	S = saturation population P_n = future population P_1, P_2 and P_3 = population at time t_1, t_2 and t_3 t_n = time since the beginning of logistic growth $t_2 - t_1 = t_3 - t_2 = n$	$S = \frac{2 P_1 P_2 P_3 - P_2^2 (P_1 + P_3)}{P_1 P_3 - P_2^2}$ $a = \log \frac{S - P_1}{P_1}$ $b = \frac{1}{n} \log \frac{P_1 (S - P_2)}{P_2 (S - P_1)}$

* Time t is generally in years for all different methods.

1.3.3 Population Densities

Population densities are generally expressed as number of persons per hectare. Table 1.2 presents some common values.

Table 1.2 Common population densities

Areas	Persons/hectare
Residential areas	
Single-family units, large lots	10 – 30
Single-family units, small lots	30 – 70
Multiple-family units	70 – 200
Apartment houses	200 – 2000
Commercial areas	30 – 60
Industrial areas	10 – 30
Total, exclusive of parks, playgrounds and cemeteries	10 – 50

Adapted from: Fair, Geyer and Okun, *Water and Wastewater Engineering*, 1966, p. 5-12.

1.3.4 Design Period

When it comes to water, civil-engineering facilities must be designed to serve the present and future needs of a specific community. The design period is defined as the number of years between the time a structure is first put into service and the time it no longer meets the needs of the community. This period is usually calculated using a financial analysis method in which the following items are considered:

- investments;
- maintenance and operation costs (labour, electricity, chemicals, etc.;)
- interest rates;
- possibility of expansion or adding elements to the system (for example, it is easy to replace a pump or increase pumping-station capacity, but difficult to increase water intake or enlarge a tunnel);
- future technological innovations (the design period can then be chosen assuming that the future will bring better and cheaper techniques).

Table 1.3 shows the design periods of structures used for drinking water and sewage treatment. There are basically three types of design periods: between 5 and 20 years, for installations that wear out rapidly and are easily replaceable; between 20 and 30 years, for installations that are more difficult and costly to replace; and 50 years or more, for installations that are extremely difficult and costly to replace or to expand.

Table 1.3 Design periods of some facilities

Facilities	Characteristics	Design period (years)
Drinking water		
Dam or tunnel	Extension difficult and costly	50 – 75
Water intake and collection pipe	Extension relatively difficult	25 – 50
Water-treatment plant and distribution network	If population-growth and interest rates are:	
	low	50
	high	25
Pipes more than 300 mm in diameter	High replacement cost	25 – 30
Pipes less than 300 mm in diameter	Construction and doubling easy	15 – 20
Pumping station:		
- structure	Easily extended	15 – 20
- pumps	Capacity easily increased; rapid wear	5 – 10
Wastewater		
Secondary pipes smaller than 375 mm in diameter	Construction and doubling easy	20
Collecting and intercepting sewers, outfalls	Extension difficult and costly	30 – 50
Treatment plant	If population-growth and interest rates are:	
	low	20 – 30
	high	15 – 20
Pumping station	Easily extended	10 – 20
Entrance pumps at the treatment plant	Replacement easy and cheap; rapid wear	5 – 10

CONCLUSION

Surface water, groundwater, and brackish and sea water can be used and consumed after being collected, treated in a water-treatment plant, and distributed through a distribution network. Wastewater is evacuated through the sewage network, then sent to a sewage-treatment plant. After treatment, the treated wastewater is returned to the environment, where the self-purification process completes organic and microbial purification. The water-utilization cycle ends, and water can then be collected again.

Facilities for the drinking-water network and the wastewater system must be in use for a minimum period of time called the design period. To design these facilities adequately, population growth during the design period must be projected as precisely as possible. The methods presented in this chapter allow projection of population growth using available data (past and present populations, economic characteristics of the given region, etc.).

PROBLEMS

1.1 The population of a city grows according to a decreasing growth rate of 1,5% per annum. The present population is 70 000 inhabitants, and urban planning forecasts an ultimate population of 100 000 inhabitants. Evaluate the population of this city in 10 years.

1.2 For a new city prone to growth spurts, which method would be the most appropriate to estimate the short-term future population (e.g., 5 years)?

1.3 The population of a city has a geometrical growth rate, during which its population has grown from 20 000 to 25 000 inhabitants in 10 years. During this period, what was the annual growth rate of this population? Calculate the geometrical growth constant.

1.4 Over a period of ten years, the population of a community has grown in the following manner:

- January 1, 1979: 12 100 inhabitants
- January 1, 1984: 15 078 inhabitants
- January 1, 1989: 18 790 inhabitants

Name the type of growth. Evaluate the population of this community for January 1, 1999. Calculate the annual growth rate of this population.

1.5 In 1950, 1970, and 1990, the population of a city was 30 000 (P_1), 172 000 (P_2), and 292 000 (P_3), respectively. In 1940, the population was zero. Assuming that the population growth is logistic, calculate the saturation population (S) and the population in 2010.

Chapter

2

HYDRAULIC CHARACTERISTICS OF FLOWS IN PIPES

INTRODUCTION

Hydraulics computations are closely integrated in the design and maintenance of distribution networks, sewer systems, and equipment found in those installations. This chapter will discuss the basic hydraulics concepts required to calculate flows, velocities, and water levels in sewers and water-distribution pipes, slopes of energy grade lines and hydraulic grade lines, and head losses.

2.1 FUNDAMENTAL EQUATIONS

In 1769, Chezy proposed an empirical equation to calculate the velocity of *uniform flows* in channels. Uniform flow is defined as the flow for which the longitudinal velocity is the *same* at any position in the channel. The Chezy equation is described as follows:

$$v = C_C\sqrt{R_h s} \tag{2.1}$$

where v = average longitudinal velocity of the water flow (m/s)

C_C = coefficient believed by Chezy to be constant, but, in fact, dependent on $f^{1/2}$

f = Darcy-Weisbach friction factor

R_h = hydraulic radius (m) (flow section, A/wetted perimeter, WP)

$R_h = d/4$ for circular pipes flowing full

A = cross sectional area of flow (m^2); $A = \pi d^2/4$ for a circular pipe flowing full

WP = wetted perimeter (m); length of contact line between the water and pipe wall in a perpendicular cross section of the flow ($WP = \pi d$ for a circular pipe flowing full)

d = inside diameter of the pipe (m)

s = slope of the energy grade line (m/m); for channels, it is also the slope of the water surface when the flow is uniform (fig. 2.1)

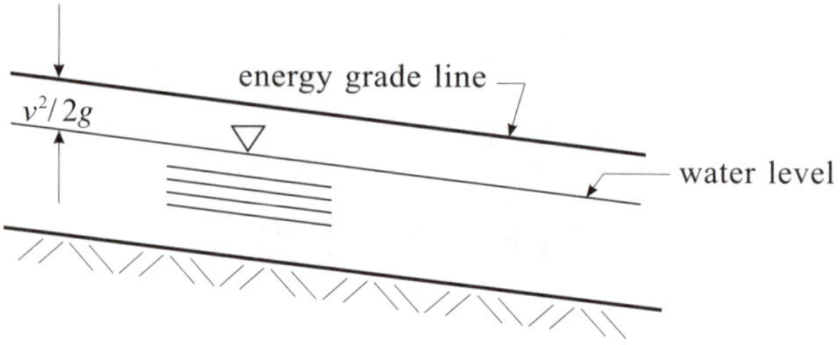

Figure 2.1 Energy grade line for a channel with a uniform flow

Figure 2.2 shows a vertical and longitudinal view of a uniform flow in a channel and in a sewer pipe. When the water level reaches the top (crown) of a circular sewer pipe and the water pressure is below 101,3 kPa (one atmosphere), the pipe is said to be flowing full. When the water pressure exceeds 101,3 kPa, the pipe is said to be under pressure. As shown in figure 2.2, the bottom of the pipe is called the *invert*.

The energy grade line (fig. 2.1) represents the energy accumulated by the water at any point in the channel or the pipe (see Bernoulli's equation, equation 2.5). The difference between the water surface and the energy grade line represents the kinetic energy, $v^2/2g$, where g is acceleration due to gravity. For some uniform flows (e.g., in channels or sewer pipes), the energy grade line, the water surface, and the invert of the channel or pipe are parallel. In other words, s is also the hydraulic slope, or hydraulic gradient.

In 1869, Ganguillet and Kutter proposed a new equation to calculate the value of the coefficient C_C of the Chezy equation. The terms of the new equation were R_h, s, and a new coefficient of roughness, n, called the Kutter roughness coefficient, or Kutter's n.

Although it is empirical, the Manning equation is the equation most frequently used when calculating the flow velocity in sewer pipes where water is subjected to atmospheric pressure (101,3 kPa). This equation is a variation of the Chezy equation where $C_C = R_h^{1/6}/n$; it then yields:

$$v = \frac{R_h^{2/3} s^{1/2}}{n} \tag{2.2}$$

where n is Manning's coefficient of roughness (different from Kutter's roughness coefficient, although numerically close).

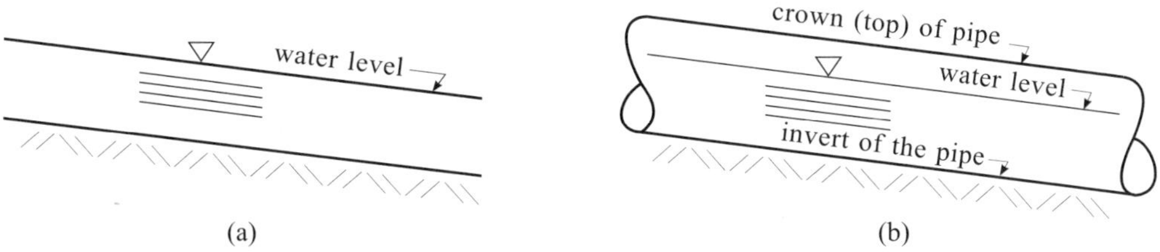

Figure 2.2 Uniform flow: a) in a channel; b) in a sewer pipe

Table 2.1 shows the most common values of Manning's coefficient of roughness for different types of material. The value of the pipes' n coefficient must be chosen with care, as it depends on the nature and number of joints, the material, and the condition of the pipes, and the nature of the water being transported. The values of n most frequently used are: 0,013 for concrete pipes; 0,010 for asbestos-cement pipes; 0,015 for old, worn-out pipes; 0,009 for plastic pipes; 0,024 for corrugated steel pipes; and 0,012 for corrugated steel pipes with interior asphalt lining.

Table 2.1 Values of Manning's coefficient of roughness, n, for different materials

Matérial	n
Pipes	
asbestos cement	0,010 – 0,015
badly corroded cast-iron or steel	up to 0,035
brick	0,013 – 0,027
clay	0,013 – 0,017
clean coated cast-iron	0,010 – 0,014
corrugated steel	0,021 – 0,027
new cast iron	0,011 – 0,015
rough concrete	0,015 – 0,027
smooth concrete	0,012 – 0,014
smooth plastic (including PVC)	0,009 – 0,013
wood	0,013 – 0,016
worn cast iron	0,013 – 0,027
welded steel	0,011 – 0,015
Channels	
asphalt	0,013 – 0,017
brick	0,012 – 0,018
concrete	0,011 – 0,020
with plants	0,030 – 0,400

The Manning nomograph (fig. 2.5), derived from equation 2.2 (Manning's equation), is used to calculate the flow velocity and flowrate in circular sewer pipes when the pipes are flowing full and the water is subjected to a pressure of 101,3 kPa (fig. 2.3a). In this case, the flow in the sewer is called the hydraulic capacity and the slope of the energy grade line is the same as the slope of the pipe. However, a sewer pipe often does not flow full (fig. 2.3b).

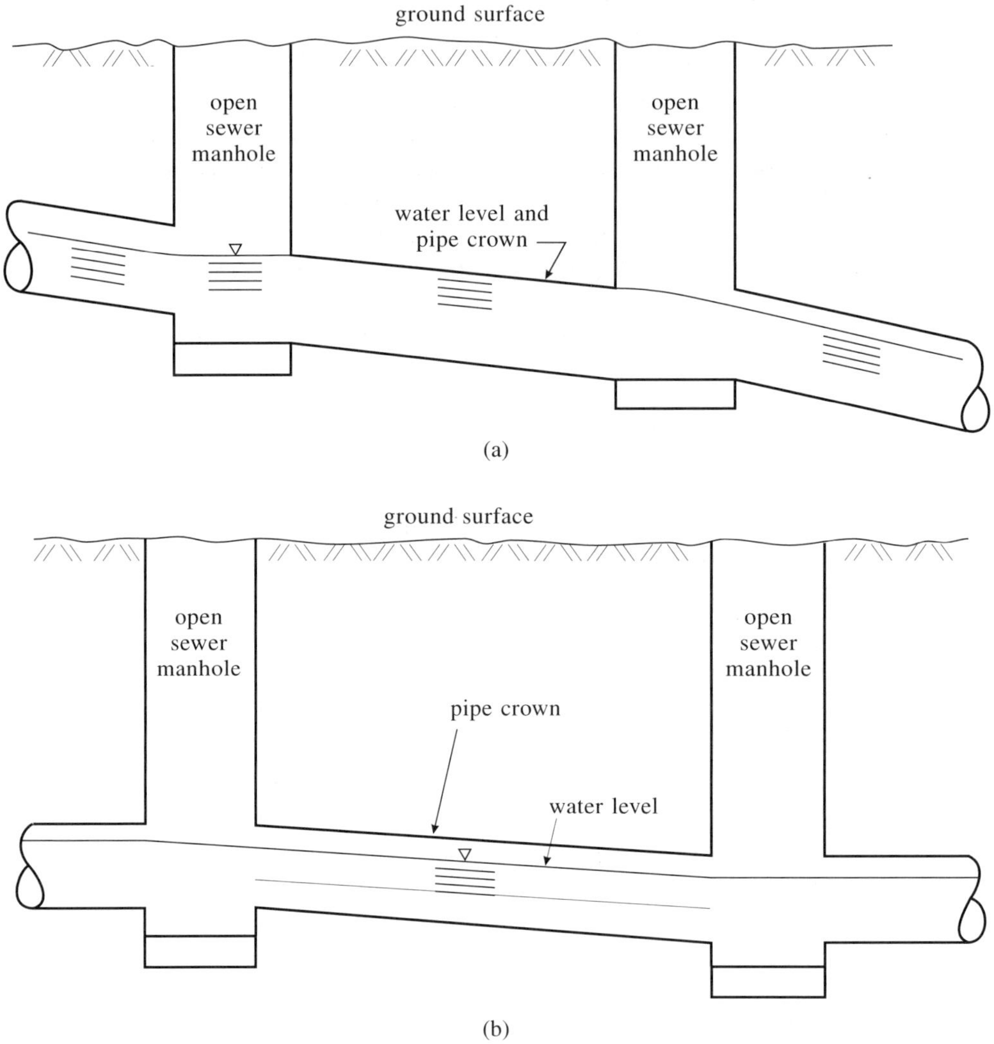

Figure 2.3 Sewer pipes with uniform flows subjected to a pressure of 101,3 kPa (atmospheric pressure): a) pipe flowing full; b) pipe flowing partly full

The Manning equation and nomograph can be used for sewer pipes flowing under pressure as long as the slope of the energy grade line is used to identify s. The Manning equation can be used for water-distribution pipes. This is not usually done, as the values of the n coefficients relative to the material of the pipes are not always known. Figure 2.4 shows a sewer pipe flowing under pressure.

The hydraulic grade line (fig. 2.4) represents the water pressure. It is expressed as the height the water would reach if a vertical tube were attached to the pipe. The difference in height between the hydraulic grade line and the energy grade line represents the kinetic energy, $v^2/2g$. As the flow velocity is uniform along the pipe (same flow, no cross-sectional area change), the energy grade line is parallel to the hydraulic grade line, except when a sudden change occurs in hydraulic conditions (e.g., at the junction between the pipe and a sewer manhole).

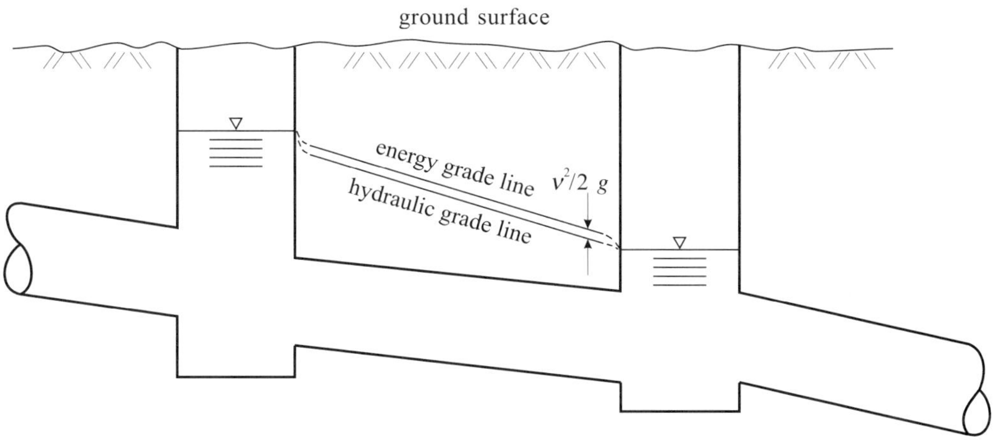

Figure 2.4 Sewer pipe flowing under pressure (uniform flow)

Figure 2.5 shows Manning's nomograph for different values of the roughness coefficient.

EXAMPLE 2.1

The slope of a reinforced concrete pipe with a nominal diameter of 600 mm (actual or inside diameter = 610 mm) is 0,002 m/m. Calculate the hydraulic capacity of this pipe when it flows full and is subjected to a pressure of 101,3 kPa (atmospheric pressure). Assume that the Manning coefficient, n, of concrete is 0,013. What is the water-flow velocity?

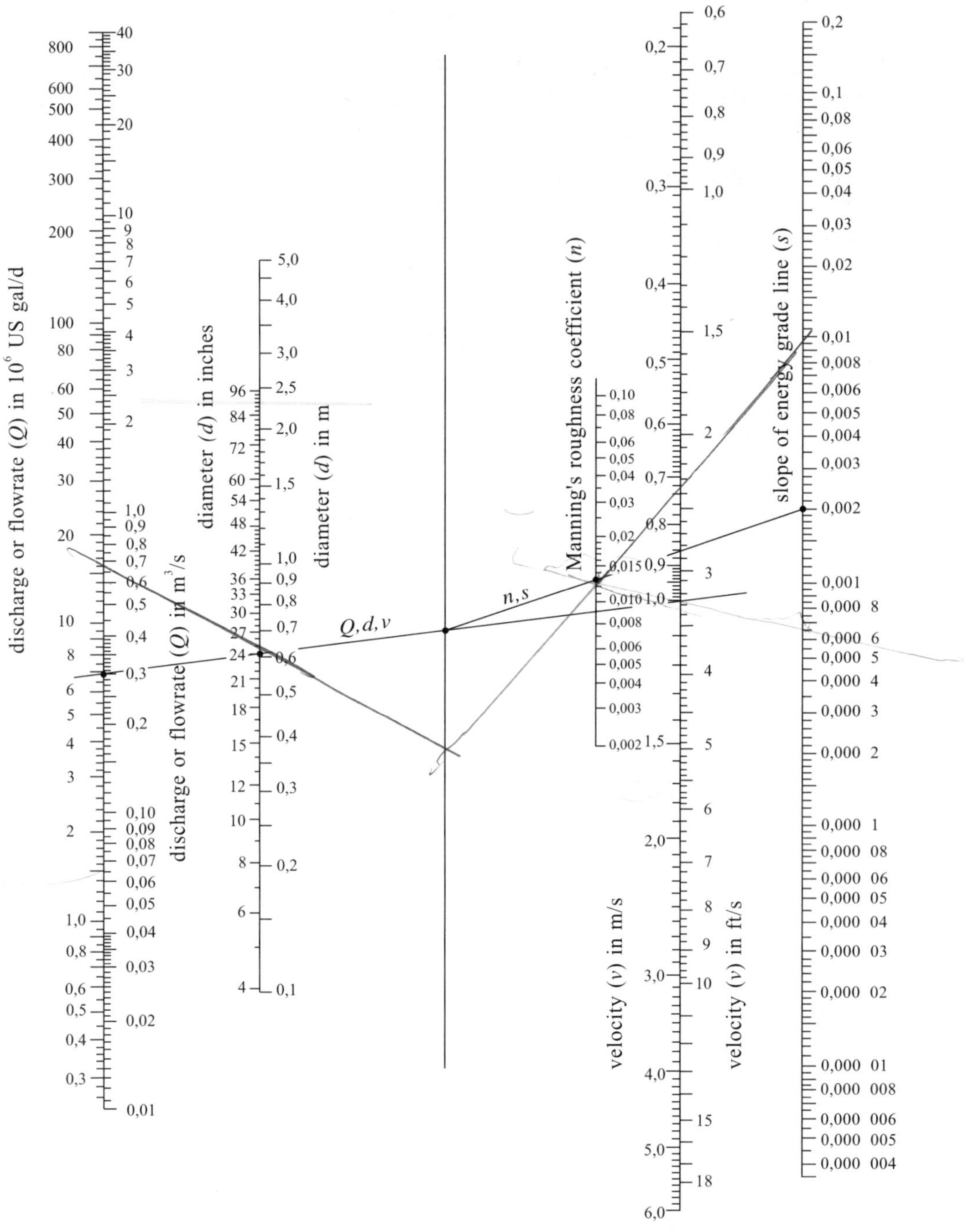

Figure 2.5 Manning's nomograph: Pipes flowing full (uniform flow)

From: American Society of Civil Engineers and Water Pollution Control Federation, *Design and Construction of Sanitary and Storm Sewers* (New York: ASCE, 1982), p. 96.

Solution

On the nomograph in figure 2.5, we draw a line from the point representing the slope (s = 0,002) to the point representing Manning's coefficient. Then, this line is extended to the vertical pivotal line in the centre of the figure. A second line is drawn from the point representing the diameter (d = 0,610 m) to the point of intersection between the first line and the pivotal line. By extending the second line, the hydraulic capacity is found for this pipe; it is 0,3 m^3/s. Its water-flow velocity is 1 m/s.

To verify the validity of these answers, the Manning equation can be used directly.

Darcy and Weisbach have proposed an equation to calculate the head losses (e.g., energy losses due to friction against the pipe walls) in circular pipes flowing under pressure. This equation is written as follows:

$$H_L = f\frac{Lv^2}{d2g} \tag{2.3}$$

where H_L = head losses due to pipe friction, pipe of length L (m)

f = friction factor; related to the Reynolds number, N_R, and the relative roughness of the pipe

N_R = Reynolds number; for a circular pipe flowing under pressure, its value is a function of the pipe diameter, d, flow velocity, v, dynamic viscosity, η, and the density of the water, ρ

L = length of the pipe (m)

g = acceleration due to gravity (9,81 m/s^2)

The Reynolds number, N_R, is a non-dimensional number. For a pipe under pressure, it is calculated using the following relation:

$$N_R = \frac{\rho_e vd}{\eta}$$

where ρ_e = density of the water (kg/m^3)

η = dynamic viscosity of the water ($N \cdot s/m^2$)

In this equation, the velocity, v, is expressed in metres per second and the diameter, d, is in metres.

Equation 2.3 can be used for water-distribution pipes. However, to solve for the values of coefficient f, Moody graphs (fig. 2.6) are used.

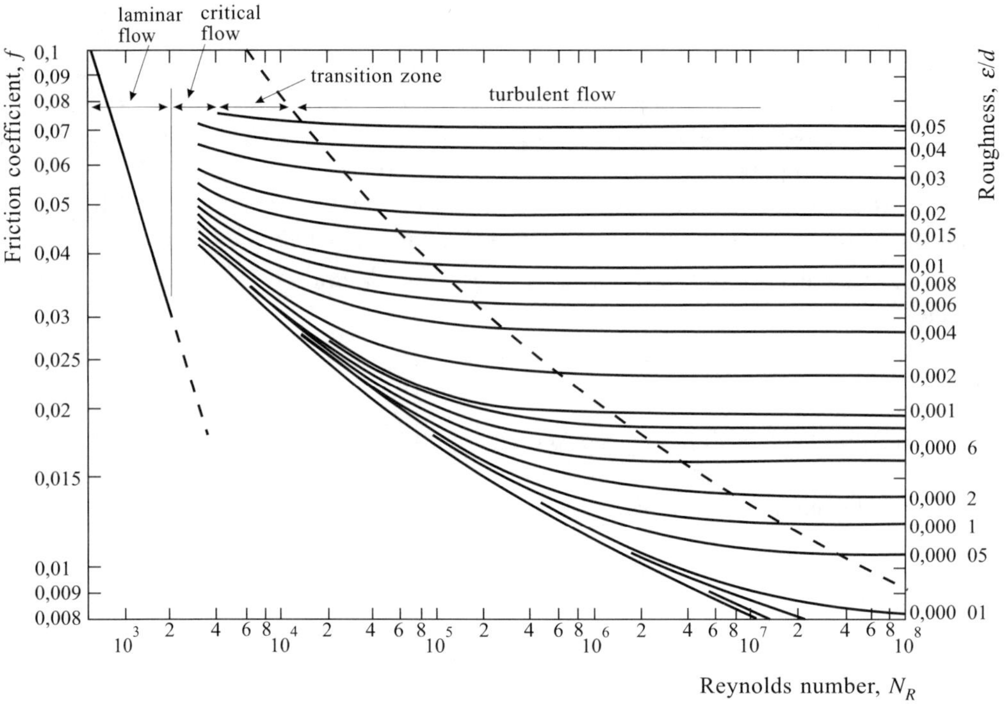

Figure 2.6 Moody graphs for flows under pressure in circular pipes (ε/d = relative roughness; ε = difference between the maximum and minimum values of the inside radius of the pipe)

The Hazen-Williams equation is most frequently used to compute flow characteristics in pipes under pressure. This equation is written as follows:

$$v = 0{,}849 C_{HW} R_h^{0{,}63} s^{0{,}54} \tag{2.4}$$

where C_{HW} = Hazen-Williams roughness coefficient.

Figure 2.7 shows the Hazen-Williams nomograph.

The Hazen-Williams equation can be used for sewer pipes. However, this is rarely done, as the values of the C_{HW} coefficients relative to the material of the pipes are seldom known.

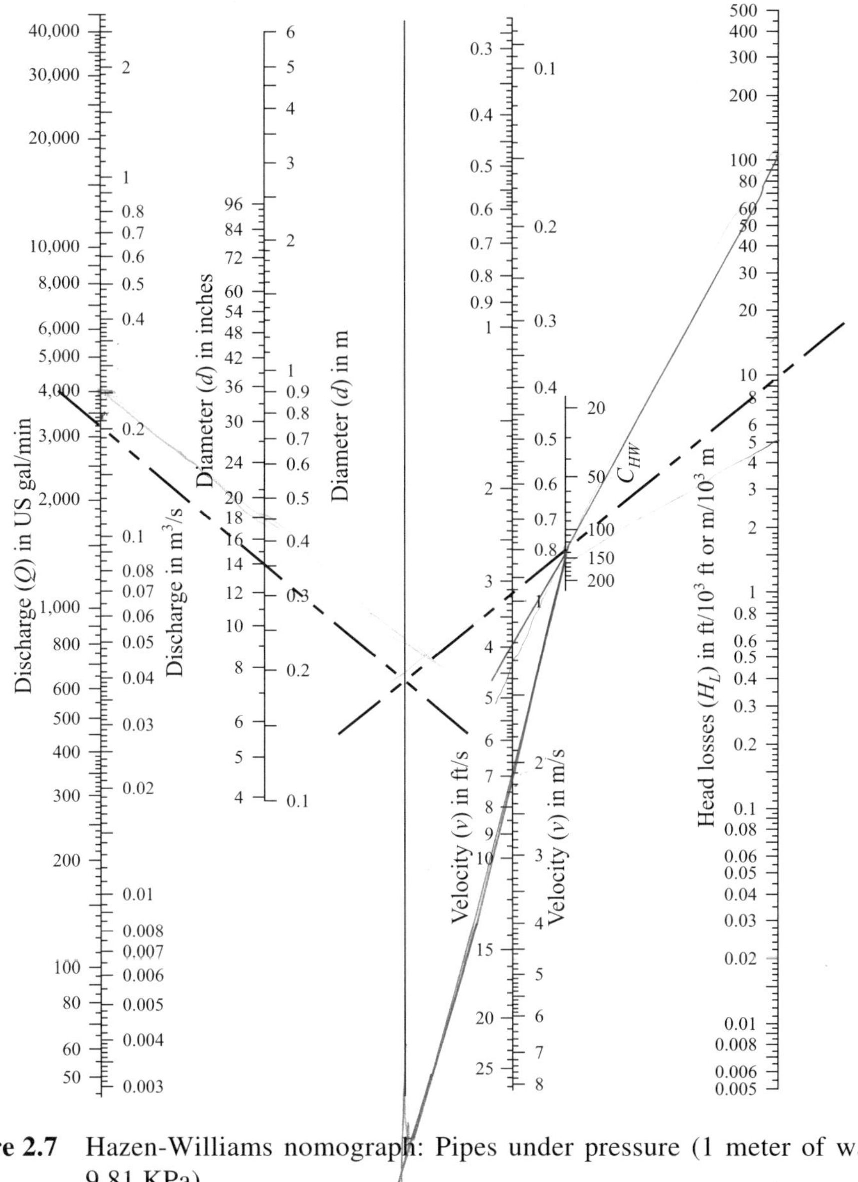

Figure 2.7 Hazen-Williams nomograph: Pipes under pressure (1 meter of water = 9,81 KPa).

From: American Society of Civil Engineers and Water Pollution Control Federation, *Design of Sanitary and Storm Sewers* (New York: ASCE, 1982, p. 98.

EXAMPLE 2.2

A horizontal water-distribution pipe, *A-B*, flowing from *A* to *B*, has a length of 2,565 m and a nominal diameter of 350 mm (actual or inside diameter = 355 mm; fig. 2.8a). Points *A* and *B* are located at the ends of the pipe. The flowrate is 200 L/s and the Hazen-Williams coefficient is 130. If the pressure at point *A* is 300 kPa, find the pressure at point *B*.

Figure 2.8 shows the pressure energy and the kinetic energy. These points will be addressed below.

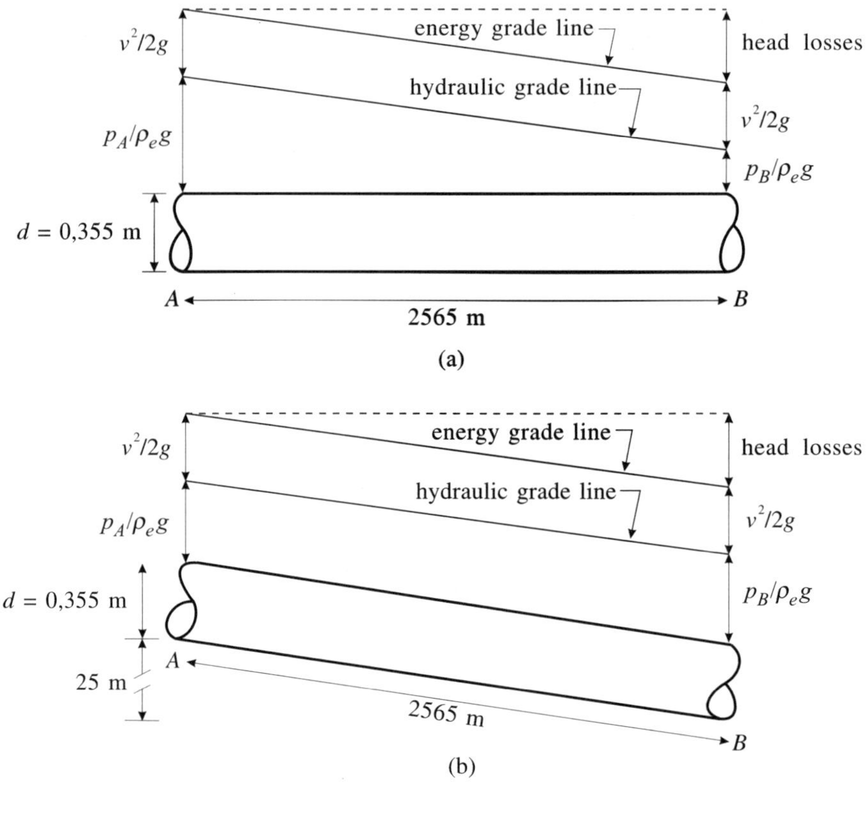

Figure 2.8 (examples 2.2 and 2.3) Pipes flowing under pressure: a) horizontal pipe; b) inclined pipe ($p/\rho_e g$ = pressure energy; $v^2/2g$ = kinetic energy)

Solution

On Hazen-Williams nomograph one finds that head losses are 10 meters of water per 1000 meters of pipe length. It then yields:

$$
\begin{aligned}
H_{tot} &= \frac{10}{1000} \times 2565 \\
&= 25{,}65 \text{ (m of water)} \\
&= 25{,}65 \times 9{,}81 \\
&= 251{,}6 \text{ kPa}
\end{aligned}
$$

However, as the pressure variations are caused only by friction between points *A* and *B* (end points), the pressure at point *B* is obtained by subtracting the head losses in the pipe from the pressure at point *A*. It yields:

$$
\begin{aligned}
p_B &= p_A - H_{tot} \\
&= 300 - 251{,}6 \\
&= 48{,}4 \text{ kPa}
\end{aligned}
$$

EXAMPLE 2.3

Find the pressure at point *B* (end of pipe) in example 2.2, if it is located 25 m lower than point *A* (fig. 2.8b).

Solution

The pressure at point *B* is increased by the pressure caused by the 25 m high water column located over point *B*. It then yields:

$$
\begin{aligned}
p_B &= p_A - H_{tot} + \text{pressure caused by water column} \\
&= (300 - 251{,}6) + (25 \times 9{,}81) \\
&= 293{,}65 \text{ kPa}
\end{aligned}
$$

In a hydraulic system, the Bernoulli equation expresses the energy conservation between two points, *A* and *B*. It states that for two given points in this system, the difference between the various forms of energy (potential energy or elevation, kinetic energy, and pressure energy) is the sum of the various head losses. In fact, in urban hydraulics, the only forms of energy being considered are the potential energy or elevation (height of points *A* and *B* in relation to a reference point), kinetic energy, and pressure energy; variations in other forms of energy are usually negligible. The Bernoulli equation is written as follows:

$$
\left(\frac{p_A}{\rho_e g} + \frac{v_A^2}{2g} + h_A\right) - \left(\frac{p_B}{\rho_e g} + \frac{v_B^2}{2g} + h_B\right) = \text{head losses (or energy losses)} \qquad (2.5)
$$

where $p/\rho_e g$ = pressure energy (m)
$v^2/2g$ = kinetic energy (m)
h = potential energy or elevation

These terms are often expressed as water height (m) or in kPa. The Bernoulli equation is more general than the Hazen-Williams equation, which is used when the kinetic energy difference between points *A* and *B* is negligible. However, the flow velocity often varies. Variation may occur, for example, in the cross-sectional area of flow, because of an additional inflow of water, or in the sewer pipe collecting wastewater from another sewer pipe. Figure 2.9 shows a pipe in which the velocity of water flow varies. It is important to note that the hydraulic grade line was not plotted for cases where head losses occurred at the exit of the upstream sewer manhole and at the entrance of the downstream sewer manhole.

2.1.1 Hydraulic Grade Line and Slope of Energy Grade Line

In an open channel–that is, a channel subjected to a pressure of 101,3 kPa, and in which the flowrate and cross-sectional area of flow are uniform–the energy grade line is parallel to the water surface and to the bottom of the channel. A sewer pipe that does not flow under pressure is also considered to be a channel subjected to a pressure of 101,3 kPa. Between two consecutive sewer manholes, the diameter, flowrate, and cross section of flow are usually uniform. In this case, the energy grade line, hydraulic grade line, the crown (top) and bottom of the pipe are parallel (figs. 2.2b and 2.3). In equations 2.1 to 2.5, the slope, *s*, is *always* the slope of the energy grade line. It is therefore a mistake to talk about the slope of the hydraulic grade line or the slope of the pipe, even though these slopes are often the same as the slope of the energy grade line.

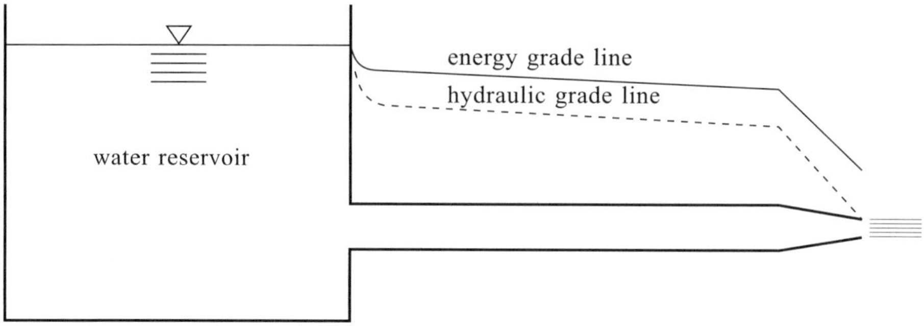

Figure 2.9 Pipe in which the water-flow velocity varies

2.2 HEAD LOSSES

2.2.1 Head Losses Due to Friction in Pipes

Head losses due to pipe friction are an important factor to take into account when designing sewer and water-distribution pipes. Minor head losses (e.g., head losses occurring when water enters or exits a structure) and head losses due to friction allow us to determine the minimum gradient of sewer pipes or the pressure drop in water-distribution pipes. The gradient of the sewer pipes ensures that sewage flows uniformly under gravity.

2.2.2 Head Losses Due to Various Structures

When the level of the hydraulic grade line is calculated, minor head losses occurring in structures (e.g., sewer manholes, valves), in alignment or in section variations of sewer pipes must be considered. Although such head losses have little influence when the gradient of the sewer pipe is high, they cannot be ignored when a pipe is long and has a large diameter, a very weak slope, and a relatively high water-flow velocity. For water-distribution networks, structural head losses must be considered to determine precisely the residual pressures.

2.2.3 Computation of Minor Head Losses (Pressure Pipes)

For appropriate sites, minor head losses are calculated using the following general equation:

$$H_m = K_L \frac{v^2}{2g} \tag{2.6}$$

where v = water-flow velocity of the studied site
K_L = coefficient relative to the structure

Table 2.2 shows the values of K_L for certain types of structures.

Table 2.2 Some values of coefficient K_L

Water Inlet	K_L	K_L	Water Outlet
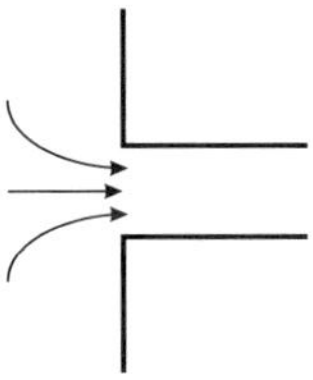	0,5	1,0	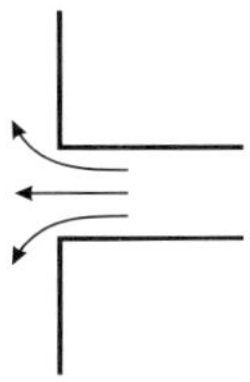
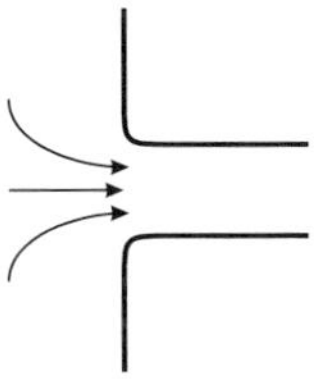	0,2	1,0	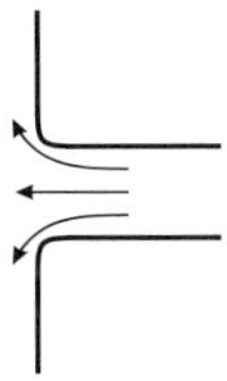
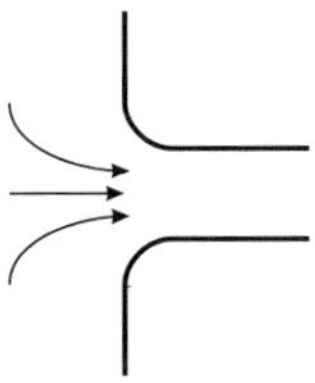	0,04	1,0	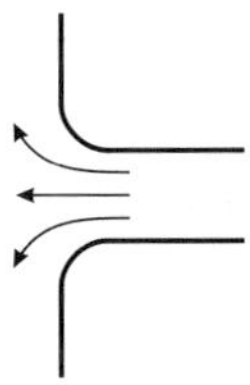

As a general rule, for a given pipe, minor head losses under 5% of total head losses are not taken into account, as calculations are accurate only within approximately 5%. For this reason, engineers do not consider minor head losses when the number of obstacles, n, in the pipe of length L satisfies the relation $n \leq L/1000\ d$.

2.2.4 Length of Equivalent Pipe (Pressure Pipes)

For studies such as the balancing of water-distribution networks, it is often useful to replace different structures with equivalent pipes. For example, a valve can be replaced by a pipe that has the same diameter as the pipe on which it was installed, and with a length inducing the same head losses as the valve itself.

Manufacturers often suggest equivalent lengths for their products, and these can also easily be obtained from the Darcy-Weisbach equation (equation 2.3); therefore:

$$H_m = f\frac{L_{equ}v^2}{d2g}$$

where H_m = minor head losses calculated using the appropriate K_L coefficient
L_{equ} = length of the equivalent pipe

From equation 2.6 :

$$K_L\frac{v^2}{2g} = f\frac{L_{equ}v^2}{d2g}$$

hence

$$L_{equ} = \frac{K_L d}{f}$$

A simple equation can also be used to calculate the length of the equivalent pipe. Let us modify the Hazen-Williams equation (equation 2.4) in the following manner:

$$Q = Av = 0{,}849C_{HW}AR_h^{0,63}s^{0,54}$$

In this equation, R_h is replaced by $d/4$, s by H_L/L_{equ}, H_L by $K_Lv^2/2g$, and it yields:

$$L_{equ} = 0{,}00773K_LQ^{0,148}C_{HW}^{1,852}d^{0,8703} \tag{2.7}$$

where L_{equ} = length of the equivalent pipe, of diameter d (m)
Q = flowrate (m^3/s)
d = diameter of the pipe (m)

Therefore, for a pipe of diameter d carrying a pressured flowrate Q, equation 2.7 allows us to calculate the length of a pipe with the same diameter that would have the same head loss as a structure or obstacle of known coefficient K_L.

EXAMPLE 2.4

A storm-water collecting sewer discharges its rainwater into a river through a sewer manhole and a reinforced-concrete outfall (Manning coefficient, n, of 0,013). To drain a new area, construction of a second collecting sewer is planned and the sewer manhole is to be rebuilt.

Figure 2.10 shows the submerged outfall. (The water then flows under pressure.) The outfall drains 6,58 m^3/s of rainwater in the river, through a drop manhole. The reinforced concrete outfall has a diameter of 1,5 m, a length of 18 m, and a slope of 1%. The level of the manhole bottom and the level of the sewers invert, where they are connected to the sewer manhole, are 26,6 m and 28,38 m, respectively. If the minor head losses are negligible, find the water level in the sewer hole.

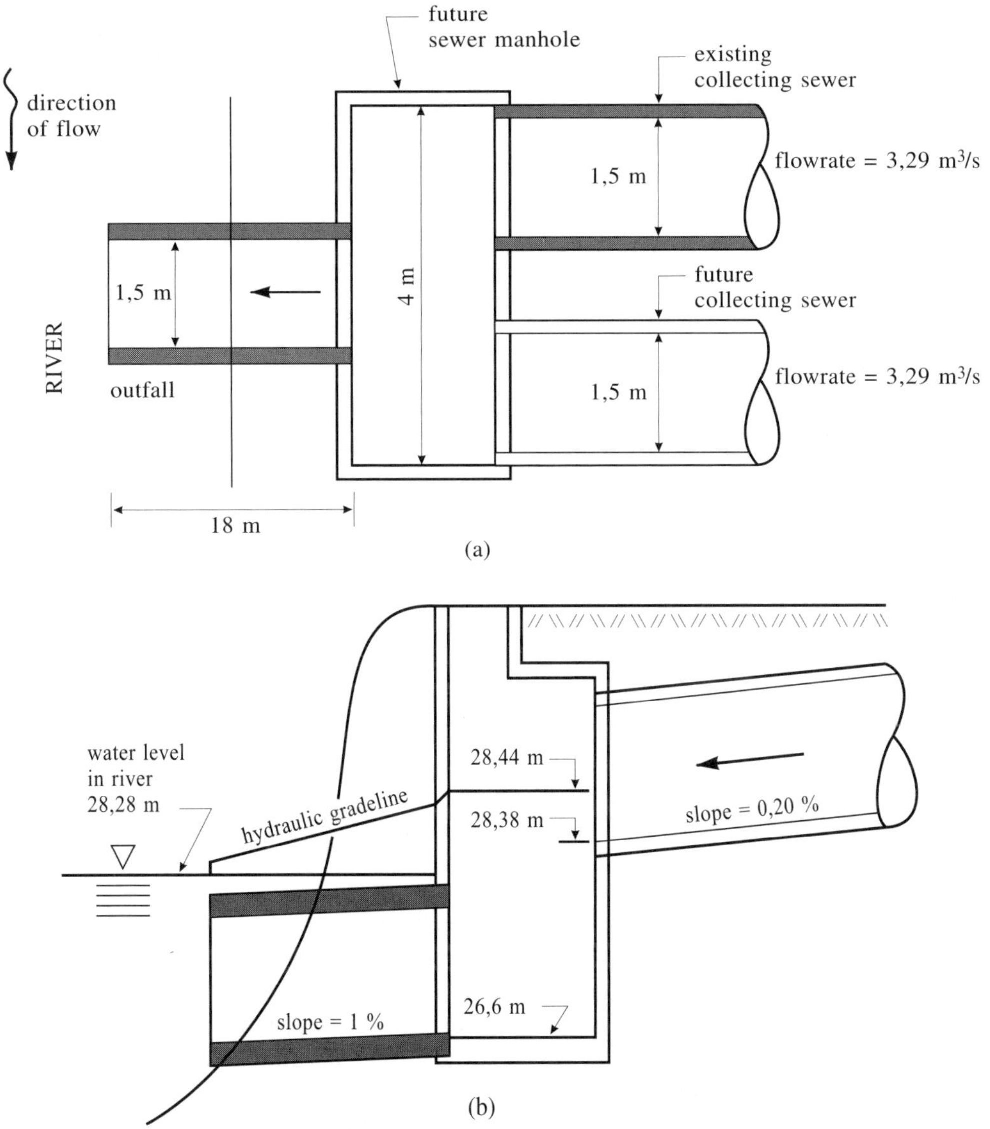

Figure 2.10 (example 2.4) Submerged sewer outfall: a) plan; b) elevation (+ = sum of minor losses at both ends of the outfall)

Solution

As the entire flow goes through the outfall, the flow velocity is calculated using the equation $Q = Av$; it yields:

$$v = \frac{Q}{A} = \frac{6{,}58}{\pi d^2/4} = \frac{6{,}58}{\pi(1{,}5)^2/4}$$
$$= 3{,}72 \text{ m/s}$$

Using the Manning equation (equation 2.2), the slope of the energy grade line is calculated. In this case, the slope is also the slope of the hydraulic grade line (the velocity of the water is the same throughout the outfall since the cross-sectional area does not change) but differs from the slope of the pipe. It then yields:

$$v = \frac{R_h^{2/3} s^{1/2}}{n}$$

therefore

$$s^{1/2} = \frac{vn}{R_h^{2/3}} = \frac{3{,}72 \times 0{,}013}{(1{,}5/4)^{2/3}} = 0{,}093$$

and

$$s = 0{,}0087 \text{ m/m}$$

As s represents the head loss per unit of length of the outfall, the total head loss is 0,16 m (0,0087 × 18). Therefore, the water level in the sewer hole is:

$$28{,}28 + 0{,}16 = 28{,}44 \text{ m}$$

If minor losses were taken into account, the hydraulic gradeline would drop at each end of the outfall as in figure 2.4.

For a zero water flow, the water level in the sewer manhole is the same as the water level in the river: 28,28 m.

2.3 COMPUTATION OF FLOWS WHEN PIPES ARE NOT FLOWING FULL

In general, a sewer pipe is designed to allow for the maximum projected flowrate when the pipe flows full. As the maximum flowrate is seldom reached, it is therefore important to study what happens in normal flow conditions and in conditions where flowrates are average or minimum. This is why, after designing a pipe in terms of a maximum flowrate, the engineer must ensure that minimum velocities of 0,6 to 0,9 m/s are maintained at all times (see section 8.4.8 of this book for minimum velocities). This verification is very important when designing large-diameter trunks and intercepting sewers or in the case of head pipes of networks where cleaning and maintenance rather than maximum flows dictate diameter choices (art. 8.4.2). In this latter case, the maximum flow is often much lower than the capacity of the pipe.

The hydraulic conditions in circular pipes not flowing full can be determined from the curves found in figure 2.11. As the Kutter roughness coefficient is practically identical to the Manning coefficient, these curves can be used as if they were plotted based on the Manning coefficient. Moreover, as these curves are almost completely unaffected by the variation in the Manning coefficient, they can be used for different values of this coefficient, especially those between 0,009 and 0,015 – more common in pipes for urban hydraulics. Finally, it has been observed that the Manning coefficient varies as a function of the height of the water in the pipes. This is usually not considered when calculating hydraulic conditions when sewer pipes are not flowing full.

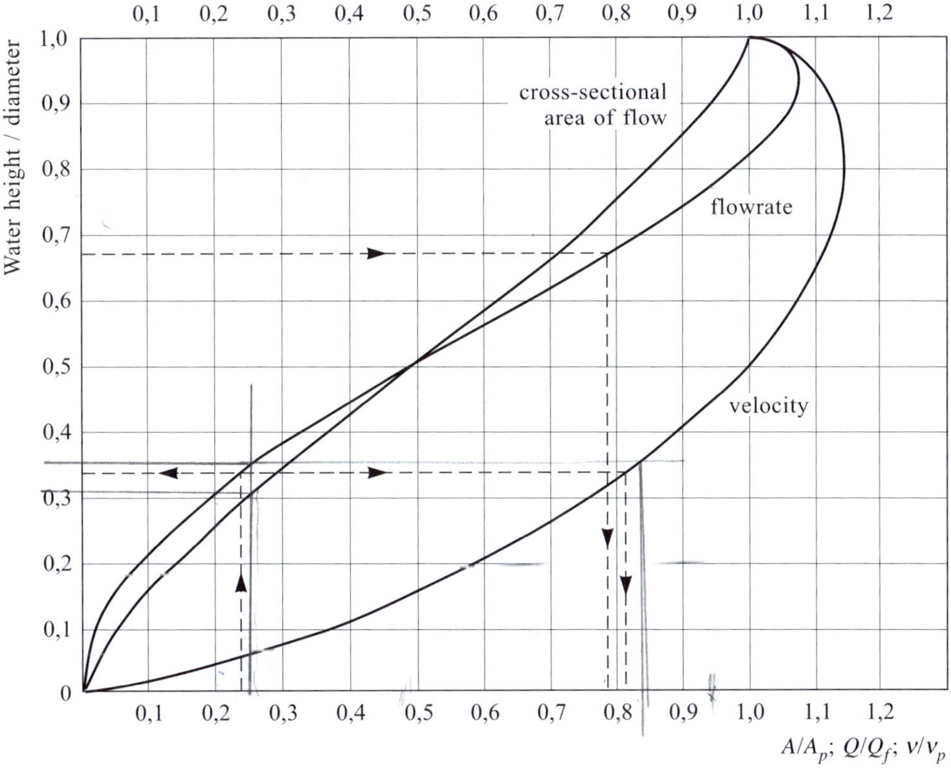

Figure 2.11 Flow conditions (uniform) in circular sewer pipes not flowing full (index f means "flowing full"); n (Kutter) = 0,015; the Manning coefficient, n, does not vary with the water height in the pipe

Adapted from: Swan and Horton, *Hydraulic Diagrams for the Discharge of Conduits and Canals* (McGraw-Hill, 1922), p. 37.

A
2xPIxD
Z/1000

EXAMPLE 2.5

In a circular sewer pipe of diameter d, the height of the water is $2d/3$ when the flowrate, $Q_{2/3}$, is 0,3 m^3/s. When the pipe carries the minimum flowrate, Q_{min}, of 0,09 m^3/s, the flow velocity, v_{min}, is 0,6 m/s (fig. 2.12). Assuming that the Manning roughness coefficient is 0,015, calculate the diameter, d, and the slope, s, of this pipe.

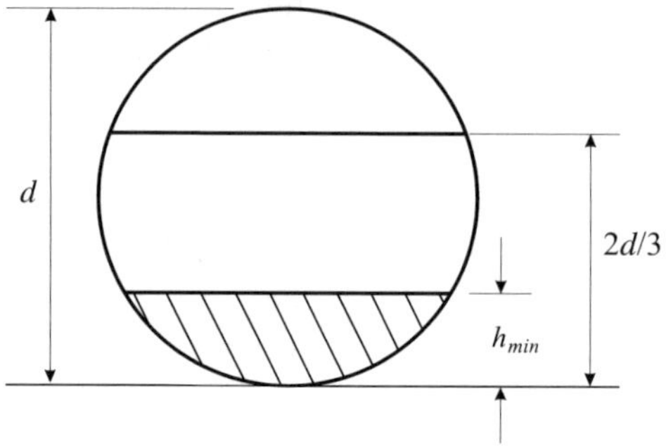

Figure 2.12 (example 2.5) Sewer pipe not flowing full

Solution

As stated in the question, when the flowrate is $Q_{2/3}$, we know that:

$$\frac{h}{d} = \frac{2}{3}$$

Using the curves in figure 2.11, we find:

$$\frac{Q_{2d/3}}{Q_f} = 0{,}79$$

hence

$$Q_f = \frac{Q_{2d/3}}{0{,}79} = \frac{0{,}3}{0{,}79}$$

$$= 0{,}38 \text{ m}^3/\text{s}$$

and

$$\frac{Q_{min}}{Q_f} = \frac{0{,}09}{0{,}38}$$

$$= 0{,}24$$

However, for $Q_{min}/Q_f = 0{,}24$, it yields (fig. 2.11):

$$\frac{h_{min}}{d} = 0{,}34$$

and

$$\frac{v_{min}}{v_f} = 0{,}82$$

where

$$v_f = \frac{0{,}60}{0{,}82}$$

$$= 0{,}73 \text{ m/s} \quad \text{(water velocity when the pipe flows full)}$$

As a result, since $Q_f = 0{,}38 \text{ m}^3/\text{s}$ and $v_p = 0{,}73$ m/s, it yields:

$$A = \frac{Q_f}{v_f}$$

$$= 0{,}52 \text{ m}^2$$

and

$$d = \left(\frac{4A}{\pi}\right)^{1/2}$$

$$= 0{,}81 \text{ m (810 mm)}$$

A choice has to be made between the two commercial diameters that are the closest to the calculated diameter, 750 and 900 mm. If a 750 mm diameter is chosen, we observe (fig. 2.5) that for the pipe to flow full when the flowrate is $0{,}38 \text{ m}^3/\text{s}$, the slope s must be 0,0015; the velocity is then 0,86 m/s. These last values should be reinserted into the above problem to solve for actual flow conditions such as h_{min} and v_{min}.

EXAMPLE 2.6

In a circular sewer pipe of diameter d, the height of the water is $3d/4$ ($0{,}75d$) when the flowrate, $Q_{0,75}$, is $0{,}14 \text{ m}^3/\text{s}$. When the pipe carries the minimum flowrate, Q_{min}, of $0{,}03 \text{ m}^3/\text{s}$, the flow velocity, v_{min}, is 0,60 m/s (fig. 2.13). Assuming that the Manning coefficient, n, is 0,015, calculate the diameter, d, the slope, s, and the hydraulic capacity, Q_f, of this pipe.

Solution

Using the curves found in figure 2.11, we find for $h/d = 0{,}75$:

$$\frac{Q_{0,75d}}{Q_f} = 0{,}915$$

hence

$$Q_f = \frac{0{,}14}{0{,}915}$$

$$= 0{,}153 \text{ m}^3/\text{s} \quad \text{(flowrate when pipe flows full)}$$

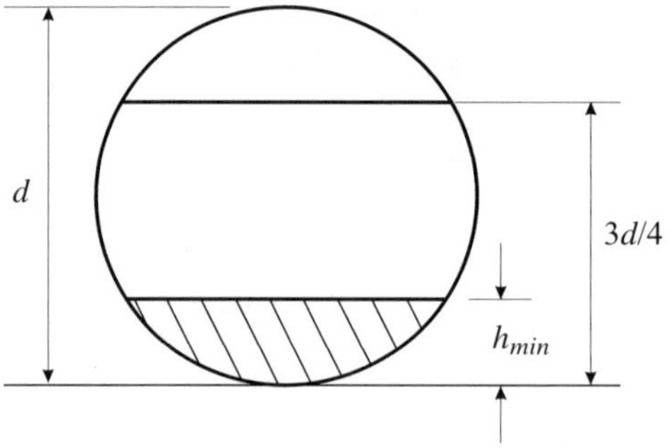

Figure 2.13 (example 2.6) Sewer pipe not flowing full

and

$$\frac{Q_{min}}{Q_f} = \frac{0{,}03}{0{,}153} = 0{,}20$$

For $Q_{min}/Q_f = 0{,}20$, we find (fig. 2.11):

$$\frac{h_{min}}{d} = 0{,}31$$

and, for $h_{min}/d = 0{,}31$, we have:

$$\frac{v_{min}}{v_f} = 0{,}77$$

However, $v_{min} = 0{,}60$ m/s; then:

$$v_f = \frac{v_{min}}{0{,}77} = \frac{0{,}60}{0{,}77} = 0{,}78 \text{ m/s}$$ (water velocity when pipe flows full)

Therefore:

$$A = \frac{Q_f}{v_f} = \frac{0{,}153}{0{,}78} = 0{,}20 \text{ m}^2$$

and

$$d = \left(\frac{4A}{\pi}\right)^{1/2} = 0{,}5 \text{ m (500 mm)}$$

There are no concrete pipes of 500 mm diameter currently available on the market. The choice is then between a pipe with a smaller diameter (455 mm) and one with a larger diameter (535 mm). In this example, d = 535 mm is chosen so that h_{min} = 0,31d = 163 mm. We then find (fig. 2.5) that for this pipe to flow full when the flowrate is 0,153 m^3/s, the slope, s, must be 0,0017; the velocity will then be 0,71 m/s. Here again, the problem should be reconsidered using these values in order to solve for actual flow conditions.

CONCLUSION

Most hydraulic computations are based on a few easy-to-use formulas. In urban hydraulics, the main problems to be solved are summarized as follows.

Sewer pipes. In sewer pipes, the water flow is uniform (the longitudinal water velocity is the same anywhere in the pipe) but not permanent (the velocity varies with time). To obtain the flow velocity, the Manning equation (equation 2.2) is used:

$$v = \frac{R_h^{2/3} s^{1/2}}{n}$$

In this equation, the slope of the energy grade line, s, is equal to the slope of the pipe itself and to the slope of the hydraulic grade line, whether the pipe flows full or not. In other words, these slopes are equal when the hydraulic grade line is located on or below the crown of the pipe – when the hydraulic pressure is 101,3 kPa (1 atmosphere).

When the water flows at a pressure higher than 101,3 kPa, the slope, s, is the slope of the energy grade line, which is the same as the slope of the hydraulic grade line if the flow is uniform. However, it is not equal to the slope of the pipe, except in special cases.

Water-distribution pipes. In water-distribution pipes, the flow is uniform but not permanent. The pressure exerted on the water is greater than 101,3 kPa. To calculate head losses, the Hazen-Williams equation is used (equation 2.4):

$$v = 0{,}849 C_{HW} R_h^{0,63} s^{0,54}$$

In this equation, the slope of the energy grade line, s, is also equal to the slope of the hydraulic grade line. When the flow velocity ($v = Q/A$) is known, the value of s for a given pipe corresponds to energy losses due to friction (head losses) per unit of pipe length.

More complex hydraulic problems. To solve these problems, the Bernoulli equation (equation 2.5) is used. It is more comprehensive and states that for two given points, A and B, of a hydraulic system, the difference between the sum of the various forms of energy represents the sum of the various head losses, therefore:

$$\left(\frac{p_A}{\rho_e g} + \frac{v_A^2}{2g} + h_A\right) - \left(\frac{p_B}{\rho_e g} + \frac{v_B^2}{2g} + h_B\right) = \text{(head losses or energy losses)}$$

Minor head losses. When water enters or exits a structure (reservoir, sewer manhole, valve, etc.) or clears an obstacle, the water sustains a head loss. When the water flow is turbulent, the value of the head loss is proportional to the square of the flow velocity (equation 2.6); in other words:

$$H_m = K_L \frac{v^2}{2g}$$

The value of K_L is a function of the nature of the obstacle and the characteristics of the surfaces over which the water flows.

Length of equivalent pipe (pressure pipes). It is often useful to replace certain structures by an equivalent pipe lengths. To determine this replacement, equation 2.7 is used:

$$L_{equ} = 0{,}0077 K_L Q^{0,148} C_{HW}^{1,852} d^{0,87}$$

For a pipe of diameter d carrying a pressured flowrate Q, this equation predicts the length of a pipe, of identical diameter, capable of producing the same head loss caused by a structure or obstacle of known coefficient K_L.

Inside diameter of pipes. In practice, when studying the flows in pipes, it is important to remember that the diameter used for computations is the inside diameter of the pipe and not the nominal or commercial diameter (see chapter 8).

PROBLEMS

2.1 For a circular concrete sewer pipe with an actual (inside) diameter of 380 mm, the Manning coefficient is 0,015. Calculate the slope of this pipe if the desired minimum flow velocity is 0,75 m/s when the pipe is flowing full.

2.2 A circular sewer pipe, with an actual (inside) diameter of 455 mm, is designed to carry a maximum flowrate of 0,175 m^3/s when the pipe is flowing full. Find the slope of this pipe. Calculate the depth and velocity of the water flow when the flowrate is only 1/5 of the maximum flowrate (the Manning coefficient is 0,013).

2.3 A city wants to supply 0,250 m^3/s of water to another city using a distribution pipe 3000 m long. The Hazen-Williams coefficient of the pipe is 130. Calculate the minimum diameter of this pipe if the total head losses due to friction cannot exceed 150 kPa.

2.4 In a pressure pipe with an actual (inside) diameter of 255 mm, the pressure at point A is 345 kPa. The height of the pipe at point A is 30 m, and head losses due to friction between points A and B are 80 kPa. If the height of the pipe at point B is 18 m, calculate the pressure at that point.

2.5 Two sewer manholes are linked by a concrete pipe measuring 100 m in length and 610 mm inside diameter. The slope of the pipe is 0,00075 m/m. The upstream invert of this pipe is at an elevation of 98 m and the ground surface is at 101 m. The Manning coefficient n for concrete is 0,013.

a) Find the capacity of this pipe (in m^3/s). What is the water velocity?

b) When the water levels in the upstream and downstream sewer manholes are at 100 and 99 m, respectively, find the flowrate (in m^3/s) and the water velocity (in m/s) through this pipe. Minor head losses may be ignored.

c) If the water level in the downstream sewer manhole is controlled and kept at 99 m, what is the highest flowrate (in m^3/s) that can be carried through this pipe without having an overflow at the upstream sewer manhole? What is the water velocity?

d) Calculate the flowrate (in m^3/s) in the pipe if the water level is at mid-diameter and the water flow is uniform.

Chapter 3

VOLUMES AND FLOWS OF DRINKING WATER

INTRODUCTION

In general, volumes and flows of drinking water vary with local conditions and the lifestyles of inhabitants. For a given community, *total water consumption* includes water distributed for domestic use, water supplied to businesses and industries, and water used for public services (street cleaning and fire fighting).

In this chapter, we discuss how to evaluate the total annual water consumption of a typical American city. From average flows and peak factors (FP), we also demonstrate how to evaluate the monthly, weekly, daily, and hourly maximum and minimum flows. These flows are instrumental in the design of water-treatment plants and water-distribution networks. They are also used to evaluate the performance of these systems and networks under various conditions.

3.1 DEFINITIONS

In order to understand the issues and problems related to the production and distribution of water for consumption, it is important to know the vocabulary used by water-management professionals. These professionals are responsible for meeting the needs of all consumers, who, in turn, are billed at various rates for these services. This is why water distributors keep a close tab on different water volumes supplied for domestic, commercial, and industrial uses in their annual budgets.

Drinking water. This is the total water supplied by municipal water-treatment plants mainly for domestic consumption, but also to businesses, shops, and some industries, for fire fighting, street cleaning, and other purposes. (Industries with their own water supply are not included in this category.)

Total consumption per person. This is the ratio of the total daily volume of water produced to the number of inhabitants. This consumption is often expressed in litres per person per day [L/(person·d)]:

$$\text{total consumption per person} = \frac{\text{total daily production}}{\text{number of inhabitants}}$$

Often, large consumers, such as industrial companies, are excluded from this equation.

Domestic consumption. This is all water supplied for residential use, including personal hygiene, laundry, lawn watering, and the filling of swimming pools. To this consumption is often added the water supplied to small businesses (convenience stores, hardware stores, gas stations, churches, etc.). Domestic consumption may vary widely depending on the region or country. Often, public consumption is included with domestic consumption.

Commercial consumption. This is the water supplied to stores and shopping centres.

Industrial consumption. This is the water supplied to industries excluding those with their own water supply.

Public consumption. This is the water supplied to various municipal services: fire fighting, street cleaning, public buildings, public pools, parks, sewage treatment plants, and so on. Cities do not collect taxes for this water use.

Water losses. Water losses are due to flaws in the distribution network. Losses occurring outside of city limits are attributed to the respective type of consumption, depending on the zones where the losses occurred. If they occurred on an industrial property, they are considered industrial consumption; if they occurred on a store property, they are considered commercial consumption, etc.

Total consumption. This is the sum of all of the different types of consumption, including water losses.

Equivalent population. It is often useful to replace a large water consumer, such as an industry, by an equivalent population. To this end, water consumption of this consumer, usually for a single day, is divided by the total daily consumption per person.

3.2 AVERAGE AND SUGGESTED WATER CONSUMPTION

Water-consumption volumes, generally expressed as volumes per day, vary considerably depending on the region or country. The following information will enable the reader to compare those volume variations as well as suggested and actual consumption.

Total consumption per person in Quebec municipalities. In Quebec, probably because of wasteful habits and important losses in the distribution networks, total consumption per person is very high. In cities, it varies from 300 to 820 L/(person·d).

In 1974, a committee (AQTE-MISE-À-JOUR) set up by Quebec's Association of Water Technologies (Association québécoise des techniques de l'eau) found that total consumption per person was 554 L/(person·d) for cities where a water meter was installed in each household. For cities without meters, this consumption rose to 820 L/(person·d).

Domestic consumption. Domestic consumption varies according to whether a country is developing or industrialized and whether an area is urban or rural.

Domestic consumption in developing countries. Although there is a lack of data for developing countries, it is believed that the domestic consumption is only a few L/(person·d), especially in regions where water is transported home on foot. According to the World Health Organization (*Public Standpost Water Supplies, a Design Manual*, Technical Paper Series, no. 14 [1979], p. 16), water consumption at public fountains in small cities and villages varies from 20 to 40 L/(person·d).

Domestic consumption in rural North America. For rural areas, the US Joint Committee on Rural Sanitation (*Water Supply of Rural Zoning and Small Communities* [Geneva: World Health Organization, 1961], p. 48) noted the following consumption rates:

- home with a manual water pump: 38 L/(person·d);
- home with tap in the kitchen sink: 57 L/(person·d);
- home with running water (hot and cold), kitchen, laundry room, shower, and toilet: 190 L/(person·d).

Domestic consumption in urban North America. In North America, urban domestic consumption varies from 90 to 330 L/(person·d), with an average of about 200 L/(person·d). These numbers represent between 30% and 40% of total urban consumption. Here, it is important to remember that industrial and commercial consumption and water losses added to domestic consumption form total consumption.

Domestic consumption is highest in large cities, where it approaches a saturation limit. In residential suburbs, overuse of lawn watering is a waste of water, and cities must impose by-laws to limit this nonessential use.

Commercial consumption. To assess existing networks, commercial consumption rates of every establishment are measured with meters. When new networks are planned, water quantities for future businesses are estimated using the commercial consumption of existing businesses. Many authors suggest the following commercial consumption rates for places of business: 225 L/(employee·d) or 2,5 to 6,1 L/(m^2·d); for office buildings, 40 to 55 L/(employee·d).

Industrial consumption. Water consumption in industrial zones varies considerably depending on the types of industries present. This type of consumption is much harder to evaluate than domestic consumption. Therefore, it is easier to make a survey of water consumption for each industrial user. When data are not available (or if the industry is not yet operational), an approximation is obtained by multiplying an evaluation of the flowrate per unit of production (table 3.1) by the current and future expected volumes of production for each industry.

Table 3.1 Water consumption according to type of industry

Industry	Consumption (L/1000 L or L/1000 kg)
Canning	
Beans (L/1000 kg)	60 000–100 000
Fruits and vegetables (L/1000 kg)	7 000– 35 000
Meat (L/1000 kg live animals)	40 000– 80 000
Chemicals	
Gasoline (L/1000 L)	25 000–125 000
Oil refinery (L/1000 L)	1 000– 6 000
Soap factory (L/1000 kg)	1 000– 2 500
Steel plant (L/1000 kg)	40 000– 50 000
Food and beverages	
Bakery (L/1000 kg)	2 000– 4 500
Brewery (L/1000 L)	10 000– 20 000
Dairy products (L/1000 kg)	15 000– 25 000
Distillery (L/1000 L)	5 000– 10 000
Slaughterhouse (L/1000 kg live animals)	15 000– 25 000
Sugar refinery (L/1000 kg)	6 000– 25 000
Pulp and paper	
Paper (L/1000 kg)	10 000–100 000
Pulp (L/1000 kg)	150 000–450 000
Tannery (L/1000 kg finished product)	1 000– 5 000
Textiles	
Cotton (L/1000 kg)	80 000–300 000
Dyeing (L/1000 L or L/1000 kg)	600 000–900 000
Wool (L/1000 kg)	40 000– 80 000

Adapted in part from: McGauhey, P.H., *Engineering Management Quality* (New York: McGraw-Hill, 1968, 295 p.).

Special consumption. To estimate and calculate the flowrates and volumes of water for special cases, water-requirement data for different users can be used. Table 3.2 lists water consumption for different users, and table 3.3 gives a general overview of water quantities required for some plumbing apparatuses.

Table 3.2 Water consumption for different users

Establishment - Purpose	Consumption (L/person·d) or L/(Unit·d)
Airport (per passenger)	8–12
Cinema or theatre	
Drive-in (per car)	20
Two shows per day (per seat)	10–15
Health etablishments	
General hospital (per bed)	600
General hospital (per employees)	40
Hotels and boarding houses	
Hotel room (per guest)	200–400
Motel with kitchen	400–600
Room and board (per boarder)	200
Laundromat (per washing machine)	1000–3000
Office building (per employees)	60
Residence	
Apartment building (per person)	200–250
Private house (per person)	200–300
Restaurant	
Bar (per seat)	38
Medium-sized restaurant (per seat)	150
Medium-sized restaurant; open 24 h (per seat)	200
Tavern (per seat)	80
School (per student)	
Boarding school	300
With cafeteria	40–60
With cafeteria, showers, and gymnasium	60–80
Service station	
Per automobile served	40
Per pair of pumps	2000
Shopping centre	
Per employee	40
Per parking space	5
Small and medium-sized business	
No cafeteria or showers (per employee)	70
No cafeteria, with showers (per employee)	140
Store	
Per customer	8
Per employee	40
Vacation Centre	
Campgrounds (per vacationer)	100
Day camps, without meals (per vacationer)	65
Holiday camps; common toilets and baths (per vacationer)	160–200
Labour camp (per worker)	140–200
Recreational-vehicule camps; private toilets and baths (per unit: 2,5 persons)	500–600

Adapted in part from: Metcalf and Eddy Inc., *Wastewater Engineering : Treatment, Disposal and Reuse* (New York: McGraw-Hill, 1979, 920 p.).

Table 3.3 Water quantities required for different types of plumbing apparatuses

Plumbing Apparatus	Quantity of water required
Bathtub	90–110 L/use
Continuous-flow drinking fountain	4–5 L/min
Fire hose (38 mm; nozzle of 13 mm; 20 m of head)	140–160 L/min
Garden hose (16 mm, 8 m of head)	10–12 L/min
Garden hose (19 mm, 8 m of head)	16–20 L/min
Home-type automatic washing machine	130–200 L/use
Home-type dishwasher	15–30 L/min
Lawn sprinkler	6–8 L/min
Shower*	0,2–0,5 L/s
Toilet (flush)	90–110 L/min
Toilet (tank)	13–25 L

* The lowest values represent the most recent models on the market.

Adapted from: Metcalf and Eddy, *Wastewater Engineering: Treatment, Disposal, Reuse* (New York: McGraw-Hill, 1979), p. 17 and "Preparing Collection Systems for Water Conservation," *Water Environment Technology*, vol. 5, no. 8 (Aug. 1993): 54.

Public consumption. Public consumption varies between 45 and 68 L/(person·d), which is about 10% of average total consumption.

Losses. Losses are on the order of 20% of average total consumption. This proportion is often higher in cities where the distribution network is old or in poor condition.

Total consumption. This is the sum of all the different types of consumption, including water losses. In American cities, total consumption is around 565 L/(person·d). In Canada, because of slightly lower domestic consumption and, more importantly, much lower industrial consumption, the total consumption is about 500 L/(person·d).

In European and developing countries, water consumption is much lower than in the United States and Canada. This phenomenon is attributed to the following factors:

- scarcity of water;
- lower percentage of homes using specialized equipment requiring water use;
- more rational use of water by consumers.

In the largest cities of the world, total consumption varies greatly, from as little as 132 L/(person·d) to as much as 675 L/(person·d).

3.3 WATER CONSUMPTION FOR DIFFERENT PURPOSES

In a city, water is used for domestic, industrial, commercial, and public purposes. In Canadian and American cities, if water losses are taken into account, we obtain the distribution presented in table 3.4.

Table 3.5 shows the water-consumption distribution for Quebec municipalities. This consumption was measured by AQTE-MISE-À-JOUR in 1974. Table 3.5 also offers a comparison for the city of Sainte-Foy. These data were determined by Demard in 1976 ("La demande en eau des résidences: le mythe des 100 gallons," *Eau du Québec*, vol. 9, no. 4 [1976]: 16–21).

Table 3.4 Distribution of total water consumption for typical Canadian and American cities

Consumption	**Average consumption**		
	Variation [L/(person·d)]	**Average [L/(person·d)]**	**Percentage calculated from averages (%)**
Domestic	150–480	220	36,7
Industrial and commercial	40–400	260	43,3
Losses	40–160	90	15,0
Public	20–80	30	5,0
Total		600	100,0

Adapted from: Tchobanoglous and Schroeder, *Water Quality* (Don Mills, ON: Addison Wesley, 1985), p. 7.

Table 3.5 Distribution of total water consumption in Quebec municipalities

Consumption	**AQTE***				**Sainte-Foy****			
	Without meter [L/(person·d) and %]		**With meter [L/(person·d) and %]**		**Without meter [L/(person·d) and %]**		**With meter [L/(person·d) and %]**	
Commercial	68	8,2	46	8,1	68	8,2	46	8,1
Domestic	473	57,1	318	57,0	291	35,1	195	34,9
Industrial	100	12,1	68	12,1	100	12,1	68	12,1
Losses	164	19,8	109	19,5	346	41,8	232	41,6
Public	23	2,8	18	3,3	23	2,8	18	3,3
Total	828	100,0	559	100,0	828	100,0	559	100,0

* AQTE-MISE-À-JOUR committee, «Rapport final sur l'inventaire de la qualité des services d'aqueduc et d'égout des municipalités du Québec», *Eau du Québec*, vol. 7, no. 3, 1974, p. 9-16 (in French).

** Demard, H., «La demande en eau des résidences: le mythe des 100 gallons», *Eau du Québec*, vol. 9, no. 4, 1976, p. 16-21 (in French).

3.4 FACTORS AFFECTING CONSUMPTION FLOWRATES

Many factor influence water-consumption volumes and flowrates. The main ones are the distribution network, the size of the community, the number of industries, the quality of drinking water, the water pressure at the consumer's tap, the installation of water meters, the population characteristics, the climate, and the rate of human activity.

Distribution network. The quality of the distribution network influences the volume of water losses, and thus affects a city's water-consumption volume.

Size of the community. Larger cities have a higher water consumption rate because water is used for a wider range of purposes (hotels attract tourists, businesses bring workers from satellite cities and suburbs; etc.).

Number of industries. Some industries use great quantities of water to produce their goods. Fortunately, new pollution laws urge industries to recover matter contained in their effluents and recycle used water, which tends to reduce industrial consumption.

Quality of drinking water. Water consumption tends to increase when the physical qualities of drinking water (colour, turbidity, and taste) are improved.

Water pressure. When water pressure is too high, the network sustains more damage, which increases water losses.

Installation of water meters. The installation of water meters is generally followed by consumers being billed for the volume of water used. Consumers are therefore urged to eliminate waste by reducing unnecessary practices (lawn watering, car washing, etc.).Table 3.6 shows the influence of water-meter installation in some Quebec municipalities. Brooks and Peters (*La gestion de la demande d'eau: les options qui s'offrent aux Canadiens* [Conseil des sciences du Canada, 1988], p. 22) state that in the urban sector of the United States, "the installation of water meters, without any increase in the price, brought permanent reductions in the water consumption in the order of 10 to 40 percent." Moreover, municipalities are urged to reduce water losses by promptly replacing broken pipes, repairing flawed equipment, or installing water meters.

Table 3.6 Influence of meter installation on water consumption in Quebec municipalities

Network	Consumption [L/(person·d)]		
	min.	avg.	max.
without meters	375	550	950
with meters	275	380	470
consumption reduction (%)	27	31	50

Population characteristics. Water consumption also varies with certain population characteristics, such as culture, standards of living, and professional activities.

Climate. The climate, and specially the temperature, influences water consumption. When it is hot outside, the use of swimming pools and lawn watering increases water consumption. During the winter months, however, some consumers let water run from the tap to avoid freezing of pipes that are badly insulated or too close to the ground surface. This practice also increases water consumption.

Rate of human activity. Water consumption fluctuates according to the month of the year, the day of the week, and the time of day. These fluctuations reflect the rate of human activity.

In some communities, for example, water consumption increases on Monday morning, as many people do their laundry on that day. In other communities, water consumption is much higher on the weekend because of an influx of tourists for cultural or sport events. Water consumption decreases during major national or international events (sport competitions, political speeches, famous trials, etc.) as a large number of people follow the coverage on television. Water consumption is high in the early morning, when many people are washing and showering, and in the early evening, when people are cooking supper. Finally, in the suburbs, water consumption is higher in the spring, because of the filling of swimming pools, and in summer, because of lawn watering (when this activity is authorized).

3.5 COMPUTATION OF MAXIMUM FLOWS (PEAK FLOWS) AND MINIMUM FLOWS

Water consumption is not constant; as we have seen, it fluctuates depending on the time of day, day of the week, the week of the month, and month of the year. For certain periods, water flows to be supplied are very far from the suggested averages. Therefore, equipment must be designed to meet both maximum and minimum consumption rates. For example, during periods of maximum flow, *head losses* must not be too high and *residual pressures* must be sufficient. Moreover, the pumping stations, treatment plants and reservoirs capacities must be appropriate; during periods of minimum flow, reservoirs must be filled in order to have sufficient quantities of water for supply during the periods of maximum consumption.

The following section discusses information and means available to engineers to compute total peak flows.

Goodrich empirical formula. This formula enables small residential municipalities to calculate the percentage by which the average flow of a certain period t (days) of a year must be multiplied to obtain the maximum flow of an identical period t of that year. The values for t may vary from 2/24 to 365 d. The formula is written as follows:

$$p = 180t^{-0,10}$$

where p = maximum percentage (%)
t = period studied (d)

Therefore, for t = 1, p = 180%: the highest daily flowrate during the year is 1,8 times the average daily flowrate for that year. However, during the year, the highest weekly flowrate (t = 7) is 1,48 times the average weekly flowrate for that year. Finally, the highest monthly flowrate (t = 30) is 1,28 times the average monthly flowrate for that year. The Goodrich formula cannot be used if commercial or industrial flowrates are significant or if the network sustains important water losses.

EXAMPLE 3.1

The average daily flowrate of a city is 51,5 L/s (4 449 600 L/d or 4 450 m^3/d). Calculate the maximum weekly flowrate for that year.

Solution

According to the Goodrich equation, the peak factor for a seven-day period (one week) is as follows:

$$\begin{aligned} p &= 180t^{-0,10} \\ &= 180\,(7)^{-0,10} \\ &= 148\ \% \end{aligned}$$

The maximum weekly flowrate, $(Q_{week})_{max}$, is therefore:

$$\begin{aligned} (Q_{week})_{max} &= 51{,}5 \times 1{,}48 \\ &= 76{,}3\ \text{L/s}\ (6593\ \text{m}^3/\text{d})\ \text{(this flowrate will apply for 1 week)} \end{aligned}$$

Most of the time, maximum and minimum flowrates are calculated by multiplying the values of the average flowrates (annual, monthly, weekly, and daily) by *peak factors* (FP), which are based on many years of observations. In general, the peak factor is the value by which the average flowrate is multiplied in order to obtain the maximum or minimum flowrate.

Monthly peak flowrates. In the northern hemisphere, maximum consumption occurs in July and August, mainly because of the use of swimming pools, lawn watering, and air conditioning. In some cold-climate communities where pipes are not buried deep enough, water consumption is high in December and January because some consumers let the water run to avoid freezing of pipes.

Weekly peak flowrates. During the year, there are seven consecutive days during which consumption is at its highest point. This week usually occurs during the month of maximum consumption, but there are exceptions to this rule.

Daily peak flowrates, minimum daily flowrates. Monday is generally the day when consumption is at its maximum. It is traditionally the day for laundry and also when most people return to work after the weekend. The day of maximum consumption usually occurs during the week of maximum consumption, but, this is not necessarily the case. Minimum consumption occurs on Sunday, as many activities cease for that day.

Hourly peak flowrates, minimum hourly flowrates. Consumption also varies depending on the time of day. It is generally higher than average between 7:00 a.m. and 8:00 p.m., and lower than average between 10:00 p.m. and 7:00 a.m. The minimum flowrate usually occurs between 3:00 and 4:00 a.m., when it reaches only 40% to 50% of the average daily flowrate. The maximum flowrate most often occurs in the evening, between 6:00 p.m. and 7:00 p.m., though some cities reach their hourly peak flowrate in the morning. In fact, the curve representing the consumption variation as a function of time is unique for each city, and reflects inhabitants' life styles.

Figures 3.1 and 3.2 show hourly variations in water consumption during the day of maximum consumption for two cities in the Montreal region. In general, the maximum hourly consumption in a year occurs on the day in that year when daily consumption is maximum.

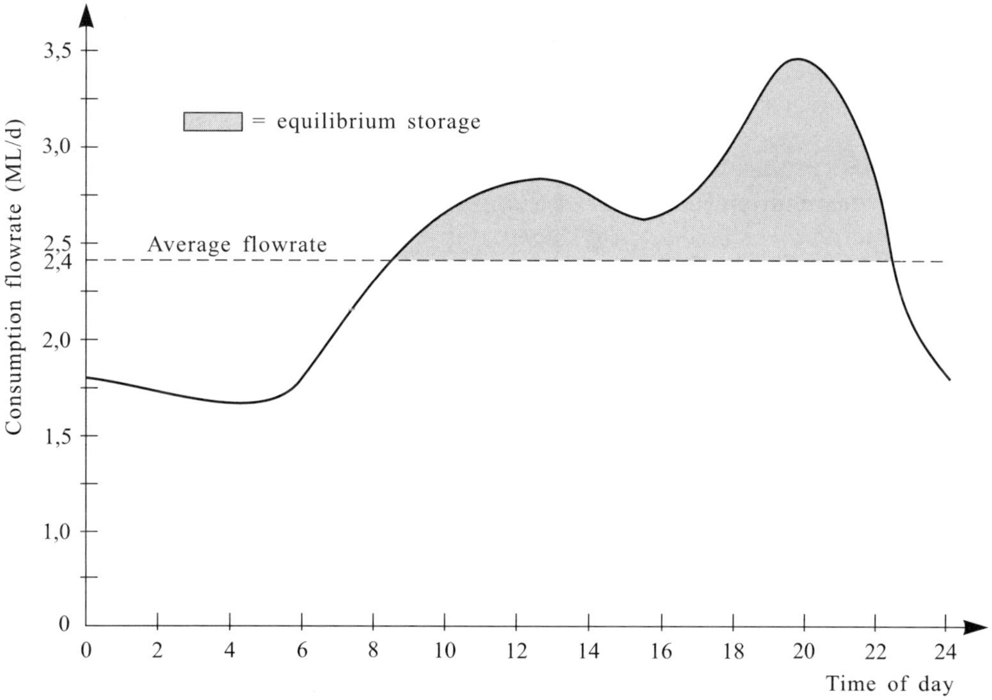

Figure 3.1 Water-consumption variation during the day of maximum consumption for a city in the Montreal region

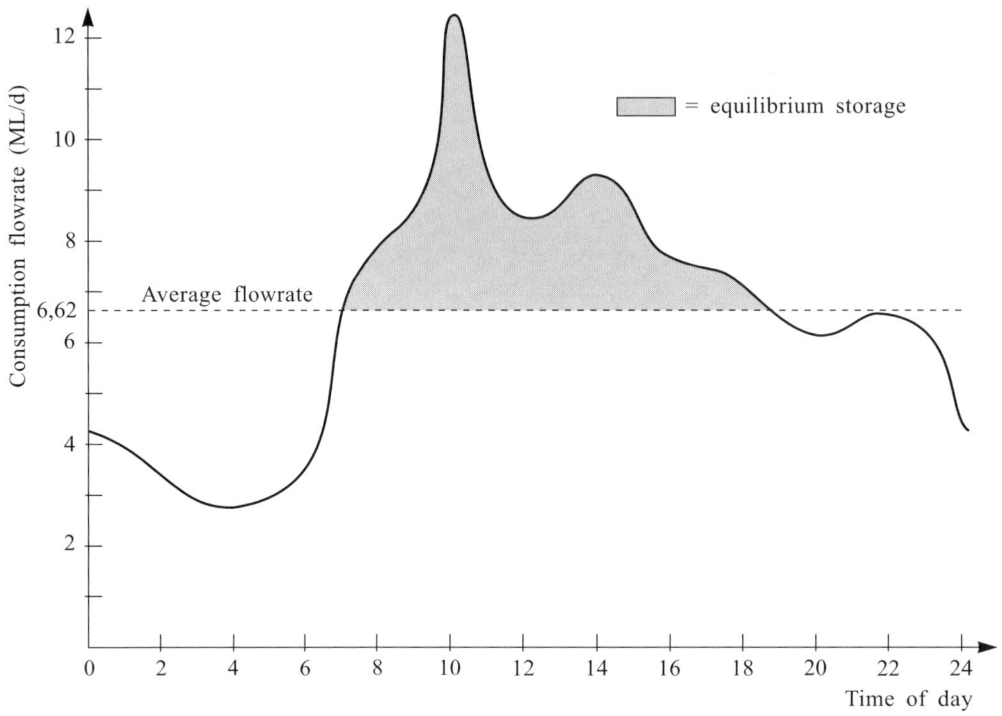

Figure 3.2 Water-consumption variation during the day of maximum consumption for a city in the Montreal region

Peak factors (FP) are recommended for calculation of maximum and minimum domestic consumption flowrates. In North America, years of observations have led engineers to discover precise peak factors for calculating maximum and minimum domestic consumption flowrates.

Maximum monthly consumption. Average daily consumption during the month of the year when monthly consumption is at its highest is about 130% of average daily consumption during the year $(Q_d)_{avg}$. The monthly peak factor to be used is thus 1,3.

Maximum daily consumption, $(Q_d)_{max}$. The consumption during the day of the year when consumption is at its highest varies between 150% and 300% of average daily consumption for the year $(Q_d)_{avg}$ $(1{,}5 \leq \text{PF} \leq 3{,}0)$, depending on the size of the community.

Maximum hourly consumption in a day. Consumption during the hour of the day when hourly consumption is at its highest is 150% of average hourly consumption for that day. The peak factor is thus 1,5.

Maximum hourly consumption for the year, $(Q_h)_{max}$. The highest hourly consumption during the year (probably during the day of the year on which daily consumption is at its highest) varies between 225% and 450% of average hourly consumption for the year. The peak factor is obtained by multiplying the maximum daily peak factor by the hourly peak factor of a day; the maximum peak factor is then 4,50 (3,0 × 1,5).

Minimum hourly consumption of the year, $(Q_h)_{\min}$. The lowest hourly consumption of the year varies between 40% and 80% of the average hourly consumption for the year ($0,4 \le PF \le 0,8$), depending on the size of the community.

EXAMPLE 3.2

In a Quebec residential municipality with 10 000 inhabitants, find the average daily consumption, $(Q_d)_{avg}$, the daily peak flowrate, $(Q_d)_{max}$, the hourly peak flowrate, $(Q_h)_{max}$, and the minimum hourly flowrate, $(Q_h)_{min}$, assuming that every home has a water meter and that there is no major leakage from the network. The following peak factors are suggested: maximum daily = 2,00; maximum hourly = 3,00; minimum hourly = 0,5.

Solution

As there are no major industries or businesses, we assume that total consumption is 445 L/(person·d) (table 3.5), which includes public and domestic consumption and water losses in the network. It then yields:

$$
\begin{aligned}
(Q_d)_{avg} &= \frac{445 \times 10\ 000}{24 \times 60 \times 60} = 51{,}5 \text{ L/s} \\
(Q_d)_{max} &= 2{,}0(Q_d)_{avg} = 103 \text{ L/s (this flowrate will apply for 24 h)} \\
(Q_h)_{max} &= 3{,}0(Q_d)_{avg} = 154{,}5 \text{ L/s (this flowrate will apply for 1 h)} \\
(Q_h)_{min} &= 0{,}5(Q_d)_{avg} = 25{,}75 \text{ L/s (this flowrate will apply for 1 h)}
\end{aligned}
$$

EXAMPLE 3.3

In a Quebec municipality with businesses and industries, there are 10 000 inhabitants. Assuming that water meters are not present in any of the homes, find the average daily consumption, $(Q_d)_{avg}$, the daily peak flowrate, $(Q_d)_{max}$, the hourly peak flowrate, $(Q_h)_{max}$, and the minimum hourly flowrate, $(Q_h)_{min}$. The peak factors are the same as those suggested in example 3.2.

Solution

In this case, we assume that the total consumption is 828 L/(person·d) (table 3.5). It then yields:

$$
\begin{aligned}
(Q_d)_{avg} &= \frac{828 \times 10\ 000}{24 \times 60 \times 60} = 96 \text{ L/s} \\
(Q_d)_{max} &= 2{,}0(Q_d)_{avg} = 192 \text{ L/s (this flowrate will apply for 24 h)} \\
(Q_h)_{max} &= 3{,}0(Q_d)_{avg} = 288 \text{ L/s (this flowrate will apply for 1 h)} \\
(Q_h)_{min} &= 0{,}5(Q_d)_{avg} = 48 \text{ L/s (this flowrate will apply for 1 h)}
\end{aligned}
$$

Domestic consumption alone (473 L/person·d) equals:

$$\frac{473 \times 10\ 000}{24 \times 60 \times 60} = 55\ \text{L/s}$$

This consumption represents 57% of the average flowrate; the remainder, 43%, is attributed to industries, businesses, public utilities, and water losses in the distribution network.

Water-waste reduction. Some commonly used equipment can be replaced to reduce water consumption. Table 3.7 shows the quantities of water that can be saved with some of this new equipment.

Table 3.7 Water conservation due to use of modern equipment

Equipment	Water-waste reduction [L/(person·d)]
Shower (11 L/min)	27
Shower (2 L/min)	52
Toilet (13 L tank)	30
Toilet (2 L tank)	74
Dishwasher (water saver)	3,8
Washing machine (water saver)	6,4

Adapted from: United States Department of Housing and Urban Development, Office of Policy Development and Research, Building Technology Division, *Residential Water Conservation Projects*, "Summary Report" (June, 1984), pp. 1-7, in D.B. Brooks and R. Peters, *La gestion de la demande en eau: les options qui s'offrent aux Canadiens* (Conseil des sciences du Canada, 1988), p. 13.

Flowrates and volumes of water needed for fire fighting. Normally, a city must be able to supply water for fire fighting. The quantity of water required is relatively low compared to that for other public uses. However, since this quantity is used during a very short period, the flowrates are quite high.

Flowrates and water volumes are determined using the formulas suggested by the Fire Underwriters Survey (see appendix F). If the standards set by this organization are followed, citizens pay lower fire-insurance premiums.

Design of reservoirs and water-distribution networks. Reservoirs and distribution networks are designed to supply consumers with sufficient quantities of water and water pressure:

- during the day of maximum consumption, $(Q_d)_{max}$, when one or many fires are burning;
- during the hour of maximum consumption, $(Q_h)_{max}$.

Variation of the peak factor. Peak factors are higher for small communities and lower for larger communities. Table 3.8 shows peak factors suggested by the Ontario Environment Ministry for communities with populations of less than 500 to those with populations of more than 150 000 inhabitants.

Table 3.8 Peak factor for water-consumption flowrates

Population	**Peak factors**		
	maximum daily of the year	maximum hourly of the year	minimum hourly of the year
under 500	3,00	4,50	0,40
500 – 1 000	2,75	4,13	0,40
1 001 – 2 000	2,50	3,75	0,45
2 001 – 3 000	2,25	3,38	0,45
3 001 – 10 000	2,00	3,00	0,50
10 001 – 25 000	1,90	2,85	0,60
25 001 – 50 000	1,80	2,70	0,65
50 001 – 75 000	1,75	2,62	0,65
75 001 – 150 000	1,65	2,48	0,70
over 150 000	1,50	2,25	0,70

From: Ontario Environment Ministry, *Guidelines for the Design of Water Storage Facilities, Water Distribution Systems, Sanitary Sewage Systems and Storm Sewers* (May 1979), p. A-1.

CONCLUSION

Before building the infrastructures needed to supply water to a community–water intakes, collection pipes, water-treatment plants, distribution networks, water reservoirs, and so on–current and future water consumption (domestic, commercial, industrial) must be evaluated for that community. The consumption variations must also be known.

During the year, the maximum daily domestic consumption, $(Q_d)_{max}$, varies from 150% to 300% of average daily domestic consumption, $(Q_d)_{avg}$. The maximum hourly domestic consumption, $(Q_h)_{max}$, varies from 225% to 450% of the average hourly domestic consumption for the year, which is considered equal to the average daily domestic consumption of the year divided by 24.

Industrial and commercial consumption must be measured on site, if the businesses and industries are already operational. If this is not the case, the consumption rates of similar types of existing businesses and industries must be used.

Fire-fighting facilities must also be available. To design these facilities, we assume that fires will occur on the day of maximum daily consumption.

PROBLEMS

3.1 In a Quebec municipality of 10 000 inhabitants, average total water consumption is 500 L/(person·d). Calculate, in L/d, L/h, and L/s, the flowrate of the maximum hourly consumption of the year. Assume that every home is equipped with a water meter.

3.2 A suburb in the province of Quebec meets the drinking-water needs of its inhabitants and of two industries. Industry A is a cannery that uses an average flowrate of 1000 m^3/d, seven days a week, from June 1 to September 15. Industry B is a wood-transformation plant that uses, on average, 2000 m^3/d, seven days a week, from mid-March to the beginning of September. Every home is equipped with a water meter, and average total daily consumption of the municipality (except for the consumption of the two industries) is 4000 m^3/d. The municipality would like to know if it can supply a third industry, which would use water on weekends. Therefore, peak flowrates for 48-hour periods must be calculated using the Goodrich equation. Let us assume that municipal water consumption is the same on week days and weekends. Assuming that for a 48-hour period, the peak factors for water consumption for industries A and B are 2,2 and 1,5, respectively, calculate (in m^3/d) the maximum quantity of water that the city must supply:

a) during weekends when both industries are operational;
b) during weekends when both industries are not operational;
c) during weekends when only industry B is not operational;
d) during weekends when only industry A is not operational.

3.3 City A has a population of 10 000 and supplies water for its citizens and for the 2000 inhabitants of neighbouring City B. From the water-treatment plant, a main pipe crosses City A and brings water to the limits of City B. This pipe is City B's sole water supply. Total average consumption is 400 L/(person·d), including 10% in water losses (average consumption + losses = total consumption). Let us assume that 20% of the water losses of City A occur on the main pipe, which stretches from the treatment plant to City B, and that these losses are split between the two cities as a function of their respective total consumption rates. Calculate the annual volume of water that City A must produce to meet the needs of City B. What is the annual amount that City A must charge City B if the cost for production and transportation of water is $1 per 4000 L?

Chapter

4

COLLECTION AND DISTRIBUTION OF DRINKING WATER

INTRODUCTION

In Quebec, as well as in many regions of North America and the world, drinking water is usually taken from surface water (lakes or rivers). This water is collected using a water intake and an intake pipe which transport water to a reservoir (raw-water reservoir) from which water is constantly drawn and sent to a treatment plant. When surface water is not available or when it is too polluted, water is drawn from the water table through wells.

This chapter discusses the means to transport water from its source to the consumer's faucet. It will not, however, address water treatments necessary to produce drinking water.

4.1 SURFACE-WATER INTAKE

A surface-water intake (fig. 4.1) is a structure that collects natural water (raw water) from the lake or river in which it is submerged. It must be installed where the raw water is of the best possible quality–that is, as far as possible from any source of pollution. Therefore, the intake must not be located downstream from a sewage outfall, even if the discharged effluent originates from a sewage treatment plant.

We evaluate the flow of a waterway–including maximum flows and, more important, the lowest flow of the dry season (minimum flows)–to determine whether the water available is sufficient for our needs and to design the volume of a future-flow equalization reservoir when hydrological surveys prove this necessary. Water characteristics vary considerably with flow variations due, for example to major precipitations or snow melting. It is also important to know the physico-chemical and bacteriological characteristics of the raw water at different times of year. This knowledge enables the engineer to determine if the water source is adequate, to plan for the treatments needed at the treatment station, and, depending on the flow variations, to determine the level at which the water intake should be installed. (Similar studies must be undertaken when raw water is drawn from a lake.)

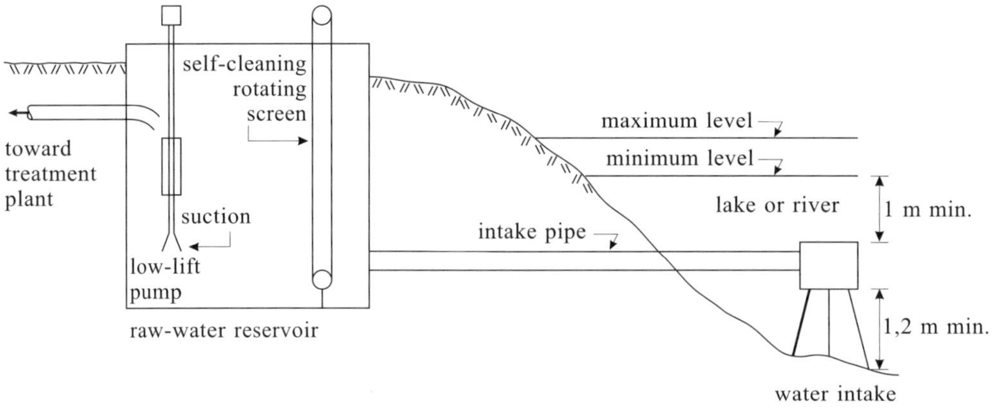

Figure 4.1 Surface-water intake, intake pipe, and raw-water reservoir

4.1.1 River Intake

When drawing water from a river, a water intake must not be installed before assessing the water-flow regime and flowrates, including maximum and minimum flows of the year.

In general, the water intake is located at a distance from the river bank to avoid stagnant and often polluted waters. In areas where water freezes, it must be ensured that frazil ice does not block the intake or the intake pipe. Actually, in regions where water is subjected to very strong turbulence, the ice sheet does not form at the surface; ice made of ice crystals, needles, or flakes mixes with the liquid mass and forms a slush. This slush, which easily sticks to grids and pipe walls, may block the flow of water to the raw-water reservoir. To avoid obstruction of the water intake by frazil, several preventive measures can be undertaken:

- The water intake can be installed in calm waters (not in rapids or immediately downstream from rapids), where a layer of ice forms during winter.
- The water intake must not be installed immediately downstream from rapids, even if a layer of ice is present, because frazil ice, which is carried by the current, can be pushed under the ice cover for a certain distance, depending on the water velocity.
- The water intake can be equipped with heating elements that warm the submerged surfaces, and therefore avoid ice built-up. Hot water can also be used to melt the ice on the surfaces.
- Air bubbles can be blown around the water intake to push ice pellets to the surface, where a sheet of ice will form.
- Cold water can be mixed with warmer water (e.g., from groundwater), which will melt ice pellets in the intake pipe.
- A very large water intake can be installed to reduce the risk of ice completely blocking the intake.
- Certain hydraulic structures (dam, breakwater, etc.) can be installed to reduce water-flow velocity in the intake zone, creating zones where water is calmer and

reducing the risk of frazil ice formation; however, the engineer must ensure that the presence of these structures does not alter the quality of the raw water.
- The intake pipe can be installed below the bottom of the river, where heat transfer from the ground helps reduce the risk of frazil ice built-up.

4.1.2 Lake Intake

Lakes behave like sedimentation tanks. Their water is often of better quality than river water, and variations in water quality are less frequent and less important than variations in river water. In deep lakes, however water quality can vary considerably according to the depth. In summer, the first metre of water, right below the surface, contains algae that cause problems at the treatment plant, such as producing a bad taste or clogging up filters. Near the bottom, the water is cold, and when dissolved oxygen is absent, certain metals in the sludge (formed by deposits of different types of matter) dissolve and water treatment may consequently be more difficult. In deep lakes, thermal stratification produces seasonal overturns: when this phenomenon occurs, usually in the fall, the water at the bottom, which is warmer and lighter, is drawn to the surface, carrying polluted sludge from the bottom. For all of these reasons, we recommend that the water intake be built so that water can be drawn at different depths.

4.2 GROUNDWATER INTAKE

When the quality and quantity of surface water are not adequate, groundwater can be used. The quantity of groundwater is evaluated by drilling and drawing flowrates equivalent to the ones needed. Then, close monitoring is kept on either the lowering of the water-table level or decreases in static pressure, depending on whether the water is at atmospheric pressure or in a confined environment. Groundwater is usually clear, but its physico-chemical characteristics vary widely depending on the location. In each case, a detailed study must be undertaken to determine the appropriate treatment for transforming groundwater into drinking water.

4.3 HYDRAULIC AND SANITARY SURVEY

When a site for the water intake is identified, a complete hydraulic and sanitary survey must be performed to learn about the quality and quantity of water available for every season of the year. When possible, the future risk of contamination must also be evaluated. To accomplish this, the following steps are undertaken:

- The sources of pollution are identified in the area: sanitary outfall, farming, pastures, septic tanks, industries, and so on.
- All data related to water management are noted: minimum and maximum flowrates, minimum water depth, minimum and maximum water levels, direction of currents, wind directions, thermal stratification of water, and so on.
- In winter, the thickness of the ice layer and the probability of frazil ice formation are evaluated.

– The raw water is sampled for one year. This sampling is used to learn about the average and extreme values of the bacteriological and physico-chemical characteristics. After that period, the engineer knows if the raw water is adequate and can determine the required treatments.

4.4 ESTIMATION OF ICE THICKNESS ON SURFACE WATER

To evaluate the average ice thickness on a river or a lake, we use simple weather data such as air temperature and the duration of the freezing period. Other factors are also taken into account, such as water depth, water turbulence, wind velocity, presence of snow, thermal conductivity and heat capacity of water and ice, and latent heat of fusion of the ice. In his 1964 book on glaciology, Lliboutry suggests that the following equation be used to obtain a preliminary evaluation of the ice thickness:

$$d = 3{,}6A\sqrt{G} \qquad (4.1)$$

where d = average thickness of the ice layer (cm)
A = coefficient that takes into account the thermal characteristics of the ice, the presence of snow, the amount of sunlight, etc.
G = freezing index in number of negative-degree-days (°C–days).

The freezing index (G) of a year, at a specific site, gives an indication of the freezing intensity and duration at this site. Its computation is made taking into account the average daily temperature of the air 1,5 m above ground when this temperature is below 0 °C, as well as the number of days during which this situation occurs. More specifically freezing index (G) is the sum of the number of degrees below 0 °C, calculated on a daily basis for the entire freezing season. For example, the contribution to the value of the freezing index of three days with average air temperature of -12, -10 and -9 °C respectively is 31 °C–days. For more detail on the computation of freezing index, see Andersland and Ledenyi (1994).

Table 4.1 gives the different values of coefficient A suggested by Robert (1986).

Table 4.2 lists the freezing indices for many regions of Quebec. Environment Canada has published a document called *Climatic Norms in Canada*, that contains the freezing indices for every region of Canada for the years 1951–1980.

Table 4.1 Values of coefficient A (equation 4.1)

River or lake	A
Lake without snow	0,8
Lake covered with snow	0,5 - 0,7
River covered with snow	0,4 - 0,5
Small stream with rapid flow	0,2 - 0,4

Table 4.2 Freezing indices for many regions of Quebec

Municipality	Freezing index					
	(F° - day)			(C° - day)		
	max.	min.	avg.	max.	min.	avg.
Chicoutimi	3143	1708	2377	1746	949	1321
Coaticook	2331	1253	1885	1295	696	1047
Drummondville	2466	1242	1750	1370	690	972
Fort George (James Bay)	5237	4662	4718	2909	2590	2621
Gaspé	2441	1598	2008	1356	888	1116
Granby	1588	1060	1344	882	589	747
Huntingdon (U.S. border)	1996	1094	1702	1109	608	946
Joliette	2015	1370	1754	1119	761	974
Kénogami	3429	1974	2665	1905	1097	1481
Mont-Laurier	2933	1892	2292	1629	1051	1273
Montreal (Dorval)	2063	983	1518	1146	546	843
Montreal (McGill U.)	1854	903	1315	1030	502	731
Nicolet	2564	1296	1864	1424	720	1036
Noranda	3146	2398	2816	1748	1332	1564
Québec	2544	1324	1832	1413	736	1018
Rimouski	2421	1255	1888	1345	697	1049
Roberval	3327	2049	2663	1848	1138	1479
Sainte-Agathe-des-Monts	2547	1709	2170	1415	949	1206
Saint-Hyacinthe	2314	1134	1688	1286	630	938
Saint-Jérôme	2899	1393	2069	1611	774	1149
Shawinigan Falls	2831	1573	2074	1573	874	1152
Sherbrooke	1828	1079	1466	1016	599	814
Tadoussac	2439	1241	1762	1355	689	979
Témiscaminguc	2441	1661	2059	1356	923	1144
Thetford Mines	2671	1630	2051	1484	906	1139
Val d'Or	3895	2228	3016	2164	1238	1676
Valleyfield	2107	910	1505	1171	506	836
Victoriaville	2560	1394	1901	1422	774	1056

Adapted from: Canada Department of Transport, Air Services, *Engineering Planning, Design and Construction Manual,* Section 1, "Design and Evaluation of Flexible and Rigid Pavements" (1969), pp. 4-8.

N.B. For design purposes, Andersland and Ledenyi suggest the use of the average of the three largest annual values (coldest years) of the index during the last 30 years.

In the Montreal region, the maximum thickness of the ice layer of a lake that is covered with snow is thus:

$$d = 3{,}6\ (0{,}6)\ \sqrt{1146} = 73 \text{ cm}$$

4.5 WATER-INTAKE DESIGN

In its simplest form, the water intake is shaped like a parallelepiped. Of the six surfaces, the ones facing the current and the bottom of the lake or stream are made of a solid material to avoid entrance of debris and various materials. Water can penetrate only from the other 4 faces. The water intake must constantly be covered with at least one metre of water (if not, the top must also be closed to avoid floating debris being drawn into the intake pipe), and the bottom of the intake should be located at least 1,2 m (preferably 1,8 m) from the bottom of the lake or stream. The intake faces, where the water enters, are covered with a grid made of 2,0-2,5 cm diameter crossbars, forming square openings 10 to 20 cm in size. The water velocity through the grid openings must not exceed 15 cm/s, to keep suspended and floating matter, as well as ice, from being drawn into the intake pipe or to avoid the formation of a suction vortex over the water intake. The design is made so that this velocity will be 15 cm/s at the end of the design period for the water intake–that is, 30 to 50 years after its construction.

4.6 INTAKE PIPE DESIGN

The intake pipe links the water intake to the raw-water reservoir. This reservoir is a basin usually built on the bank of the lake or stream, and its water is pumped (low-lift pumps) to the treatment plant. The water flows naturally, without pumping, from the water intake to the raw-water reservoir. Water is drawn by gravity only: the principle of connected reservoirs applies here. To design the intake pipe, the engineer uses the maximum daily flowrate at the end of the design period and the minimum daily flowrate at the beginning of the period. This design period is the same for the water intake. The diameter of the intake pipe must enable the flow velocities to remain, at all times, within the limits set by table 4.3.

Table 4.3 Flow velocities of water in intake pipes

Criterion	Velocities (m/s)	
	min.	max.
Objective	0,9	1,22
Limit	0,6	1,85

If the flow velocities are too low, deposits (mostly sand) may accumulate in the intake pipe and clog it. On the other hand, flow velocities that are too high may cause premature aging of the pipe and lead to excessive head losses due to friction. Head losses may require additional excavation and involve extra pumping costs. Because of these constraints, two intake pipes are often built side by side, to be used separately or together. This way, flow velocities can be maintained between the maximum and minimum velocities (table 4.3), at all times, for the duration of the design period. The slope of the pipe must also be constant to avoid sedimentation at low points and air bubbles entrapping at high points.

4.7 RAW-WATER RESERVOIR DESIGN

The intake pipe leads to the raw-water reservoir (usually through the side of the reservoir, not the bottom), where the raw water is screened before being pumped to the treatment plant. The raw-water reservoir is also a flow-regulating basin that feeds the treatment plant with a constant flow via low-lift pumps.

The depth of the raw-water reservoir depends on the maximum and minimum water levels of the river or lake where the water intake is installed. It also depends on head losses in the water intake and intake pipe, and on the space required to install the low-lift pumps. The depth of the raw-water reservoir is usually greater than 2,5 m. To plan for maintenance of the raw-water reservoir without hindering the operations of the treatment plant, it is advisable to divide the reservoir into two parts that can be operated independently. Moreover, to limit turbulence, the cross section of the well's vertical surface that receives the intake pipe should be at least four to five times larger than the pipe's cross-sectional area.

4.8 SCREENS AT THE RAW-WATER RESERVOIR ENTRANCE

The screens installed at the entrance to the raw-water reservoir filter some suspended solids. The screen openings vary from 0,6 to 1,0 cm in size. The accumulation of suspended solids on the screens causes an increase in head losses, which then leads to a larger difference between the screens' downstream and upstream water levels. When head losses are too high, the screens must be cleaned. In small reservoirs, the screens are simply removed and cleaned by hand. In large reservoirs, rotating self-cleaning screens are installed; the cleaning is then done automatically and often continuously. The water velocity through the screen openings should be less than 60 cm/s in the case of a rotating screen and less than 18 cm/s in the case of a fixed screen.

4.9 LOW-LIFT PUMPS

Vertical turbine motor pumps are widely used to pump water from the raw-water reservoir to the treatment plant. Electrical motors are installed on the concrete slab covering the raw-water reservoir. The pumps are designed according to the water level in the treatment units (flocculators, sedimentation basins, and filters). The maximum flowrate of the pumped water is equal to the maximum daily consumption (generally at the end of the design period).

EXAMPLE 4.1

Using the following data related to a raw water reservoir (fig. 4.2), calculate:

a) the level of the concrete slab covering the raw-water reservoir;
b) the total head losses in the intake pipe;
c) the minimum water level in the reservoir;
d) the level of the concrete slab located at the bottom of the reservoir.

Design period (life of equipment):	30 years
Average daily flowrate in 30 years:	20 000 m^3/d
Peak factor related to the maximum daily flowrate, $(Q_d)_{max}/(Q_d)_{avg}$:	1,6
Peak factor related to the maximum hourly flowrate, $(Q_h)_{max}/(Q_h)_{avg}$:	2,65
Maximum water level of the river:	54 m
Minimum water level of the river:	53 m
Inside diameter of the intake pipe:	0,51 m
Hazen-Williams coefficient of the pipe:	130
Length of the concrete pressure intake pipe*: (45,8 m must be added to the length of this pipe to take minor head losses into account in the computation of total head losses)	300 m
Space between the maximum water level and the concrete slab located on top of the raw-water reservoir:	0,3 m
Minimum free space separating the bottom of the well and the suction (water entrance) of the low-lift pumps:	0,3 m
Minimum height of water in the raw water reservoir over the suction of the pumps:	1,2 m

Solution

When the low-lift pumps are turned off, the water level of the raw-water reservoir becomes the same as the water level of the river, therefore, at its maximum: 54 m.

a) Level of the concrete slab covering the raw water reservoir

$$54 + 0{,}3 = 54{,}3 \text{ m}$$

b) Total head losses (minor head losses and head losses due to friction)

– *Average design flowrate = maximum daily flowrate at the end of the design period*

$$\left(Q_d\right)_{max} = 20\,000 \times 1{,}6$$

$$= 32\,000 \text{ m}^3/\text{d} \ (370 \text{ L/s})$$

* Concrete pressure pipes are steel pipes covered with concrete on the inside and outside.

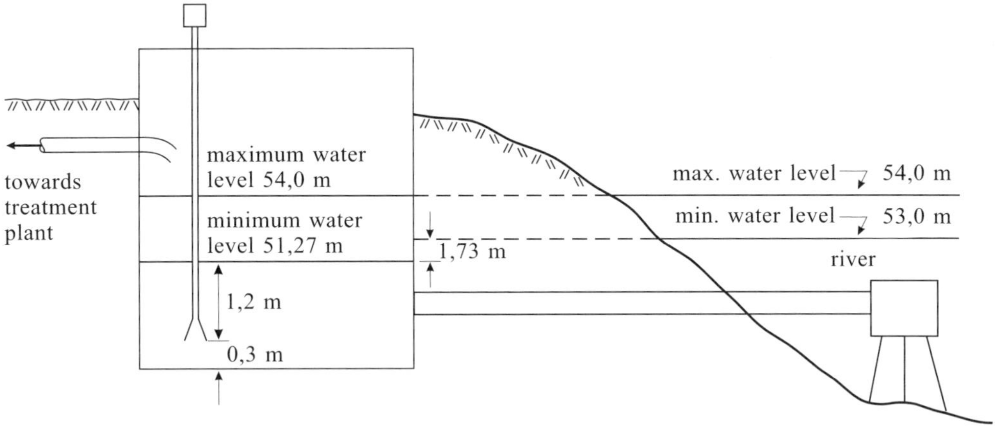

Figure 4.2 (example 4.1) Raw-water reservoir, intake pipe, and water intake

– *Total head losses*

On the Hazen-Williams nomograph (fig. 2.7), for C_{HW} = 130, Q = 370 L/s and d = 510 mm, we find s = 5/1,000; it then yields:

$$H_{tot} = 5/1000 \times 345{,}80$$
$$= 1{,}73 \text{ m}$$

c) **Minimum water level in the raw-water reservoir**

This level is the difference between the minimum water level in the river and the head losses when the flowrate is maximum; therefore:

$$53{,}00 - 1{,}73 = 51{,}27 \text{ m}$$

d) **Level of the concrete slab at the bottom of the raw-water reservoir**

This level is obtained by taking into account the minimum water level in the reservoir and the installation requirements regarding the suction of the pumps with respect to the water level and the level of the concrete slab at the bottom of the reservoir. It yields:

$$\text{level of the bottom concrete slab} = 51{,}27 - (0{,}3 + 1{,}2)$$
$$= 49{,}77 \text{ m}$$

4.10 WATER-TREATMENT PLANT AND DISTRIBUTION RESERVOIR

Treatment requirements vary depending on the quality of the water source. For example, for groundwater of relatively good quality, simple disinfection is sufficient to produce drinking water that meets current standards. However, lake or river water often requires extensive treatment: coagulation-flocculation, sedimentation, filtration, and disinfection.

The different units of the treatment plant are designed to produce water at the rate of the maximum daily consumption of the last year of the design period. This flowrate varies from 150% to 300% of the average daily flowrate in that year (see chapter 3). Looking closely at consumption variations during the day when the maximum daily consumption occurs, we note that instantaneous consumption for many hours (about 14 hours in the example shown in figure 4.3) is greater than the average consumption for that day. During these hours, the water requirements are met by drawing water from the clear-water reservoirs (also called the service or distribution reservoirs) one of them being located at the treatment plant.

The distribution reservoir is therefore a basin of clear (treated) water that fills up during daily low-consumption periods and empties during daily high-consumption periods. This type of reservoir ensures a better and more constant use of the treatment plant while providing high quality drinking water at a relatively low cost.

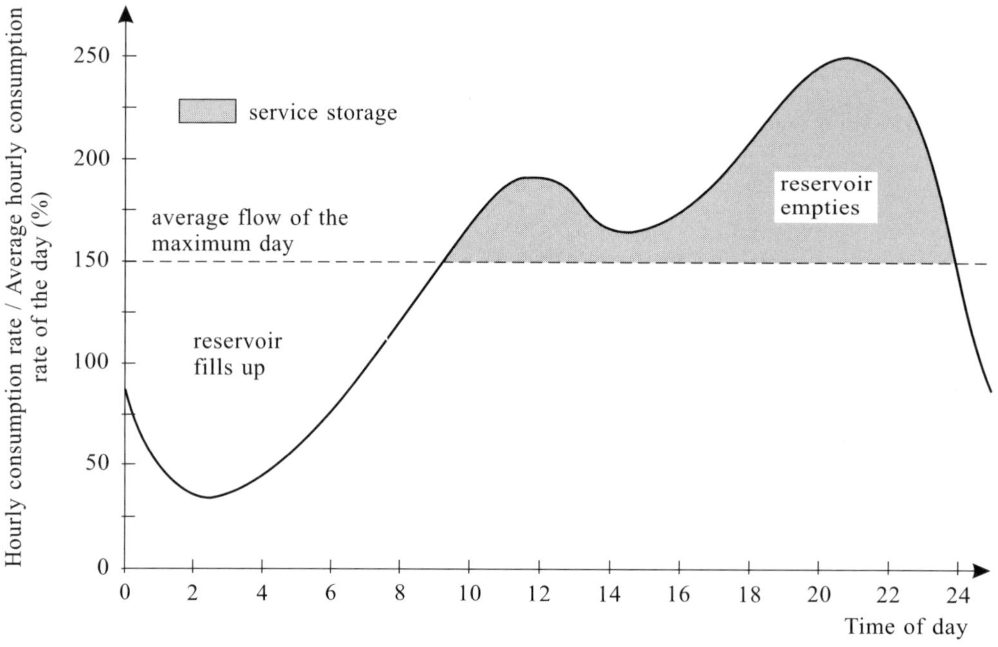

Figure 4.3 Hydrogram of maximum daily consumption

4.11 HIGH-LIFT PUMPING STATION

Water is supplied to the distribution network through high-lift pumping stations. These pumps, which are integrated into the station, draw water from the distribution reservoir. Unlike a treatment plant, a pumping station must meet an instantaneous demand. The pressures used at pumping stations depend on the topography of the area, water demand, the structure of the distribution network, and the diameter of the pipes. In fairly flat terrain, pressures used vary between 300 and 600 kPa (30 and 60 m of water).

To ensure adequate pressure at consumers' faucets, other high-lift pumping stations may be integrated into the distribution network. The pumps can be installed in the network or be coupled with reservoirs such as elevated water reservoirs (figure 4.4 where the numbers written along the pipes are their diameters.). Depending on the topography of the area, the pressure may be too high when the water reaches the consumer (for example, for people living at the bottom of a valley or at the foot of a hill). This problem can be solved by installing a pressure-reducing valve at the appropriate location on the network.

When the topography of a region shows strong differences in elevation, it may be necessary to create independent pressure zones. In other words, pressure-reducing valves are installed between the zones if the pressure needs to be reduced (for water coming from a higher zone), and booster pumping stations are installed to increase water pressure (for water coming from a lower zone). This technique would create different pressure levels, and therefore the pressure would never be too high or too low (figure 4.5 shows different layouts for pressure levels).

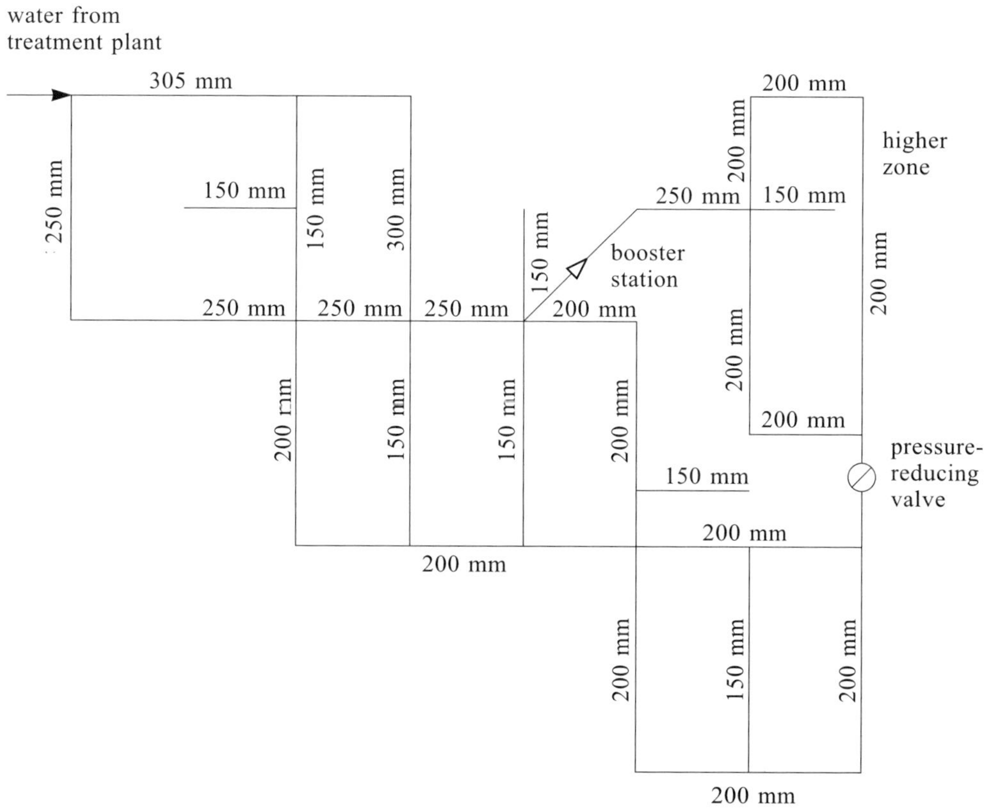

Figure 4.4 Network with a booster pumping station and a pressure-reducing valve (diameters are nominal)

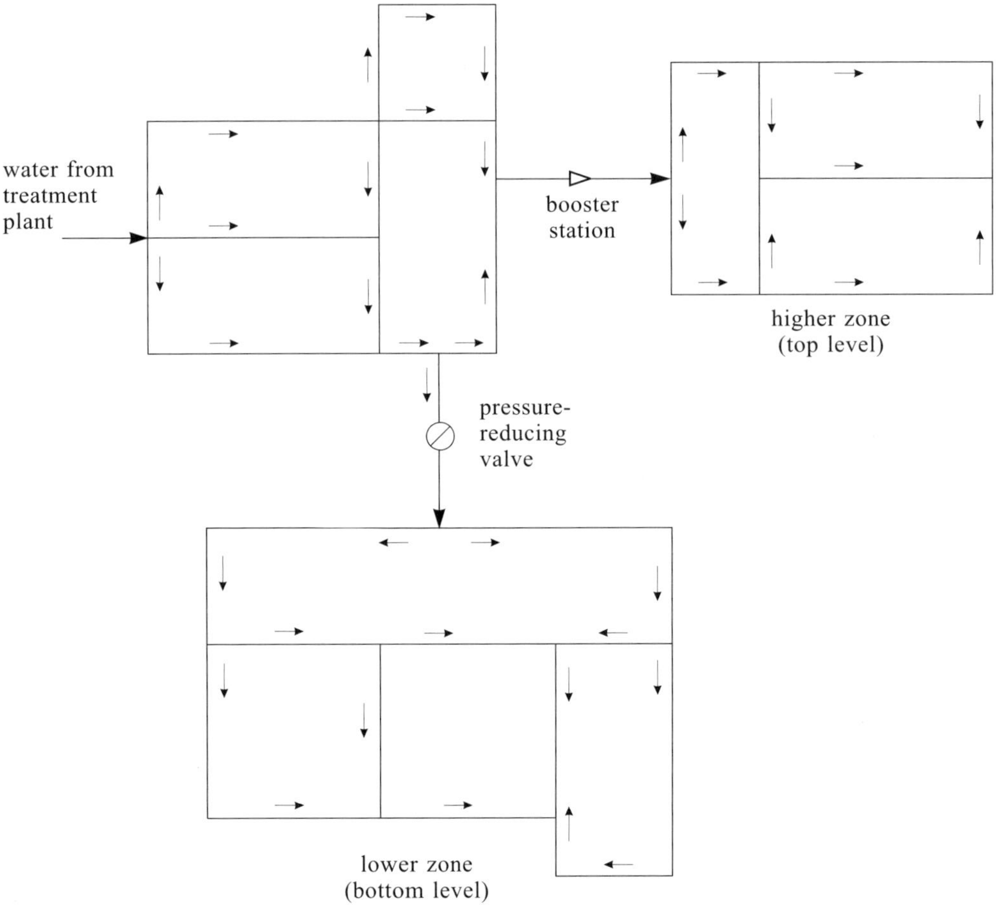

Figure 4.5 Pressure levels

4.12 WATER DISTRIBUTION NETWORKS

The drinking-water distribution network is an important infrastructure that distributes water in adequate quantities to meet current and future consumer needs. It also provides the water necessary for fire fighting.

4.12.1 Structure of the Water-Distribution Network

In cities, water-distribution pipes are usually buried under the pavement, and the system often incorporates loops (looped network), which provide better service to users. In the looped network, the pressure is more balanced, and when pipes must be repaired, the number of users affected is reduced to a minimum, as water reaches its destination by different routes. Figure 4.6a shows a looped network; the arrows pointing in opposite directions show that the flow direction in some pipes varies depending on demand. In the looped network, the flow velocity is rarely zero: this is an asset since it ensures good-quality water (see section 4.12.13).

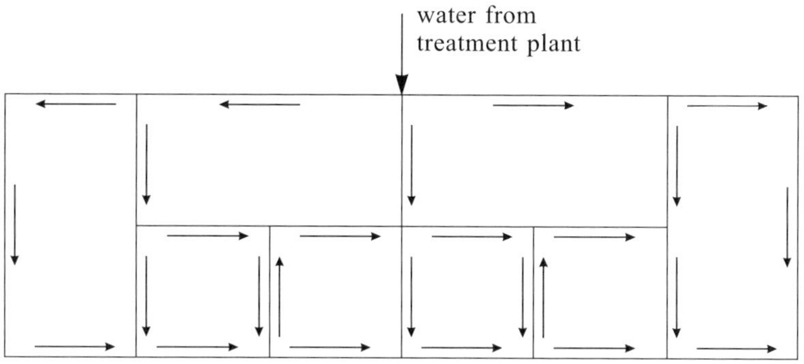

(a)

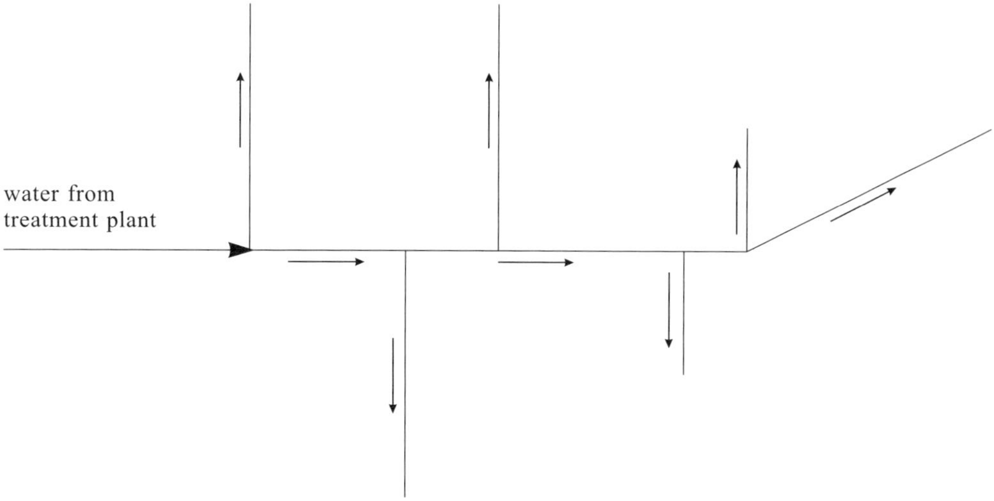

(b)

Figure 4.6 Water-distribution network: a) looped or grid network; b) branching network

When the network does not incorporate loops, it is said to be a branching network (fig. 4.6b). Here, it is easy to see that a breakdown in the network leads to the interruption of service for all users located downstream. Moreover, for any given pipe, the water always flows in the same direction, which is not always the case in the looped network (fig. 4.6a). Most often, however, the network is partly looped and partly branched: city centres and highly populated areas are connected to the looped part of the network, and peripheral districts are connected to the branching part of the network.

4.12.2 Distribution Networks

For water to be distributed in adequate quantities and pressures to meet local needs, it can be transported by gravity; by one or many high-lift pumping stations coupled with one or many reservoirs; or by high-lift pumping stations alone.

Networks with water supplied by gravity. A network distributing water by gravity can be used when the water supply is located at a higher altitude than the targeted population. This type of network is ideal, as it does not need pumping stations to ensure adequate quantities of water at appropriate pressures. There are no energy costs related to pumping, and service quality is not a function of the pumping stations. However, even if the topography allows adequate pressure during normal consumption periods, it may not be sufficient in other circumstances like peak and future demands or for fire fighting; emergency pumps must then be used.

Networks with high-lift pumping stations coupled with reservoirs. To ensure adequate service, high-lift pumping stations and reservoirs distributed along the network are used (figure 4.7). During high-consumption periods, reservoirs make up the difference between the water demand and the production capacity of the water-treatment plant. They are also needed for fire fighting or for drinking-water supply when the treatment plant is shut down (repairs, maintenance, etc.) or when a pipe is broken (primary or arterial main or secondary pipe). Finally, reservoirs and high-lift pumping stations work together to maintain adequate pressures at all times for the entire network. During low-consumption periods (most often at night), reservoirs are filled, and high-lift pumping stations alone ensure desirable pressures for the entire network.

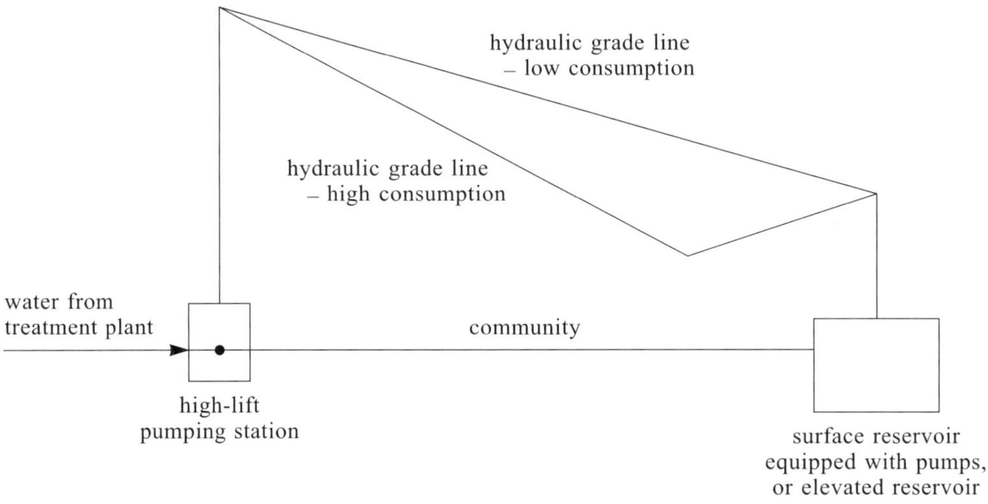

Figure 4.7 Water distribution with high-lift pumping stations coupled with reservoirs

These systems have many advantages:

- low pressure variation in the network, especially during high consumption periods (as water originates from different sources);
- water storage for fire fighting;
- relatively uniform pumping flowrates (enabling optimum operation of the pumps);
- drinking-water supplied at least partly by gravity in case of breakdown at the treatment plant.

Networks with high-lift pumping stations only. Here, the drinking-water supply depends solely on high-lift pumping stations (figure 4.8). To deal with major variations (for example, during the day) in water demand, the pumping stations are designed to be very flexible, with many pumps of different capacities, pumps with variable speeds, and pumps equipped with gasoline-powered motors.

These types of systems, however, have many disadvantages:

- Often, many high-lift pumping stations must be built, and the costs of construction and operation are high.
- There is no backup water supply in case of breakdown at the pumping stations.
- Strong fluctuations in pressure related to variation in water demand may occur if there is an insufficient number of pumping stations.
- Even for relatively constant consumption, there are great pressure variations in the network when a limited number of pumping stations are installed.
- Many pipe breakdowns resulting in water leakage are caused by important and frequent pressure variations.

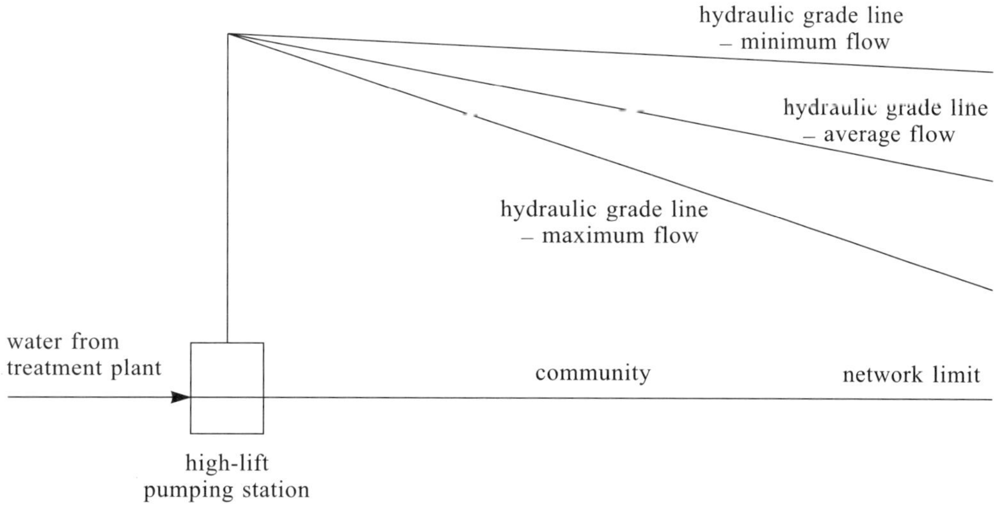

Figure 4.8 Water distribution with high-lift pumping stations only

4.12.3 Distribution Reservoir

A water-distribution network usually has many reservoirs, including those at the treatment plant. These reservoirs have the following functions:

- to ensure uniform constant operation of the plant under a constant supply of raw water: when demand is high, clear water is partly supplied by the reservoir as the treatment-plant production capacity cannot meet demand;
- to supply at least part of the water necessary for fire fighting;
- to ensure water supply, at least in part, when breakdowns at the plant or in arterial or secondary pipes occur, or during power failures;
- to reduce pressure variations in the network during high-demand periods, through multiple water sources; during low-demand periods, the reservoirs are filled up.

A distribution network is a very complex system that can meet the population's needs only by storing clear water in reservoirs. This water is drawn when special needs arise (consumption variation during the day and fire fighting) or when operational problems occur (main pipe breakdown, maintenance of the treatment plant, etc.). Depending on the situation, storage is referred to as service storage, fire-fighting storage, emergency storage, or production storage. The locations of these reservoirs must be well planned.

Service storage or reserve. During the day, when demand exceeds the production capacity of the water-treatment plant, high-lift pumps draw water from the service reservoir, R_{equ}, to balance supply with demand. The clear-water reservoir at the treatment plant stores at least part–often, all–of the service storage for the community supplied by the treatment plant. This reservoir fills during low-demand periods and empties during high demand periods. On any given day, service storage represents about 20% of the consumption for that day. For a district, community, or city, the maximum volume of service storage is a function of the maximum daily consumption at the end of the design period of the treatment plant. To this end, two curves can be drawn with actual data (but mainly predicted future consumption data): the cumulative-consumption curve and the consumption-rate curve.

Cumulative Consumption Curve. To evaluate the volume of service storage (required on a daily basis) using this method, this procedure must be followed:

- Collect consumption data for the day when consumption is maximum, $(Q_d)_{max}$; these data are kept at the treatment plant.
- Organize the data into tables (table 4.4).
- Plot a cumulative-consumption curve for that day (figure 4.9, broken line).
- On the same graph, plot a straight line from the point representing consumption at time zero hour to the point representing consumption at time 24 hours; the slope of this line represents the average consumption rate for that specific day.
- Plot two straight lines parallel to the line plotted above: one line is tangential to the upper part of the cumulative-consumption curve, the other line is tangential to the lower part of that curve.

- Measure the vertical distance between the two parallel lines; this distance represents, according to the scale, the volume of service storage for that specific day. This storage is shared among the different reservoirs of the municipality.

Consumption rate curve. To evaluate the volume of service storage using this method, this procedure must be followed:

- Plot a curve of the hourly consumption rate for the day when consumption is maximum (fig. 4.10).
- On the same graph, plot a horizontal line at a level equal to the average consumption rate of that day.
- Evaluate (using planimetry, for example) the surface contained between the upper part of the curve and the horizontal line plotted above: this surface is the volume of service storage for that specific day. This storage is shared among the different reservoirs of the municipality.

Water storage or reserve for fire fighting. A certain volume of water must be set aside for fire fighting; this is often called the fire-fighting storage, R_{fire}. The Fire Underwriters Survey in Canada (guidelines are found in appendix F)–the National Board of Fire Underwriters in the United States–ultimately sets guidelines that a municipality must follow, stipulating:

- necessary flowrate to fight one or many fires;
- duration of a conflagration that must be extinguished;
- minimum pressure required in pipes and fire hydrants.

In general, provisions are made for one major conflagration occurring at the time or a number of smaller fires requiring together the same flowrate or volume of water when the municipality needs to meet its maximum daily consumption (not the maximum hourly consumption).

This storage is also shared among the different reservoirs of the municipality.

The Fire Underwriters Survey also deals with other aspects of fire protection, such as the number of fire engines required, fire-hydrant distribution, the minimum number of fire fighters, and so on.

Emergency water storage or reserve. It is recommended that a certain volume of water (*emergency storage in reservoir*, R_{emer}) be stored in order to meet the demand during exceptional events, such as breakdown of pumps or main pipes.

Emergency storage is also shared among the different reservoirs of the municipality. The quantity stored must be able to meet the average daily consumption during a period varying from 2 to 14 hours, depending on the size of the network.

Production storage. The *production storage in reservoir*, R_{prod}, is the water reserve necessary to meet the demand when operational problems occur at the treatment plant. This reserve corresponds to about four hours of production at the treatment plant and is located on site.

Table 4.4 Hourly consumption on the day when consumption is maximum

Hourly consumption (L/s)	**Hourly consumption (m^3/h)**	**Cumulative consumption (m^3)**
140	504,0	504,0
130	468,0	972,0
125	450,0	1 422,0
124	446,4	1 868,4
125	450,0	2 318,4
131	471,6	2 790,0
229	824,4	3 614,4
328	1 180,8	4 795,2
355	1 278,0	6 073,2
372	1 339,2	7 412,4
380	1 368,0	8 780,4
400	1 440,0	10 220,4
405	1 458,0	11 678,4
402	1 447,2	13 125,6
390	1 404,0	14 529,6
400	1 440,0	15 969,6
420	1 512,0	17 481,6
462	1 663,0	19 144,6
589	2 120,4	21 265,0
525	1 890,0	23 155,0
318	1 144,8	24 299,8
162	583,2	24 883,0
156	561,6	25 444,6
145	522,0	25 966,6
	25 966,6	

* Service storage required in reservoir, R_{equ}, to balance supply with demand during the day.

Note: Average hourly consumption (= average hourly production): 25 966,6/24 = 1 081,94 m^3/h.

Time of day (h)	**Hourly consumption (% of average hourly consumption)**	**Average hourly consumption minus hourly consumption: Water volume in reservoir**	
		decreases by (m^3)	**increases by (m^3)**
1	47		577,94
2	43		613,94
3	42		631,94
4	41		635,54
5	42		631,94
6	44		610,34
7	76		257,54
8	109	98,86	
9	118	196,06	
10	124	257,26	
11	126	286,06	
12	133	358,06	
13	135	376,06	
14	134	365,26	
15	130	322,06	
16	133	358,06	
17	140	430,06	
18	154	581,06	
19	196	1 038,46	
20	175	808,06	
21	106	62,86	
22	54		498,74
23	52		520,34
24	48		559,94
Total		5 538,24*	5 538,20*

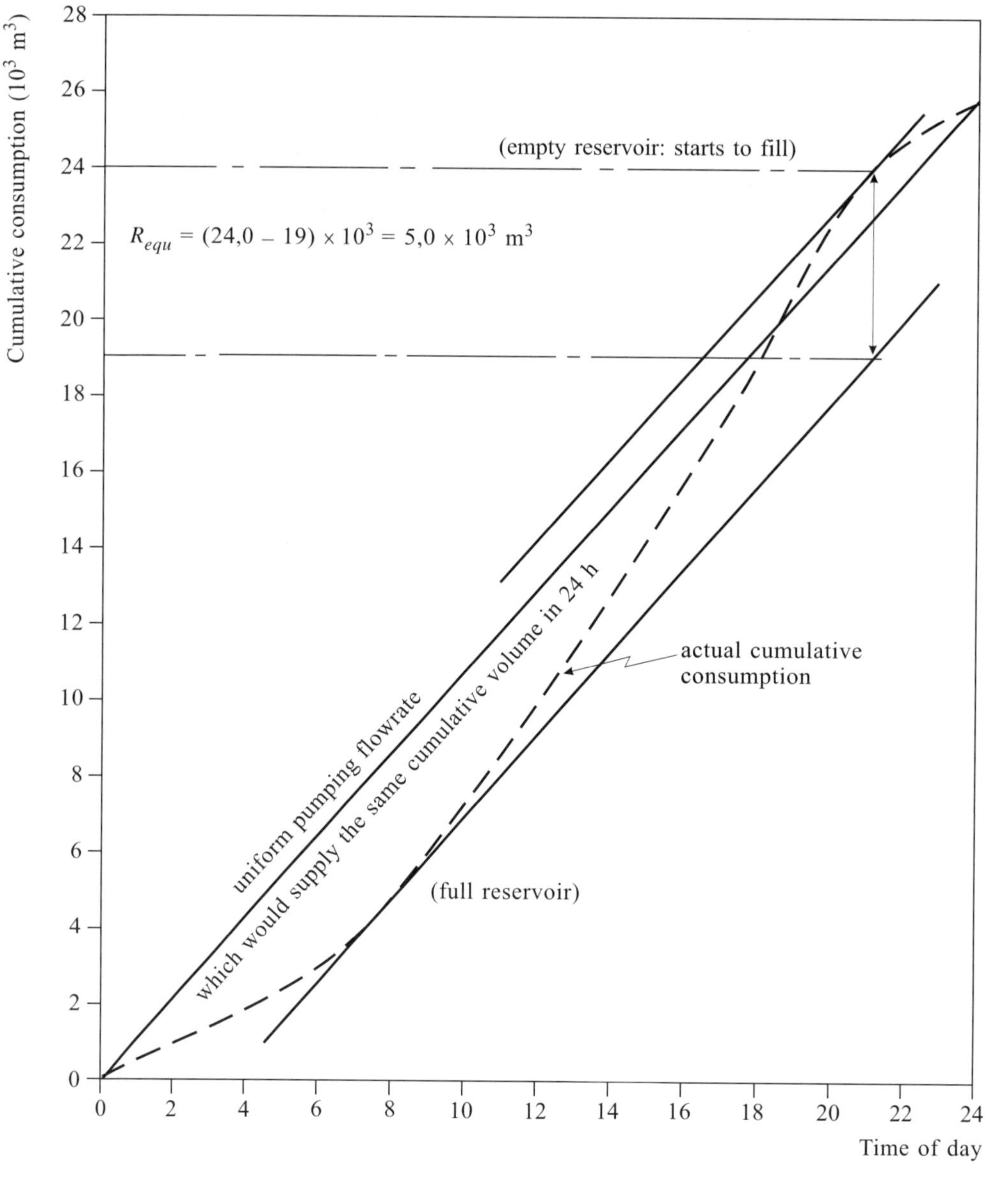

Figure 4.9 Cumulative consumption during the day

Computation of the total storage or reserve. The probability of having a major fire during the day when consumption is maximum, and at the same time sustaining a breakdown at the treatment plant and on one of the main network pipes, is very low. It is not economically feasible to store, at all times, a volume of water equal to the sum of the water reserves discussed above. To determine the volume of total

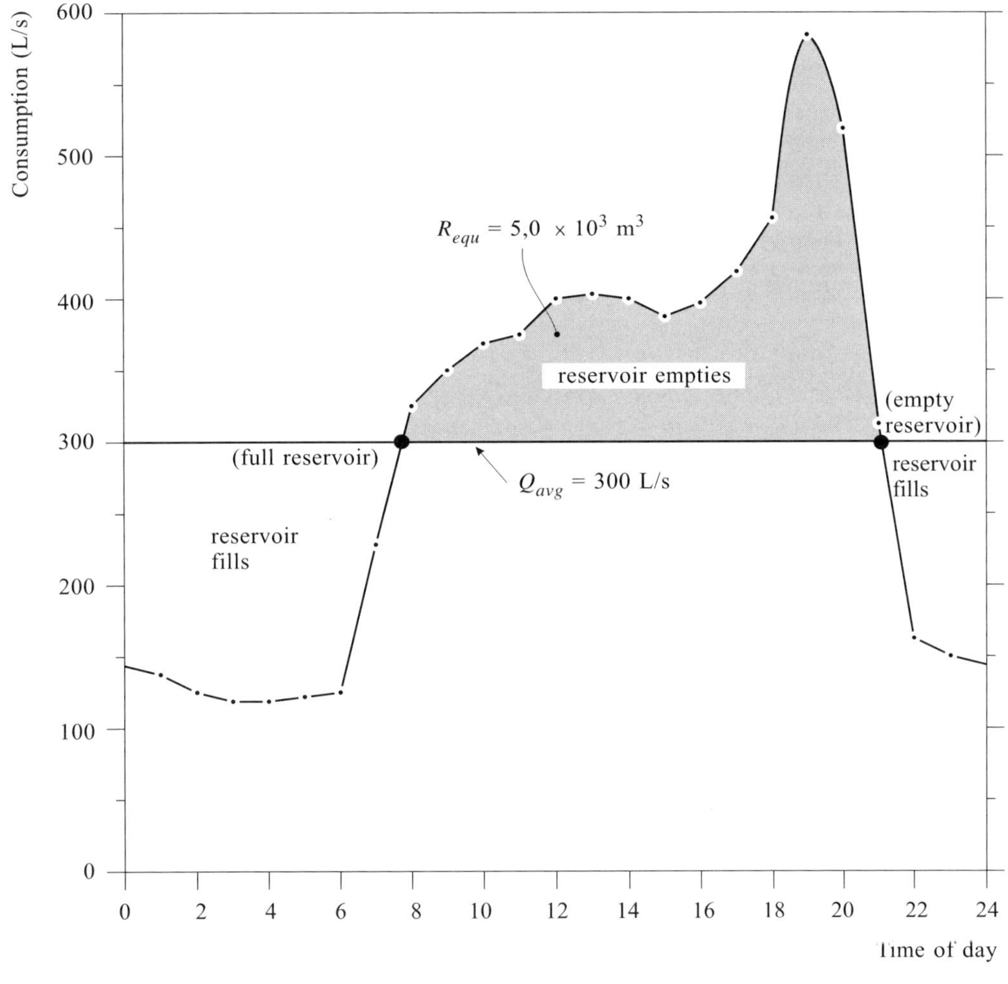

Figure 4.10 Hourly consumption during a day

storage, the following three equations are often used to determine the maximum storage in reservoir, R_{max}, the recommended storage, R_{recom}, and the minimum storage, R_{min}:

$$R_{max} = R_{equ} + R_{fire} + R_{emer} + R_{prod}$$

$$R_{recom} = R_{equ} + R_{fire} + R_{emer}$$

$$R_{min} = R_{equ} + R_{fire}$$

According to many authors, the sum of the emergency and production reserves represents about one third of the sum of the service and emergency reserves. However, local authorities (municipalities, Fire Underwriters Survey, Quebec Ministry of the Environment) ultimately determine the volume of the total reserve.

Locations and types of reservoirs. There are two types of water reservoir: elevated reservoirs and surface reservoirs (built at ground level). Nowadays, the use of elevated reservoirs is controversial, as we shall see below.

Elevated reservoirs. These reservoirs are built on a mountainside or on a specially designed structure (fig. 4.11). The water is pumped up by high-lift pumping stations during low consumption periods, and flows down by gravity during high consumption periods. Often, the same pipe is used to fill up and to empty the reservoir, as shown in figure 4.11. Elevated reservoirs have the following advantages:

- Power failures and pump breakdowns do not hinder the distribution of stored water.
- The pumps filling the reservoir work intermittently.
- A strategic choice for the site of the elevated reservoir reduces spatial pressure variations in the network.

However, elevated reservoirs also have the following disadvantages:

- The reservoir must be installed to ensure the required pressures in the network, even when the reservoir is filled at the lowest water level; the energy required to fill it up is therefore wasted.
- The maximum storage volume is limited, because, economically, installation is limited to reservoirs available on the market.
- The water pressure generated by the reservoir cannot be modified after installation, as it cannot be easily modified for higher or lower pressure;
- During low-consumption periods, more specifically at the beginning of the reservoir's design period, the stored water may not be used for long periods

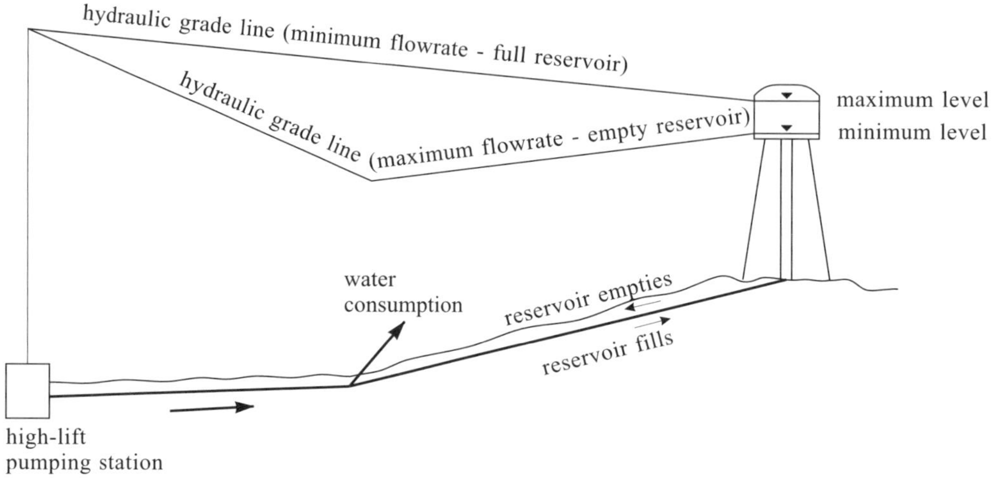

Figure 4.11 Use of an elevated reservoir: Maximum and minimum consumption flowrates

(sometimes many months); as a matter of fact, the pressure built by high-lift pumping stations (like those at the treatment plant) at the base of the reservoir may be too high for the elevated reservoir to empty. This problem may cause poor water quality in the reservoir.

– Part of the water stored in the reservoir may become unusable under certain conditions; for example, if the pressure in the vicinity of the reservoir has to be maintained higher than the design pressure because consumption is exceeding projections (and head losses are higher), the minimum water level in the reservoir will be higher than the minimum level planned at the design stage; the water volume between those two minimum levels cannot, to all intents and purposes, be used.

To summarize, the problems caused by the use of elevated reservoirs originate from their dual roles. As well as being used for water storage, the reservoirs create the required water pressures in the network. As pressures are a function of the water levels in the reservoirs, engineers have no real control over the pressures created by these reservoirs. On the other hand, operation of high-pressure pumps can easily be modified.

Surface reservoirs. Water can also be stored in reservoirs at ground level. These reservoirs can be built either slightly above the surface or completely in-ground. High-lift pumping stations draw water from these reservoirs and distribute it through the network under pressure. Surface reservoirs have the following advantages:

– Construction and operation costs are lower than for elevated reservoirs.
– Water quality is better, as there is less risk that the water will remain in the reservoir for long periods of time.
– The water pressures are a function of the chosen pumps and not of the water level in the reservoirs.
– Aesthetically, the reservoir blends easily with the environment.
– It is easy to determine the location of reservoirs.
– There is a wide choice of sizes and volumes for the reservoirs, as the engineer is not limited to reservoirs available on the market.

Special cases. When topography allows, there are some advantages to installing a surface reservoir in a zone of higher pressure level (fig. 4.12). The surface reservoir will then behave as an elevated reservoir for an adjacent district, located at a lower pressure level, while still meeting certain needs in the district where it is located (fire-fighting storage, peak hourly consumption, etc.).

This type of arrangement can ensure:

– a constant water supply to the higher pressure level, because of the reservoir and the high-lift pumping station;
– fire protection in both pressure levels;
– water supply by gravity to the lower pressure level, especially in case of emergency.

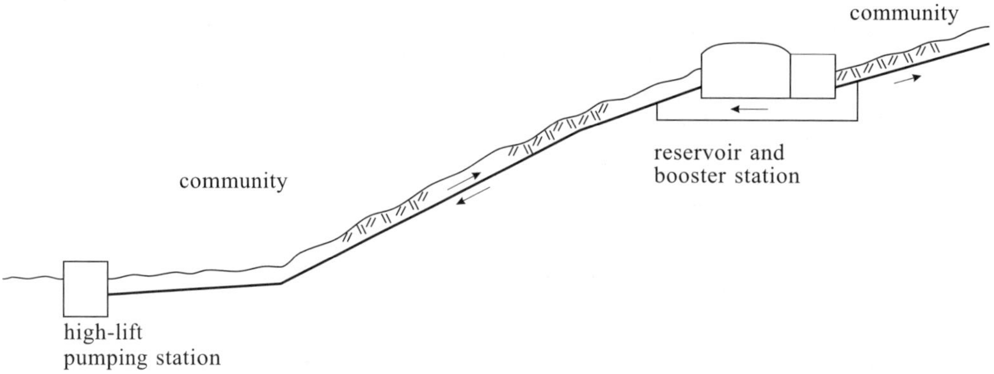

Figure 4.12 Reservoir: special case of a surface reservoir used as an elevated reservoir

4.12.4 Pipes Available on the Market

Table 4.5 shows the diameters of pipes available in Canada and the United States.

Pipes are made of different materials, each of which has advantages and disadvantages. *Plastic* pipes are light and easy to handle, and friction due to water flow is low. However, they have a tendency to deform when an external load is applied and with temperature variations. Moreover, they cannot be thawed by use of an electrical current, as can metal pipes. Pipe sections 6,1 m long are connected with mechanical joints and push-on joints (see section 4.12.10). Plastic pipes for sewage systems are produced in 4,0 m long sections.

For small-diameter pipes, *ductile cast-iron pipes (also simply called ductile-iron pipes), lined with cement mortar on the inside*, have very good rigidity. Their inner lining ensures a stable roughness coefficient for the entire duration of the design period, while providing interior protection against corrosion of the cast iron.

Ductile cast-iron pipes are easily thawed by running an electrical current through them from fire hydrant to fire hydrant. Pipes of nominal diameter smaller than 250 mm are sold in 6 m long sections; for diameters between 300 and 600 mm, the pipes are sold in 5,5 m long sections. Three types of joint are used to link cast-iron pipes: flanged joints, mechanical joints, and push-on joints.

Ductile cast-iron pipes without inner cement lining are rarely used nowadays. Until recently, however, most pipes with a nominal diameter smaller than 600 mm were made of cast-iron without the inner lining.

Stainless-steel pipes are never buried underground, where they can break because of their lack of flexibility. They are used in water-treatment plants and valve chambers mostly because of their very high strength and easy installation, as a wide variety of fittings are readily available on the market. Their cost is relatively low, they are light, and they are available in the same diameters as ductile cast-iron pipes. Welded joints are used to join stainless-steel pipes, and equipment is attached with flanged joints.

Table 4.5 Nominal diameters of water-distribution pipes available on the market

Nominal diameter*		Materials						
(mm)	(inches)	Ductile cast-iron**	Ductile cast- iron with interior cement lining	Plastic	Concrete pressure pipe	Copper	Asbestos-cement	Stainless steel
20	3/4			X		X		X
25	1			X		X		X
32	1 1/4			X		X		X
37	1 1/2			X		X		X
50	2			X		X		X
75	3	X		X				X
100	4	X	X	X			X	X
125	5			X			X	X
150	6	X	X	X			X	X
200	8	X	X	X			X	X
250	10	X	X	X			X	X
300	12	X	X	X			X	X
350	14	X	X	X	X		X	X
400	16	X	X	X	X		X	X
450	18	X	X	X	X		X	X
500	20	X	X	X	X		X	X
600	24	X	X	X	X		X	X
750	30	X	X	X	X		X	X
900	36	X	X	X	X		X	X
1050	42		X		X			X
1200	48		X		X			X
1350	54		X		X			
1500	60		X		X			
1600	64		X		X			
1650	66				X			
1800	72				X			
2100	84				X			
2400	96				X			
2700	108							

* The actual inside diameter may vary plus or minus a few millimetres, depending on the type of pipe and the nominal diameter.

** Ductile iron pipes without interior coating are not used any more in water distribution networks.

For larger diameters, *concrete pressure (hyprescon)* pipes are often used. These pipes are made of welded steel lined with concrete on the inside and outside; the concrete is pre-stressed in the longitudinal direction of the pipes, using coiled steel wires. These pipes are available in diameters varying from 350 to 2700 mm and in sections of variable lengths (4,7 to 7,3 m). Flanged and push-on joints are used to join these pipes.

Asbestos-cement pipes are rarely used in Quebec because they are brittle and have a tendency to crack. Moreover, this type of pipe is difficult to use as there is a limited number of fitting parts. These pipes are joined by push-on joints.

Finally, *copper* pipes are used mostly in service-pipe or house-service connections where the minimum diameter is 20 mm. These pipes are welded or joined by compression joints.

4.12.5 Burying Depth of Pipes and Freezing Depth in Soil

In the Montreal area, water-distribution pipes must be buried at least 1,85 m under the pavement to avoid freezing. In colder regions, pipes are buried deeper-as deep as 4,25 m, as in the James Bay area (upper Quebec). Moreover, as the water flow is under pressure, these pipes do not have any specific slopes; they simply follow the ground configuration and are not required to be buried any deeper than the freezing protection requirements. The freezing depth below paved surfaces can be calculated using data from Transport Canada. The curves shown in figure 4.13 indicate variations in freezing depth as a function of the type of surface and the freezing index. The freezing indices for different regions of Quebec are shown in table 4.2.

4.12.6 Pipe Protection Against Freezing

In 1969, Transport Canada developed the following equation between the freezing depth, h, under a surface that is free of snow and the freezing index, G:

$$h = K \sqrt{G} \tag{4.2}$$

where h is in inches and G is in (°F – day). The value of the coefficient K is 1,40 for an asphalt pavement free of snow, 1,62 for a concrete pavement free of snow, and 0,78 for natural ground covered with snow. Freezing index (G) has been defined in article 4.4.

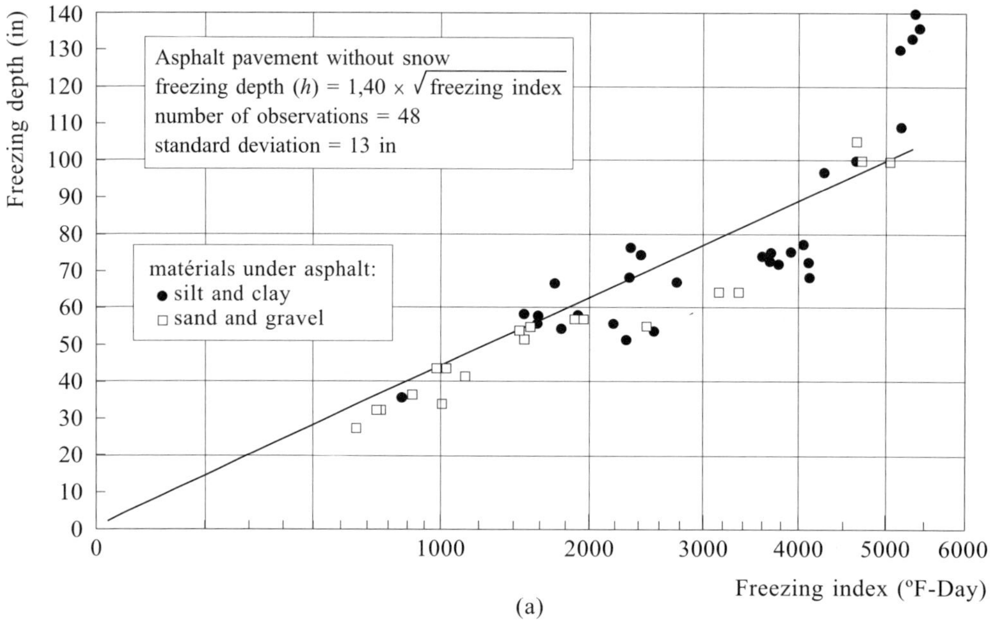

(a)

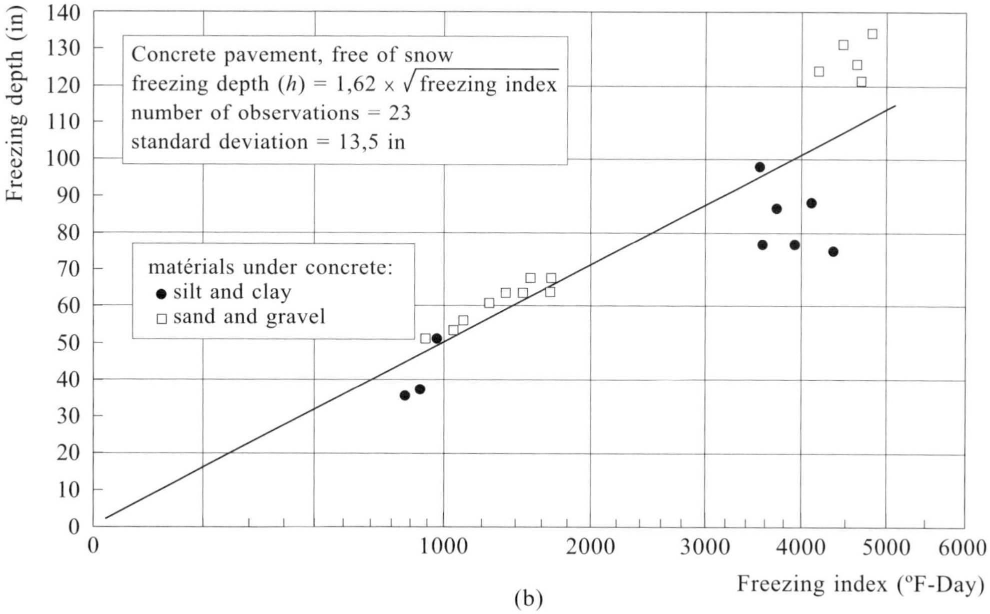

(b)

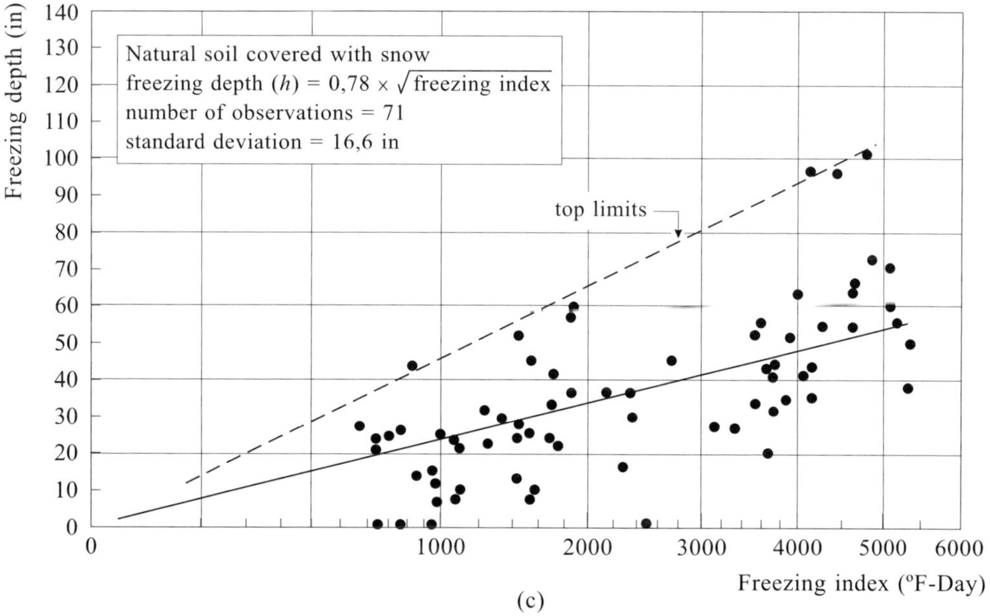

(c)

Figure 4.13 Variations in freezing depth under the surface as a function of the freezing index: a) asphalt surface without snow cover; b) concrete surface without snow cover; c) natural ground covered by snow

Adapted from: Department of Transport, Air Services, *Engineering, Planning, Design and Construction Manual,* Section 1, "Design and Evaluation of Flexible and Rigid Pavements" (Ottawa: 1969), chapter 4.

When pipes cannot be buried at an adequate depth, they have to be protected by polystyrene foam insulation. The Dow company developed the following formulas to determine the necessary width of insulation:

$$w = 3d_{out} + 2\left[h - \left(\delta + 0{,}5 + d_{out}\right)\right] \quad \text{(U.S. customary units)} \qquad (4.3a)$$

$$w = d_{out} + 2(h - \delta) - 0{,}3 \quad \text{(metric units)} \qquad (4.3b)$$

where w = width of the insulation (ft; m)

d_{out} = outside diameter of the pipe (ft; m)

h = freezing depth without insulation (ft; m)

δ = insulation depth (ft; m)

As insulation is installed 0,15 m (6 in) above the pipe, its burying depth (δ') is equal to the depth of the pipe minus 0,15 m minus the insulation thickness (fig. 4.14). The surface of the insulation must never be closer than 0,4 m (16 in) to the ground surface. The Dow table (tables 4.6a and 4.6b) enables us to determine the thickness of the insulation required as a function of the freezing index (G) and the thickness of the backfill covering the top of the insulation. Nowadays, insulated pipes and even heated pipes are readily available on the market.

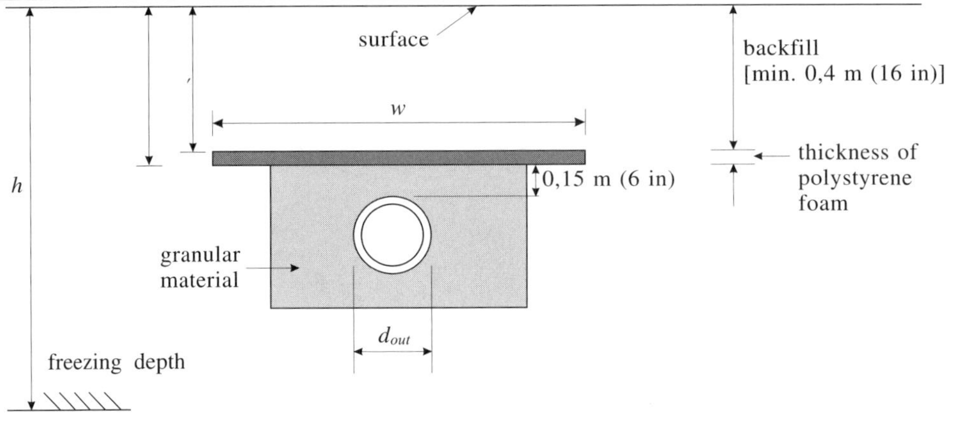

Figure 4.14 Pipe protection against freezing using polystyrene foam (Styrofoam) insulation.

From: Dow, *Insulation for Geotechnical Applications* (1997), p. 6.

Table 4.6a Thickness in inches of polystyrene foam (Styrofoam) insulation necessary to protect insufficiently buried pipes from freezing

Freezing index, *G* (°F - day) / δ'* (ft)	1500	2000	2500	3000	3500	4000	4500
2	2	2 1/2	3	3 1/2	4	4 1/2	5
3	1 1/2	2	2 1/2	3	3 1/2	4	4 1/2
4	1	1 1/2	2	2 1/2	3	3 1/2	4
5	1	1	1 1/2	2	2 1/2	3	3 1/2
6	1	1	1	1 1/2	2	2 1/2	3
7			1	1	1 1/2	2	2 1/2
8				1	1	1 1/2	2
9					1	1	1 1/2
10						1	1

* Thickness of the backfill cover over the insulation.

Table 4.6b Thickness, in millimeters, of polystyrene foam (Styrofoam) insulation necessary to protect insufficiently buried pipes from freezing

Freezing index, *G* (°C - day) / δ'* (m)	850	1125	1400	1675	1950	2225	2500
0,6	50	65	75	90	100	115	125
0,9	40	50	65	75	90	100	115
1,2	25	40	50	65	75	90	100
1,5	25	25	40	50	65	75	90
1,8	25	25	25	40	50	65	75
2,1			25	25	40	50	65
2,4				25	25	40	40
2,7					25	25	25
3,0						25	25

* Thickness of the backfill cover over the insulation

From: Dow, *Insulation for Geotechnical Applications* (1997), p. 7.

EXAMPLE 4.2

Because of a surveying error, a water-distribution pipe with an outside diameter of 325 mm will be buried under a paved (asphalt) street in Val d'Or, Quebec, with only 1,45 m of backfill cover. To avoid freezing of the pipe, the local engineer decides to install some insulation. Find or compute:

a) the maximum freezing index (G) in Val d'Or;
b) the required width of the insulation;
c) the required thickness of the insulation.

Note: 1 ft = 12 in; 1 in = 2,54 cm; 1 m = 3,28 ft.

Solution

a) Maximum freezing index (G) in Val d'Or

3895 (°F – day) (Table. 4.2); we choose 4000 (Table 4.6a).

2164 (°C – day) (Table 4.2); we choose 2225 (Table 4.6b).

b) Width of the insulation, w

- Outside diameter of the pipe, d, = 325 mm (≅ 0,33 m), yielding about 1,07 ft.
- Insulation depth, δ. According to figure 4.14, this depth is equal to the burying depth of the pipe minus 6 in (0,15 m); it yields:

$$\begin{aligned}\delta &= (1{,}45 \times 3{,}28) - (6 \div 12)\\ &= 4{,}26 \text{ ft}\end{aligned}$$

or

$$\begin{aligned}\delta &= 1{,}45 - 0{,}15\\ &= 1{,}30 \text{ m}\end{aligned}$$

To solve for the thickness of the backfill cover, δ', located over the insulation (this value is required in tables 4.6a and 4.6b), the insulation thickness must be known, but this figure is still unknown for the moment. Tables 4.6a and 4.6b can still be used by substituting the value of δ; the error is insignificant.

- Freezing depth, h (Fig. 4.13a)

$$\begin{aligned}h &= 1{,}40 \sqrt{\text{freezing index}} = 1{,}40 \sqrt{4000}\\ &= 88{,}5 \text{ in } (\cong 7{,}38 \text{ ft; } 2{,}25 \text{ m})\end{aligned}$$

- Width of the insulation (equations. 4.3a and 4.3b)

$$\begin{aligned}w &= d_{out} + 2(h - \delta) - 0{,}3\\ &= 0{,}33 + 2(2{,}25 - 1{,}3) - 0{,}3\\ &= 1{,}93 \text{ m}\end{aligned}$$

or

$$\begin{aligned}w &= 3d_{out} + 2\left[h - (\delta + 0{,}5 + d)\right]\\ &= 3 \times 1{,}07 + 2\left[7{,}38 - (4{,}26 + 0{,}5 + 1{,}07)\right]\\ &= 6{,}31 \text{ ft}\end{aligned}$$

c) Thickness of the insulation (tables 4.6a and 4.6b)

For a freezing index of 2225 (°C – day) and a backfill thickness, δ, of about 1,2 m (value that takes into account the desired thickness of the insulation), the thickness must be 90 mm. It is left to the reader to show on a graph the cross section of the insulation location in relation to the pipe and to the surface of the street. The thickness and width of the insulation could also be indicated.

4.12.7 Valves

Managing a water-distribution network requires different types of equipment. Valves are probably the most important pieces of equipment, as they provide control over the water flows of the network. There are different kinds of valves with a wide range of purposes (fig. 4.15).

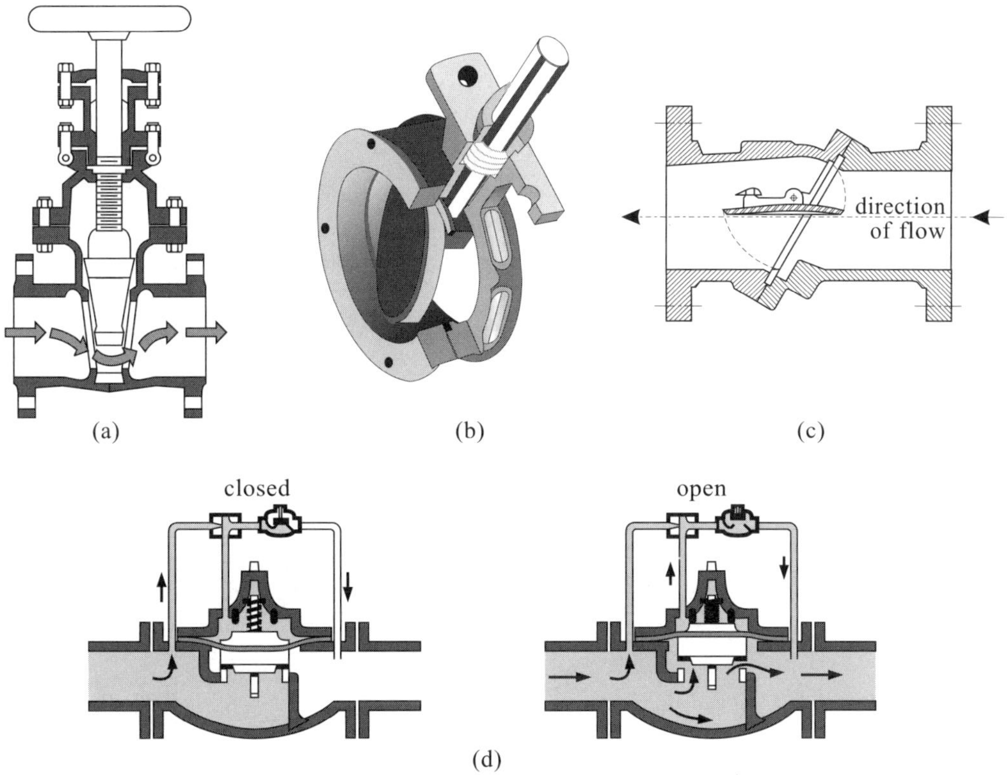

Figure 4.15 Types of valves: a) gate valve; b) butterfly valve; c) back-water (check) valve; d) pressure-reducing valve

from: b) Kamyr Valves Inc., *Metal Seated Tight Closing Butterfly Valve* (catalogue), Bulletin 2K-3; c) M. Héroux, *Dictionnaire de la robinetterie et de la tuyauterie industrielle*, (Les Publications du Québec, 1992), p. 241; d) CLA-VAL CO., California.

Isolation valves isolate parts of the network that require inspection, repairs, or maintenance: for large-diameter pipes (350 mm and over), *butterfly valves* are used; for smaller-diameter pipes, *gate valves* are used. The spacing between valves must be such that the length of pipe being shut down does not exceed 150 to 250 m, depending on budgetary constraints and the requirement of the Fire Underwriters Survey. For main pipes (> 350 mm), this may extend to 300 or 400 m. However, it should not exceed 150 m in commercial districts and 250 m for secondary pipes and small distribution mains (diameters between 150 and 300 mm). Ideally, when pipes cross and are interconnected, the number of valves should equal the number of branches minus one. In practice, for economical reasons, fewer valves are installed (fig. 4.16).

Back-water (check) valves direct water flow in one direction only. They are installed at the junction of two pressure levels, on backflow pipes of pumps and on pipes where water meters are installed.

Pressure-reducing valves control pressure either by bringing it back to recommended values or by reducing it by a predetermined value.

Altitude valves are located at the entrance to an elevated reservoir; when the water level reaches the maximum in the reservoir, the valve closes under the effect of water pressure, preventing overflowing of the reservoir.

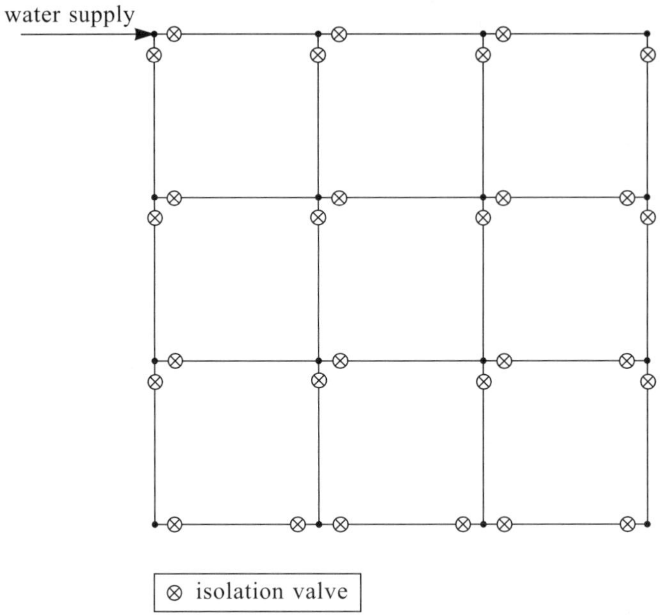

Figure 4.16 Location of isolation (gate) valves in a looped network. The distance between the valves varies from 150 to 600 m depending on district and pipe sizes.

4.12.8 Fire Hydrants

Fire hydrants (figs. 4.17 and 4.18) are installed on the distribution network and supply water for fire fighting. The guidelines for installation of fire hydrants are set mainly by the Fire Underwriters Survey. The area covered by a fire hydrant depends on the flowrate required for fire fighting; hydrants are closer together and in greater number when flowrates are high.

Fire hydrants must be equipped with at least two lateral outlets 65 mm in diameter to which a front outlet of 100 mm is added if the fire flowrate exceeds 5000 L/min or the water pressure is low. Fire hydrants must be linked to the water network by connecting pipes at least 150 mm in diameter and equipped with isolation (gate) valves.

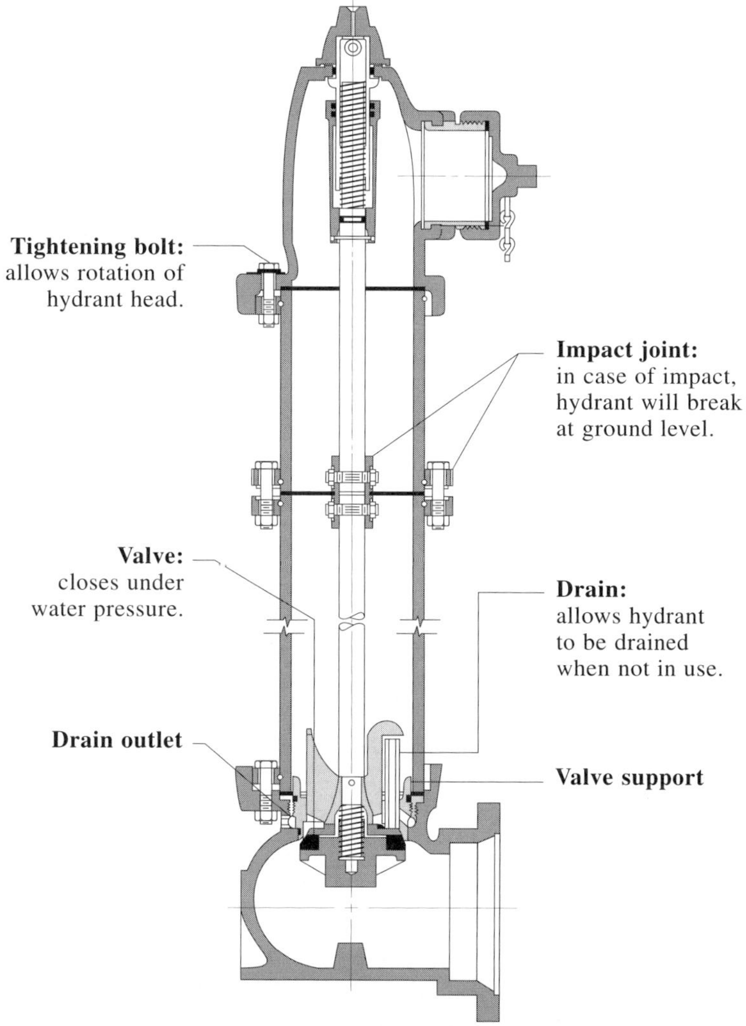

Figure 4.17 Fire hydrant

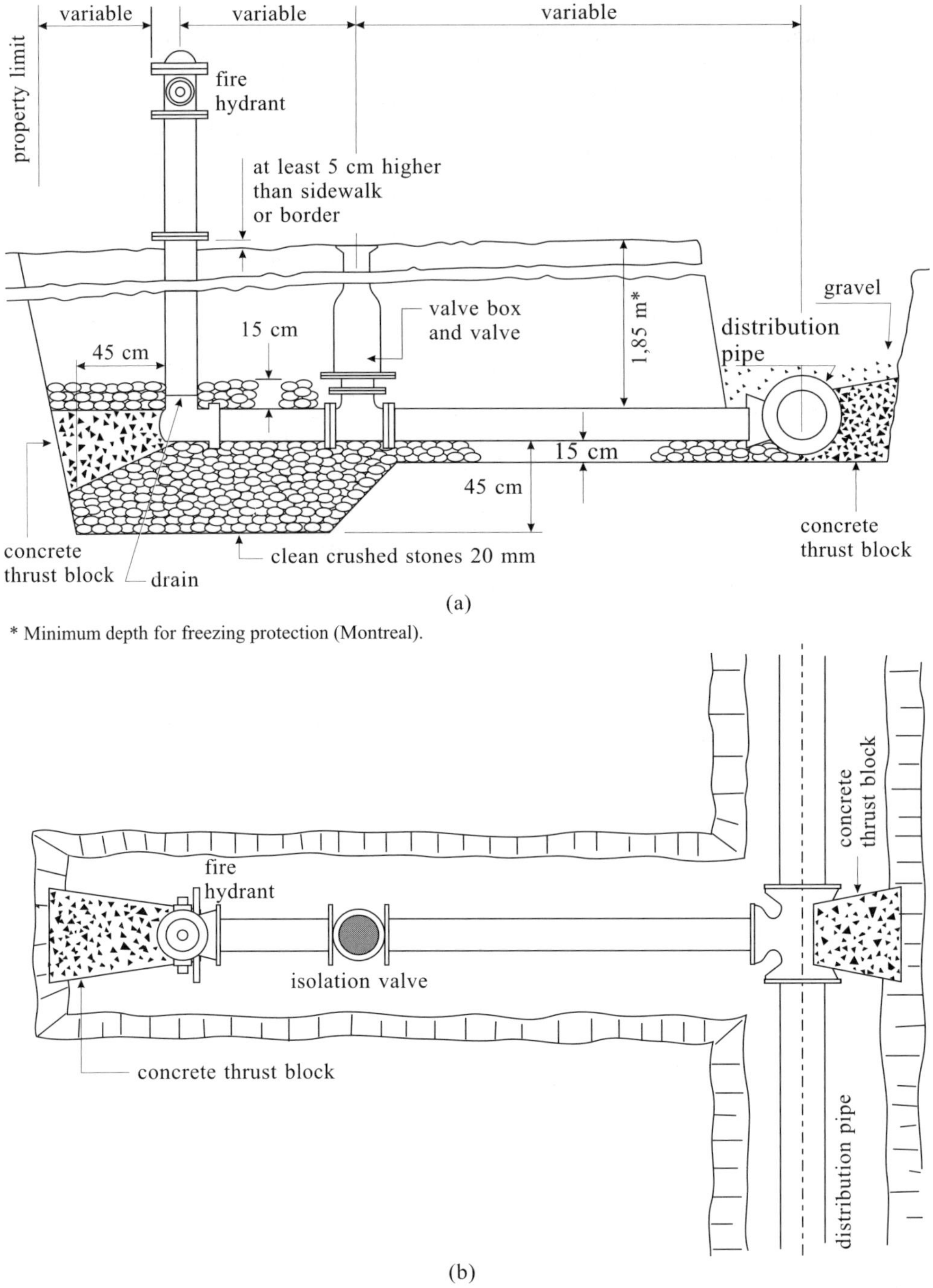

Figure 4.18 Installation of a fire hydrant: a) section; b) plan view

Normal spacing between fire hydrants on residential streets must not exceed 200 m. This is twice the length of a normal fire hose: therefore, fire fighters can attend to any conflagration occurring along the street between two fire hydrants. In high-income and commercial districts, fire hydrants are closer together (100 m). Hydrants should be located at street intersections, along streets where there is a long stretch of houses, at extremities of dead ends, and in the backyard of large establishments when fire engines can have access.

To avoid freezing, fire hydrants must be empty at all times. A cubic metre of stones is placed around the base, and the hydrant is drained of all water after being used. Hydrants must never be drained into a sewer pipe: any contact between drinking water and wastewater is prohibited by the Quebec Ministry of the Environment.

In choosing the side of the street on which to locate the hydrant, we minimize the length of the connection to the water-distribution network. For wide boulevards where two water distribution pipes are often installed, fire hydrants are installed on both sides, either for reasons of economy or to minimize traffic obstructions during a fire.

4.12.9 Service Pipes

The service pipe (fig. 4.19), often called the house-pipe connection, links a water-distribution pipe to the user's home. Service pipes are at least 20 mm in diameter and are connected to a pipe in the distribution network using a corporation cock. The user's home may be isolated by an isolation (gate) valve (often called stopping valve). In general, the service pipe is never connected to a pipe exceeding 300 mm in diameter. A gooseneck is installed on the service pipe to avoid breakage of the line in the event of ground motions (the gooseneck can be installed in the vertical or horizontal plane: in the latter case, risk of freezing is minimized).

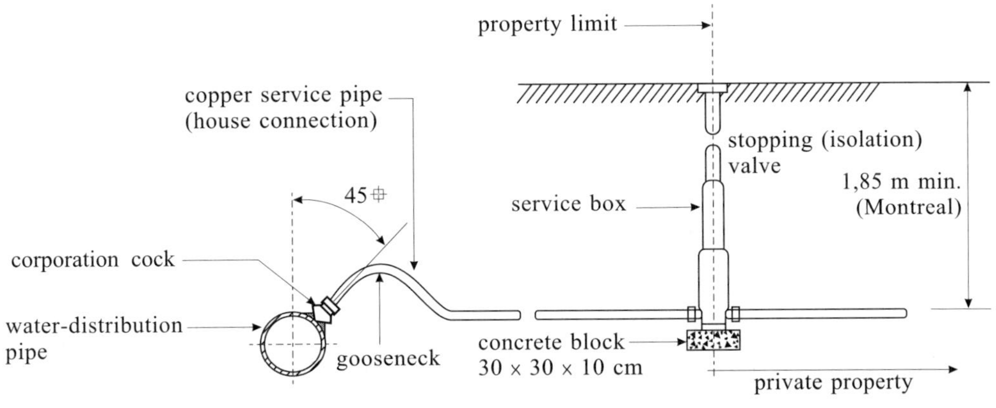

Figure 4.19 Service pipe

Table 4.7 Water requirements (in water supply units[1]) of different appliances

Appliance[2]	**Distributing valve**
Bath*	Faucets
Commercial kitchen sink*	Flush valve or faucets
Dishwasher*	
Domestic kitchen sink*	Faucets
Drinking fountain	Automatic shut-off faucet
Exterior outlet (15 m pipe)	Faucets
Laundry sink*	Faucets
Service sink*	Faucets
Service sink*	Automatic flush valve
Shower (1 head)*	Combined valve
Toilet	Flush tank
Toilet	Automatic flush valve
Urinal	Flush tank
Urinal	Automatic flush valve
Washbasin*	Faucets
Washbasin*	Automatic shut-off faucet
Washing machine*	
Washroom*	Toilet with flush tank
Washroom*	Toilet with automatic flush valve
Appliances not identified in this table, if diameter of water supply pipe is one of the four indicated.	

1. Water supply unit: comparative measurement of the water supplied to an appliance based on flowrate, utilization frequency, and duration of normal operation. For small systems, the comparison unit is based on the flowrate of a 3/8 in (9,5 mm) sink faucet over one minute: 0,189 L/s.
2. Appliances marked with an asterisk must be supplied with hot and cold water.
3. Minimum diameter is the same for hot- and cold-water supply.
4. If only hot water or cold water is used, 1/2 of the total demand is calculated.

Minimum diameter[3] of water-supply pipe (in)	**Minimum pressure (kPa)**	**Flowrate at tap (flowrate of appliance) (L/s)**	**Water demand (water-supply units)[4]**	
			private building	**public building**
1/2	55	0,25	2	4
3/4	55	0,19	-	4
1/2	100	0,19	3	-
1/2	55	0,19	3	-
3/8	100	0,05	-	1
1/2	100	0,31	-	-
1/2	55	0,25	3	-
1/2	55	0,19	-	3
3/4	100	0,95	-	5
1/2	55	0,25	2	4
3/8	55	0,19	3	5
1	100	0,95 à 2,2	6	10
1/2	55	0,19	-	3
3/4	100	0,95	-	5
3/8	55	0,12	1	2
3/8	55	0,16	-	2
1/2	100	0,19	3	-
			6	-
			8	-
3/8			1	2
1/2			2	4
3/4			3	6
1			6	10

NOTE: For a fire pipe, the pressure must meet the requirements of the Fire Underwriters Survey.

From: R.R.Q. (Réglement refondu), C I-12.1, r. 1, table 6.4; D.1638-83, a. 1, 1981.

When the service pipe is supplying water to a large building or to a very small community (e.g., mobile-home park), the size of the pipe is determined by considering all of the equipment requiring water supply installed at this location. The water-supply-unit method can be used. The number of water-supply units is obtained for each appliance (table 4.7), and the units are totalled for the entire building or small community in question. Then, table 4.8 is used, if necessary by interpolation, to determine the maximum flowrate (in L/s) that the service pipe must supply. Head losses in the service pipe must be taken into account when choosing its diameter to secure adequate pressure at the building or community.

Table 4.8 Water consumption: conversion of water-supply units in L/s

	Maximum flowrate to supply to connection (service) pipe	
Water demand (water-supply units)	**if predominance of flush tanks (L/s)**	**if predominance of automatic flush valves (L/s)**
5	1,4	0,25
10	1,7	0,50
20	2,2	0,88
30	2,6	1,3
40	2,9	1,5
50	3,2	1,8
60	3,4	2,0
88	4,0	2,5
124	4,7	3,0
160	5,1	3,5
236	6,2	4,5
300	6,8	5,4
400	8,0	6,7
470	8,5	7,4
500	9,0	7,8
600	9,9	9,0
660	10,2	9,6
700	10,7	10,1
800	11,5	11,2
850	11,9	11,7
900	12,4	12,3
1000	13,1	13,1
1060	13,6	13,6
1280	15,3	15,3
1510	17,0	17,0
1990	20,4	20,4
2480	23,8	23,8
2990	27,2	27,2

4.12.10 Pipe Joints

Manufacturers build pipes in sections, which must be joined to make conduits and distribution networks. There are three main types of joints: mechanical, push-on, and flanged (fig. 4.20). Mechanical and push-on joints are used to connect underground pipes, and the more rigid flanged joints connect pipe sections inside buildings and structures (wastewater-treatment plants, water-treatment plants, pumping stations, etc.). Contractors will often make the decision regarding the choice between mechanical and push-on joints. Flexible ball joints (fig. 4.20d) allow the deflection of adjoining sections by up to 15°: some of those joints are also extensible.

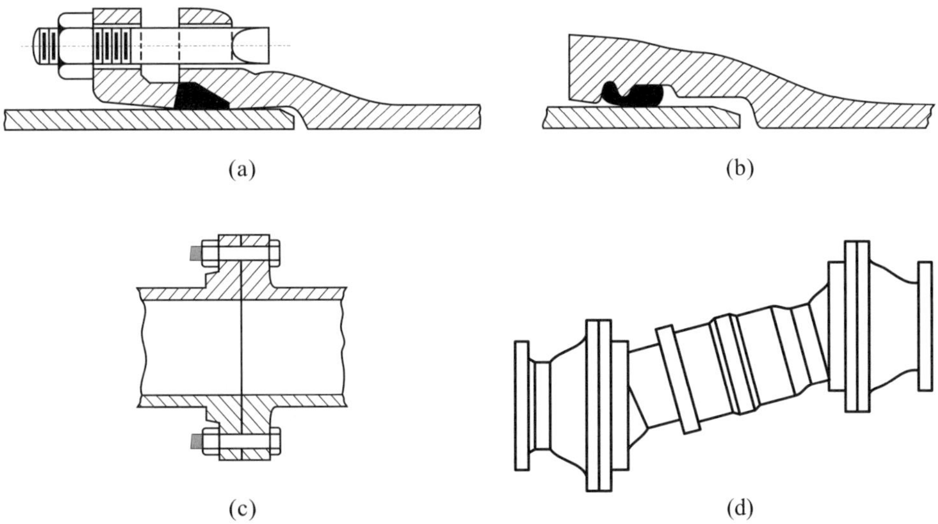

Figure 4.20 Types of joints: a) mechanical; b) push-on; c) flanged; d) flexible-expansion joints.

From: a) and b) Bureau de normalisation du Québec, *Devis normalisé, travaux de construction - clauses techniques générales - conduites d'eau et égouts* (MQ 1809-300, 1987), p. 103; d) EBAA-IRON, Texas.

4.12.11 Air Valves and Vacuum Valves

Air valves (fig. 4.21a), or vents, are installed at high points of the water network to evacuate *small quantities of air* accumulating after, for example, degasification of air or oxygen. However, air valves are not installed on pipes of 300 mm in diameter or less, as these pipes evacuate accumulated air through service pipes and domestic appliances. Air valves also allow air to enter when a vacuum is created in the pipes. Air valves then act as vacuum valves, preventing negative pressure, which could lead to the crushing of pipes or the suction of wastewater.

Another type of air valve (fig. 4.21b) is also used. The air and vacuum valve allows *large amounts of air* to escape from or penetrate into the pipes. They can be useful, for example, when a pipe is emptied or filled during repairs.

Finally, these two types of air valve can be combined to make a universal air valve (fig. 4.21c) which allows the entry or the evacuation of small or large quantities of air.

4.12.12 Minimum and Maximum Pressures

To ensure consumer satisfaction, the water-distribution network must supply water at a desirable minimum pressure when water is flowing. In other words, the water must reach the top floors of buildings. The pressure must also allow the proper and efficient use of domestic appliances (washing machines, dishwashers, etc.) and meet the needs of fire fighting. Thus, a minimum pressure is usually recommended. At the entrance to buildings, minimum pressure is 150 kPa (≅20 lb/in^2 or 14 m of water), the same pressure required for fire hydrants. However, it seems that dishwashers are more efficient when minimum pressure is increased by 50% (225 kPa or 30 lb/in^2 or 21 m of water). Practically, it is sufficient and accepted that these pressures be maintained in the distribution pipes laid in the streets rather than at the bottom of the buildings. Finally, automatic sprinklers require a minimum pressure of 275 kPa (36,6 lb/in^2). Maximum pressure at the entrance to buildings should not exceed 500 kPa in order to protect welded pipes.

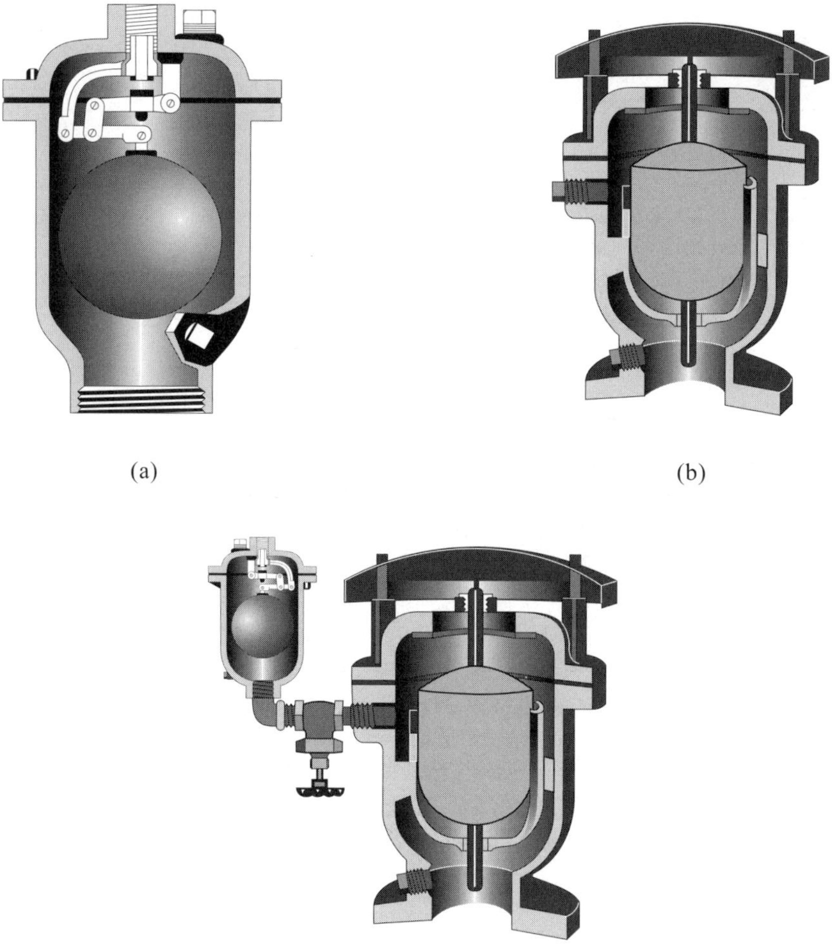

(a) (b) (c)

Figure 4.21 Types of air valves: a) for small amounts of air; b) for large amounts of air; c) universal air valve

From: Les Associés Provan, *Which Air Valve Should I Use* (Ville Saint-Laurent, Quebec: APCO).

4.12.13 Flow Velocity

Distribution pipes are generally designed for a maximum flow velocity of 2,5 or 3,0 m/s, which avoids excessive head losses and delays pipe erosion. There are no requirements for minimum flow velocity; however, zero or close to zero velocities are responsible for poor quality water so a minimum flow velocity of 0,3 m/s is usually recommended during a low-consumption period (see section 5.4.1). The "water hammer" phenomenon, due to the rapid opening and closing of valves or the starting and stopping of pumps, must also be avoided. The water hammer pressure (and associated noise) occur because of rapid variations in flow velocity (see section 4.12.18).

4.12.14 Fire-Fighting Requirements

The flowrate necessary to protect a building against fire is calculated using the requirements set by the Fire Underwriters Survey (guidelines given in Appendix F). The fire flowrate is determined using the following equation: (equation taken from document mentioned above, where it is written as follows: $D = 220\,C_{fire}\sqrt{S}$):

$$Q = 220\,C_{fire}\sqrt{A} \qquad (4.4)$$

where Q = fire flowrate (L/min)
C_{fire} = coefficient (depends on the type of building)
A = total area of floors, basement not included

The flowrate is then increased or decreased depending on circumstances: building allocation (use or content), presence of automatic sprinklers, and risk of fire propagation (to neighbouring buildings, mostly; see appendix F). For a municipality with a population under 250 000, we assume that there is only one major fire at a time (or a number of smaller fires requiring the same flowrate or volume of water) and that it occurs during the period of maximum daily consumption $(Q_d)_{max}$ (see section 5.2.2).

EXAMPLE 4.3

Calculate the necessary flowrate for fire protection of a church that has a basement and a first floor of 35 m by 100 m. The exterior walls are made of wood, the roof is made of wood shingles (highly combustible), and the frame is also made of wood. There are no automatic sprinklers, and the neighbouring buildings (on three sides) are 15 m away. If a fire occurs, how much time should be allowed to control the situation?

Solution

Basic Flowrate

$$A = 35 \times 100 = 3500 \text{ m}^2 \text{ (ignore the basement)}$$

$$C_{fire} = 1,5 \text{ (wooden construction)}$$

As a result (equation 4.4):

$$\begin{aligned} Q &= 220\,C_{fire}\sqrt{A} \\ &= 220 \times 1{,}5\ \sqrt{3500} \\ &= 19\,523\ \text{L/min} \end{aligned}$$

This value is rounded off to 20 000 L/min.

For a wooden structure, the maximum flowrate must not exceed 30 000 L/min; therefore, the value of 20 000 L/min is acceptable.

Correction Related to the Building Allocation (use or content)

An increase of 15%–as we consider that the contents of the building (such as pews) are highly combustible:

$$20\,000 \times 0{,}15 = 3000\ \text{L/min}$$

It then yields:

$$Q = 20\,000 + 3000 = 23\,000\ \text{L/min}$$

This value must not be rounded off.

Correction Related to the Presence of Automatic Sprinklers

None present.

Correction Related to the Risk of Fire Propagation to Neighbouring Buildings

An increase of 45% (15% for each side where neighbouring buildings are within 10,1 to 20 m):

$$23\,000 \times 0{,}45 = 10\,350\ \text{L/min}$$

It then yields:

$$Q = 23\,000 + 0 + 10\,350 = 33\,350\ \text{L/min}$$

This value is rounded off to 33 000 L/min.

Correction Related to Wood Shingle Roof

2000 L/min must be added.

The total fire flowrate then equals:

$$Q = 33\,000 + 2000 = 35\,000\ \text{L/min}$$

The data in the table on page 11 of the Guide (see Appendix F) indicate that we must be able to fight this fire for 8,25 hours.

NOTE: This example clearly illustrates that many of the guidelines in appendix F are open to interpretation.

EXAMPLE 4.4

An area is composed entirely of two-storey duplexes built two by two. These buildings have a basement area of 70 m^2, the exterior walls are made of bricks, and the frames are made of wood. The distances between the buildings are: back, 25 m; side, 5 m; front, 35 m. The party wall of each building is a fire wall. Calculate a) the basic fire-protection flowrate for each property; b) the various corrections applied to the basic fire-protection flowrate; c) the fire flowrate; d) the duration of the fire; e) the area covered by each fire hydrant.

Solution

a) Basic fire protection flowrate for each property

$$A = 70 + 70 = 140 \text{ m}^2 \text{ (the basement is not taken in to account)}$$
$$C_{fire} = 1{,}0 \text{ (ordinary construction)}$$
$$Q = 220\, C_{fire} \sqrt{A} = 220 \times 1{,}0 \sqrt{140} \text{ (equat. 4.4)}$$
$$= 2603 \text{ L/min}$$

This value is rounded off to the nearest 1000 L/min: 3000 L/min.

b) Corrections Related to the Basic Fire Protection Flowrate

No correction required as the content is combustible.

Correction related to the presence of automatic sprinklers

None, as there are none present.

Correction related to the risk of fire propagation to neighbouring buildings

For neighbour at 25 m: increase of 10%
For neighbour at 5 m: increase of 20%
For neighbour at 35 m: increase of 5%
Total increase: 35%

The fire flowrate is therefore:

$$Q = 3000 \times 1{,}35 = 4050 \text{ L/min}$$

This value is rounded off to the nearest 1000 L/min: 4000 L/min.

c) Duration of the Fire

The table on page 11 of the Guide (appendix F) shows that the required fire flow is 1,5 hours for a flowrate of 4000 L/min.

d) Average Area Covered by Each Fire Hydrant

The table on page 10 of the Guide (appendix F) shows that for a fire flowrate of 4000 L/min, each fire hydrant covers a circular surface of 15 000 m^2. The hydrants must be located close enough to each other so that all the 15 000 m^2 circles cover the total area built with duplexes that needs protection.

4.12.15 Roughness Coefficient

In a pipe, the flowrate calculated using the Hazen-Williams equation (equation 2.4) varies as a function of the roughness coefficient, C_{HW}. For constant head losses, the flowrate of a pipe with a roughness coefficient C_{HW} = 130 is 30% higher than the flowrate of a pipe with C_{HW} = 100. Therefore, the correct value of C_{HW} is very important. Many parameters affect the value of the roughness coefficient: pipe material, pipe diameter, flow regime (laminary or turbulent), age and condition of the pipe, and water temperature.

For new pipes made of ductile cast-iron (without cement lining), a C_{HW} of 140 is generally adequate; however, this value tends to decrease with pipe wear (table 4.9a). Plastic pipes, in particular those made of PVC, have a C_{HW} of 150, which does not seem to vary as the pipe ages. Table 4.9b shows C_{HW} values of various pipe materials.

In general, we choose a roughness coefficient C_{HW} of 140 for ductile cast-iron pipes with a cement lining and for steel-concrete (hyprescon) pipes. We assume, however, that this value does not vary with the age of the pipe.

The Hazen-Williams coefficient of existing pipes can be evaluated experimentally. A given flowrate is circulated in a long length of pipe of known diameter. Head losses are measured and flow velocity calculated. The Hazen-Williams equation (eq. 2.4) is then used to compute the C_{HW} coefficient of the pipe.

Table 4.9a Values of the Hazen-Williams roughness coefficient, C_{HW}, for common pipe material

Type of pipe	C_{HW}
PVC pipe	150
Very smooth pipe	140
New ductile cast-iron or welded steel	130
Wood, concrete	120
Clay, new riveted steel	110
Old cast-iron, brick	100
Badly corroded cast-iron or steel	80
Ductile cast-iron with inside cement coating	140

From: R.W. Jeppson, *Analysis of Flow in Pipe Networks*, Ann Arbor Science, 1977, p. 41.

Table 4.9b Hazen-Williams roughness coefficient of cast-iron pipes without interior ciment lining as a function of pipe age

Real diameter (mm)	pipe (years) (in)	0	5	10	15	20	30	40	50
100	4	130	115	105	100	90	75	65	55
150	6	130	115	105	100	90	75	65	55
200	8	130	120	110	100	95	80	70	60
255	10	130	120	110	100	95	80	70	60
305	12	130	120	110	100	95	80	75	65
355	14	130	120	110	100	95	80	75	65
405	16	130	120	110	100	95	85	80	70
460	18	130	120	110	100	95	85	80	70
510	20	130	120	110	100	95	85	80	70
610	24	130	120	110	100	100	90	85	75

4.12.16 Head Losses in a Pipe

When water flows through a distribution pipe, head losses occur; these can be calculated using the Hazen-Williams equation or the Darcy-Weisbach equation.

When determining the optimum diameter of a pipe, a compromise is made between investment costs and operation costs. In fact, when we increase the diameter of a pipe, investment costs rise but head losses decrease, creating important savings in terms of pumping costs.

4.12.17 Minor Head Losses

In a water-distribution network, the use of special equipment (elbows, valves, check valves, and service pipes) creates additional head losses, called minor head losses. These head losses are described as water heights and are proportional to the square of the velocity of the water circulating around that piece of equipment; this equation is written:

$$H_m = \frac{K_L v^2}{2g} \tag{4.5}$$

where H_m = head losses due to an obstacle (m)
K_L = coefficient of the obstacle
v = velocity of water circulating around or close to the obstacle (m/s)
g = acceleration due to gravity (m/s^2)

Values of K_L for different situations are given in chapter 2.

4.12.18 Water Hammer

The water hammer is a pressure wave caused by a rapid variation in the hydraulic regime. It travels through the water and the pipe containing it. It may result from valve closure or power failure. The change in pressure can produce either a positive (above normal) or a negative (below normal) water hammer. It often causes noise and breakage. This phenomenon is important in pumping stations and in distribution networks, where it requires careful consideration, especially when velocities are high and pipes are long. To avoid water hammer problems, open pipes and open reservoirs may be installed.

4.13 THRUST BLOCKS AND RESTRAINED JOINTS

Distribution pipes are subject to strong internal pressures. The Fire Underwriters Survey requires that pipes resist pressures of 1000 kPa, and some organizations require even higher standards.

When a pipe is diverted in a new direction (vertical or horizontal), an internal force is exerted on the coupling between the two pipe sections, allowing the change of direction. If there were no external resisting force, the coupling installed where the direction change occurs would be dislocated. For this reason, a thrust block is installed behind the coupling, or special restraints are installed on the joints between the pipe sections on each side of the coupling.

4.13.1 Thrust Blocks

The thrust block (fig. 4.22, example 4.5) is a concrete mass used to transfer the thrust exerted on the coupling to the surrounding ground. The bearing surface between the thrust block and the ground must be large enough for the ground to sustain the thrust exerted on the coupling. When this surface is calculated, the height and depth of the burying of the thrust block, the unit weight of the ground, and the nature of the ground (especially the internal friction angle, ϕ; table 4.10) must be taken into account.

For silt, the friction angle varies from 26° to 30°; for a mixture of sand and stones, it varies from 32° to 36°.

Table 4.10 Internal friction angles, ϕ, for sand

Compaction of sand	Internal friction angle ϕ (°)
Very little	< 30
Little	30 to 45
Average	35 to 40
Sufficient	40 to 45
High	> 45

Adapted from: J. Verdeyen, V. Roisin, and J. Nuyens, *La mécanique des sols* (Paris: Presses universitaires de Bruxelles and Dunod, 1968), p. 194.

For all granular material is suggested that you use the internal friction angles of sand. In the absence of information concerning compaction, the smallest angle, 30°, is chosen. In the case of clay, shearing forces must be taken into account.

According to Proctor Standard, the thrust block should rest on untouched natural ground, or when possible on backfill compacted to 90% density. The height of the thrust block should be equal to or smaller than the distance separating the ground surface and the top of the thrust block.

EXAMPLE 4.5

A distribution pipe is diverted on a 90° angle, θ. The pipe has a 300 mm nominal diameter (305 mm inside diameter) and is installed in the horizontal plane.

a) Calculate the force exerted on the coupling if the pipes must sustain an internal pressure p of 1200 kPa.

b) To avoid dislocation of the coupling and pipes, a thrust block 0,6 m high and 2,25 m deep is installed (fig. 4.22). Calculate the length of the thrust block if the ground is composed of sand of 1765 kg/m^3 with a 30° internal friction angle, ϕ, and if the weight of the pipe is 61 kg/m?

Solution

a) Force, F_{res}, exerted on the coupling

A pressure p of 1200 kPa is exerted along the longitudinal axis of the two pipe sections joined by a coupling (fig. 4.22). Forces F_1 and F_2 caused by this pressure are both equal to the product of p by the area, A; this area corresponds to the inside transversal cross section of the pipe in which the force is exerted.

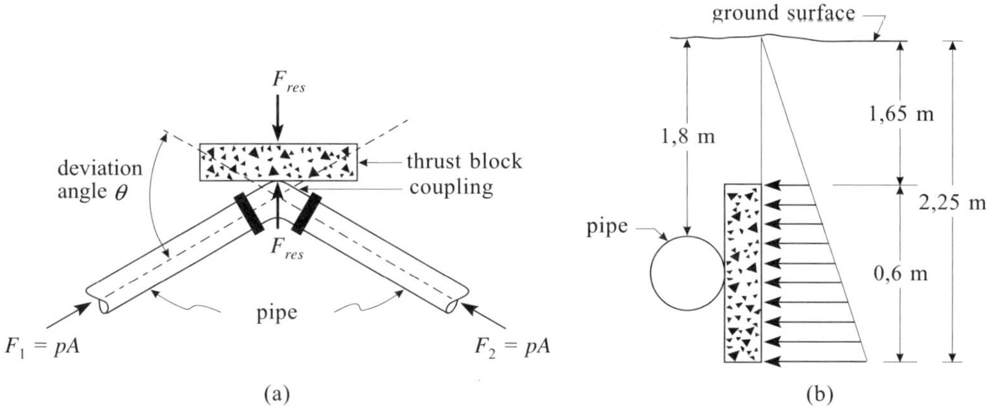

Figure 4.22 (example 4.5) Thrust block: a) plan view; b) side view

Along the x axis, the two components of forces F_1 and F_2 cancel out, but along the y axis, the two components add up and tend toward dislocation of the coupling. The net or resulting force, F_{res}, of these two forces is calculated as follows:

$$F_{res} = F_1 \cos[(180 - \theta)/2] + F_2 \cos[(180 - \theta)/2]$$

since $F_1 = F_2$

it yields

$$\begin{aligned} F_{res} &= 2F_1 \cos[(180 - \theta)/2] \\ &= 2F_1 \cos(90/2) \\ &= 2 \times 1200 \times (\pi\ d^2/4)\cos(90/2) \\ &= 2 \times 1200 \times [\pi(0{,}305)^2/4] \times 0{,}707 \\ &= 124 \text{ kN} \end{aligned}$$

b) Length of thrust block, L_b

The thrust block must have a wide enough surface for the bearing soil to sustain the force exerted on the coupling without moving.

The diagram of the forces exerted on the thrust block in the vertical plane is represented by a trapezoid (fig. 4.22b). The net force per unit length acting on the thrust block, F_b, is calculated as follows:

$$\begin{aligned} F_b &= \left(F_{top} + \frac{F_{bot} - F_{top}}{2}\right) \times h_b \\ &= \left(85{,}71 + \frac{116{,}87 - 85{,}71}{2}\right) \times 0{,}6 \ \text{(see computations below)} \\ &= 60{,}77 \text{ kN/m} \end{aligned}$$

where F_{top} = force exerted by the ground at the top of the thrust block (kPa)
F_{bot} = force exerted by the ground at the bottom of the thrust block (kPa)
h_b = height of the thrust block (m)

The length of the thrust block, L_b, can then be calculated using the following equation:

$$\begin{aligned} L_b &= F_{res} \div F_b \\ &= 124 \div 60{,}77 \\ &= 2{,}04 \text{ m} \end{aligned}$$

Let us examine how the forces exerted by the ground on the top and bottom of the thrust block are calculated.

– Force exerted by the ground at the top part of the thrust block,

F_{top} = unit weight of the ground × h of ground × g × soil coefficient (see below)
= 1765 × 1,65 × 9,81 × 3
= 85 708 N/m^2 (85,71 kPa)

where g is the acceleration of gravity.

The soil coefficient takes into account the ground's resistance to the motion due to the friction between the ground particles. This coefficient is determined using the following equation, in which angle ϕ is the internal friction angle of the ground:

$$\textbf{soil coefficient} = \frac{1 + \sin \phi}{1 - \sin \phi} \tag{4.6}$$

In this example, $\phi = 30°$, so it yields:

$$\textbf{soil coefficient} = \frac{1 + 0{,}5}{1 - 0{,}5} = 3$$

– Force exerted by the ground at the bottom part of the thrust block, F_{bot}

$$\begin{aligned} F_{bot} &= \text{unit weight of the ground} \times (h \text{ of ground} + h_b) \times g \times \text{soil coefficient} \\ &= 1765 \times (1{,}65 + 0{,}6) \times 9{,}81 \times 3 \\ &= 116\ 874\ \text{N/m}^2\ (116{,}87\ \text{kPa}) \end{aligned}$$

Often, 1 m^3 concrete blocks are used as thrust blocks.

4.13.2 Restrained Joints

Another way to avoid dislocation of pipe sections in areas of directional changes is to restrain the joints between the pipe sections over a certain length of the pipe on each side of the coupling (fig. 4.23). Pipe sections so linked act monolithically. This method seeks to use the friction between the pipe sections and the ground while taking into account the weight of the backfill, the pipe, and the water. The following equation is used:

$$L = \frac{1000\ pA\ (1 - \cos \theta)\ FS}{fg(2m_r + m_c + m_e)} \tag{4.7}$$

where L = length of the pipe on each side of the coupling (m)
p = water pressure inside the pipes (kPa)
A = inside cross-sectional area of the pipe (m^2)
θ = deviation angle (see fig. 4.22)
FS = safety coefficient (1,5 is suggested)
f = friction coefficient between the pipe and the ground (0,3 is suggested)
m_r = mass of the backfill (kg/linear metre of pipe)
m_c = mass of the pipe (kg/linear metre of pipe)
m_e = mass of water in the pipe (kg/linear metre of pipe)

According to equation 4.7, the weight of the backfill affects the friction between pipe and ground twice as much as the weight of the pipe itself or of the water it contains. This is because the backfill creates friction on all the outside surface of the pipe while the two other weights create friction mainly on the outside invert of the pipe.

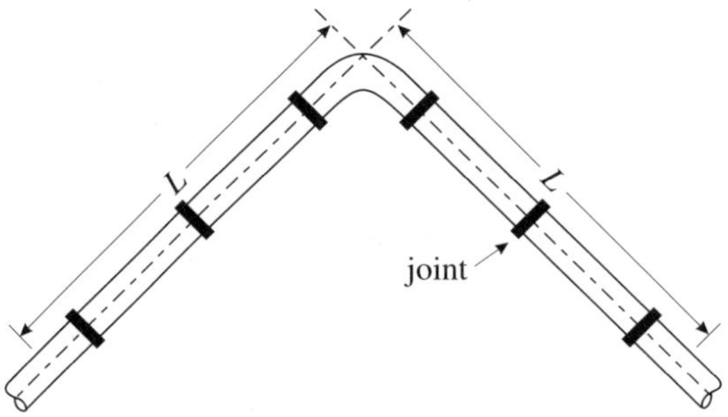

Figure 4.23 Pipe length on which the joints between the pipe sections are restrained (in the present case, six joints are restrained)

EXAMPLE 4.6

A water-distribution pipe of 305 mm inner diameter (320 mm outside diameter) and weighing 61 kg/m is buried under 1,8 m of backfill weighing 1765 kg/m^3. The pipe deviates at an angle of 60° and must sustain an internal pressure of 1200 kPa. Calculate the length of pipe (L) on each side of the coupling on which the joints between pipe sections must be restrained.

This length is determined using equation 4.7:

$$L = \frac{1000\ pA\ (1 - \cos\theta)\ FS}{fg\left(2m_r + m_c + m_e\right)}$$

$$= \frac{1000 \times 1200 \times 0{,}07(1 - 0{,}5)\,1{,}5}{0{,}3 \times 9{,}81\left[(2 \times 1017) + 61 + 73{,}1\right]}$$

$$= 9{,}9 \text{ m } \left(\text{on each side of the coupling}\right)$$

where $A = \pi\,(0{,}305)^2/4 = 0{,}07$ m^2

$\cos\theta = \cos\,(60°) = 0{,}5$

$m_r = 1765$ kg/m^3 × 1,8 m (height) × 0,320 (outside diam.) = 1017 kg/m

$m_e = 1000$ kg/m^3 × [$\pi\,(0{,}305^2/4)$] = 73,1 kg/m

CONCLUSION

Supplying water to an urban community is a demanding task for local authorities that rely on engineers to analyze problems and compute alternatives before suggesting appropriate solutions. For the water intake and intake pipe, they must plan the needs of the community for up to 50 years, without obstructions from ice or floating and suspended solids.

Every day, the drinking-water treatment system operates at the average daily flowrate for that day. This system consists of a raw-water reservoir, low-lift pumps, and the treatment plant itself. Engineers must take these conditions into account when choosing pumping equipment and when designing hydraulics related to the raw-water reservoir and the clear-water basin (volume, maximum and minimum water levels, elevation of the bottom of basins, etc.).

The water distribution network must supply water in adequate quantities and pressures to meet the needs of consumers and for fire fighting. Therefore, engineers designing such networks must determine pipe sizes, identify fire-hydrant locations, and evaluate the need for clear-water reservoirs and pumping-station installations.

Finally, pipes must be buried deep enough to avoid freezing water in winter, and equipment must be installed to evacuate accumulated air and protect the pipes of the network against negative pressures or water hammers.

PROBLEMS

4.1 Calculate the flowrate required to put out a fire in an ordinarily constructed gas station. The building is 16 m long, 6 m wide, and 8 m high (equivalent to a two-storey building). Two houses are located close to the gas station: one is 20 m from the back of the building, and the other is 10 m to the side. The gas station is equipped with automatic sprinklers (50% reduction). The content of the building is easily combustible (free burning).

4.2 A water-treatment plant is needed to meet the water requirements of a suburb in the Montreal region. This suburb contains only single-family brick houses with a maximum of two storeys (basement, first floor, and second floor). The spacing between the houses is a minimum of 3 m and a maximum of 10 m. There are 18 000 inhabitants, total consumption is 400 L/person·d, and the maximum daily and hourly consumption factors of the year are 1,8 and 2,7, respectively.

a) Compute the required capacity, in L/s, of the arterial main (pipe) between the treatment plant and the municipality, to ensure fire protection.

b) Compute the minimum theoretical diameter, in mm, of the intake pipe linking the water intake to the raw water reservoir if the water velocity must remain between 0,3 and 2,5 m/s.

c) In the river, if the water level over the intake is 100 m, draw the hydraulic grade line between the water intake and the raw-water reservoir. Assume that the total equivalent length of the intake pipe is 500 m (its actual length is 485 m), that the Hazen-Williams coefficient is 130, and that the water level in the raw-water reservoir is less than 1,5 m below the level of the river. We also assume that minor head losses, for which the equivalent length is 15 m, are uniformly distributed along the pipe. Is the diameter calculated in (b) acceptable? If not what must be the new diameter and what is the flow velocity in the pipe?

4.3 What is the diameter of the drinking-water service pipe (house pipe) connection of a public building, assuming that:

– there could be up to 100 persons in the building;

- average water consumption is 50 L/person·d;
- maximum daily consumption for the year is 1,5 times the average daily consumption of the same year;
- maximum hourly consumption for the year is 3 times the average hourly consumption of the same year;
- there are 5 automatic flush-valve toilets, 2 commercial kitchen sinks equipped with flush valves, and 5 automatic flush-valve urinals;
- the building also has 7 showers equipped with combined valves, 1 utility sink with an automatic flush valve, and 2 unidentified appliances with a water-supply-pipe diameter of 14 mm (1/2 in);
- automatic flush valves are predominant in the building;
- the value of the Hazen-Williams coefficient, C_{HW}, of the house connection is 130;
- water velocity in the service pipe must not exceed 2,2 m/s;
- the service pipe (in soft copper *K* with commercial diameters of 20, 25, 32, 37, and 50 mm) is 12 m long.

Compare the maximum flowrate calculated with water-supply units and the maximum hourly flowrate. Choose the diameter of the service pipe, and comment.

4.4 A plywood warehouse (material easily combustible but not free-burning) is 15 m high (equivalent to a three-storey building) and at the edge of a residential area. The surface area of the warehouse is equivalent to the area of a rectangle 80 m by 40 m minus the area of a smaller rectangle 40 m by 20 m, which is removed from one of its corners. There is a basement under half of the floor surface of the building. The building is made entirely of bricks and protected by an automatic sprinkler system that is without proper system supervision (50% reduction). The front of the building, the longest side, faces the street, and two-storey residential buildings are located on the three adjacent sides. One is 5 m away, the other two are 15 m away.

a) Calculate the required fire-protection flow. Make all necessary assumptions and discuss your choices.

b) How many fire hydrants are required and what type should they be if the front of the building is located parallel to the street, 5 m from the street line? The fire hydrants are located on the street line, on the same side of the street as the warehouse.

Chapter 5

DESIGN OF A DRINKING-WATER DISTRIBUTION SYSTEM

INTRODUCTION

As mentioned in chapters 3 and 4, the first task of the engineer designing a drinking-water distribution network is to evaluate the volumes of water required to meet the current and future needs of the population. As well as planning for water consumption, the engineer must also take into account domestic, industrial, and commercial needs, as well as storage for fire protection. Based on this information, the water-distribution network can then be designed.

This chapter reviews the methods engineers use to analyze and design a water-distribution network. The needs of such a network are identified, and mathematical as well as computer methods for the design of new networks or evaluation of existing ones are presented.

5.1 DEFINITIONS

A water-distribution network is composed of different types of pipes, which are classified according to their diameters and functions: arterial mains, secondary pipes, and small distribution mains.

5.1.1 Primary or Arterial Mains

Arterial mains are pipes that link the water-treatment plant or pumping station(s) to different parts of the city. They have a diameter of 350 mm or more. They are equipped with draining chambers at low points, as well as air valves (vents) and vacuum valves at high points (to allow air to enter or evacuate, depending on the situation). These pipes have no direct connections (service pipes) to houses.

5.1.2 Secondary Lines or Pipes

Secondary pipes are 250 or 300 mm in diameter. They are linked to the primary mains and form looped networks. They are located about 1000 m from one another. Houses can be connected to this secondary pipes.

5.1.3 Small Distribution Mains (Local Pipes)

Small distribution mains are the smallest pipes in the network. However, their diameter cannot be less than 150 mm to ensure water supply for fire fighting. They supply water to the inhabitants of only a few streets.

5.1.4 Nodes, Loops, and Disaggregated Flow (Consumption)

A node is the junction between two or among several interconnected pipes. The loop is a closed-circuit system, made of three pipes or more. A looped distribution network is made of many loops. When designing a network the flows drawn along the pipes are assumed to be drawn at their nearest node, where the resulting flow is called disaggregated flow.

5.2 SPECIFIC REQUIREMENTS RELATED TO NETWORK DESIGN

For a water-distribution network to perform adequately, it must be able to supply sufficient pressure, flowrate, and water volume at all times throughout the design period. The engineer must also design networks that perform well under some critical conditions, described below.

5.2.1 Maximum or Peak Hourly Consumption

First, the engineer must make sure that the network works properly during current and future peak hourly consumption. In general, as seen in chapter 3, the maximum hourly flowrate in a given year varies from 225% to 450% of the average hourly consumption for that year (section 3.5, table 3.8).

5.2.2 Maximum or Peak Daily Consumption During One or Many Fires

The engineer must then evaluate the behaviour of the network when a major fire occurs during a period of maximum daily consumption for a year (usually between 1,5 and 3,0 times the average daily flowrate for that year). When the population exceeds 250 000, the engineer must also foresee the possibility of two major fires occurring at the same time. Moreover, the engineer must study the behaviour of the network according to the conflagration site (see the guidelines for the design of water networks for fire protection of the Fire Underwriters Survey, appendix F; for questions related to numbers of fires, fire flowrates, and volume of water to be set aside for five protection, engineers should contact this organization).

5.2.3 Maximum or Peak Daily Consumption during Rupture of an Arterial Main or a Secondary Pipe

It can be useful to study the behaviour of a network when a section of a secondary or main pipe is put out of service on the day of maximum daily consumption. Of course, this service interruption is for maintenance or repairs, and is not in any way intended to justify the construction of parallel pipes. It simply allows the engineer to study the performance of the network during such events and to apply corrective measures if the quality of the service is diminished.

5.2.4 Special Cases

For a given network, certain special cases may be analyzed, for example, to ensure that a distribution reservoir can be filled during a specific period, such as night time, when consumption is at a minimum.

5.3 WATER-DISTRIBUTION NETWORK DESIGN

The design of a water-distribution network has many steps, each of which must be addressed before water is supplied to a community (fig. 5.1). These important steps are briefly described below.

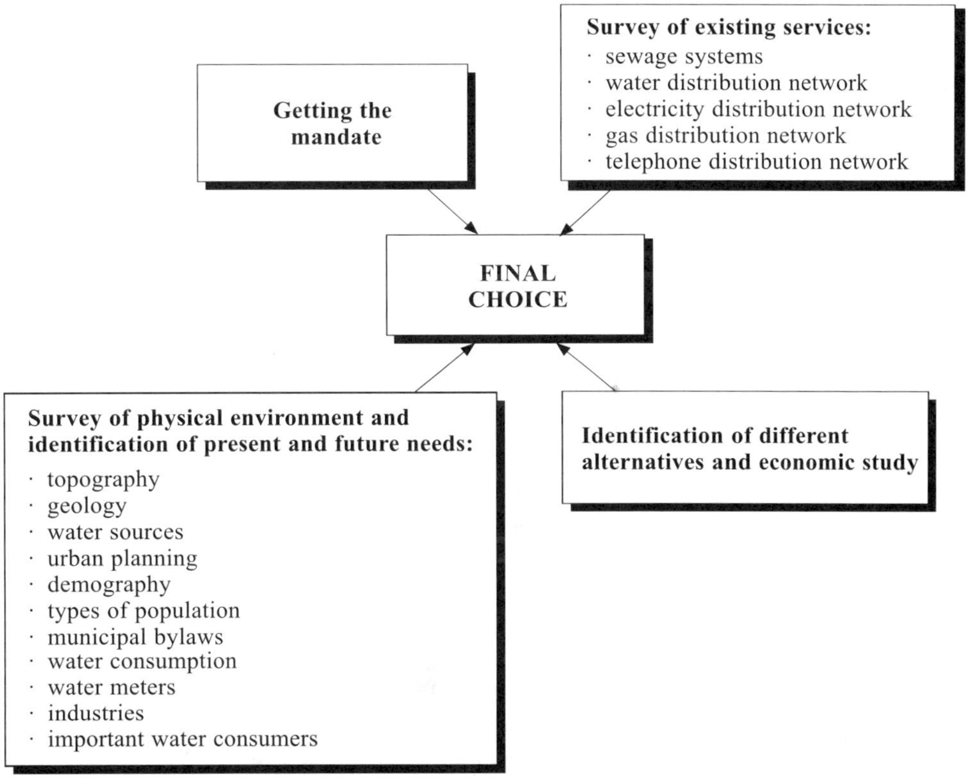

Figure 5.1 Stages of the study leading to final choices for a water distribution system

5.3.1 Obtaining the Mandate

- The engineer obtains from local authorities (usually the city) a clear, precise and detailed mandate.
- The engineer must define the targeted area (current and future). This area may exceed city limits.

5.3.2 Survey of Existing Services

The engineer must obtain from local authorities the most recent drawings of the sewage systems and the water, gas, electricity, and telephone distribution networks. These drawings must include all existing equipment: pipes, pumping stations, valves, reservoirs, and so on.

5.3.3 Survey of the Physical Environment and Identification of Current and Future Needs

- Study of the physical and geological environment.
- Identification of the current and future use of the land: zoning and demography.
- Evaluation of current and future water consumption using available data (water production over the preceding few years; water-meter readings for businesses, stores, and institutions; etc.). The engineer can then determine with maximum precision:
  - total water consumption per inhabitant;
  - domestic consumption;
  - commercial consumption;
  - industrial consumption;
  - institutional consumption;
  - water losses;
  - variations in all types of consumption;
  - consumption hydrograms for maximum daily consumption over the preceding few years.
- Establishment of criteria to study existing and future networks.
- Identification of critical conditions for which the network must be designed.
- Study of the behaviour of the network and the pressures for current conditions: for example, the results of a computer model are compared with on-site measurements.

5.3.4 Identification of Alternatives and Economic Study

- Identification of different possible sites for arterial main and secondary pipes.
- Evaluation of the size of the zones served by each node and computation of the disaggregated flow at each of these nodes for the various critical conditions.
- Simulation of network behaviour for different critical conditions.
- Economic study: construction and maintenance costs for the various possible solutions.

5.3.5 Final Choice

- Choice of the network that best suits the needs.
- Presentation of a detailed report to the organization in charge; the report should present a recommended solution, construction costs, and a construction schedule.

5.4 HYDRAULIC STUDY OF A WATER-DISTRIBUTION SYSTEM

A number of means and tools are available for engineers to make a hydraulic study of a distribution network. These are presented below. Basic knowledge of hydraulics will be outlined and equations will be presented to help you analyze and design hydraulic networks under pressure.

5.4.1 Initial Selection of Pipe Diameters

Before simulating the behaviour of a network for different water-consumption conditions, you must choose the diameter and location of the pipes. Here, you must rely on experience–which can be a great help–and apply certain principles and guidelines, as follows:

- Pipes with a diameter of less than 150 mm (6 in) are authorized only if water does not have to provide fire protection.
- In a looped network, when pipe sections of 150 mm in diameter are longer than 180 m, the intersection pipes (the pipes on which they are linked) must be at least 200 mm (8 in) in diameter to avoid excessive head losses.
- In a dead end, or when the network is only partially looped, the use of pipes with a diameter exceeding 150 mm is recommended.
- For secondary pipes, two smaller pipes located a few streets from each other may be preferable to one large pipe, and make operation of the network more flexible.
- When the network crosses a train track, river, highway, or any other such obstacle, it is preferable to have several crossings at different locations to make operation of the network more flexible.
- The loops of a network must not exceed 350 m (1200 ft) across.
- The loops of a secondary pipe network must not exceed 900 m (3000 ft) across.
- Water-flow velocity during the normal design period must vary from 0,8 to 3 m/s (3 and 10 ft/s).
- During low-consumption periods, for example at night, the minimum velocity should not fall below 0,3 m/s (section 4.12.13).
- In a network used for fire fighting, the minimum pressure is 150 kPa (section 4.12.12).
- The maximum pressure at a building entrance must not exceed 500 kPa.

Simulation of network behaviour allows the engineer to evaluate these initial choices. If the results prove that pressures or velocities are too high or too low in some areas or pipes, corrective measures can be applied to these initial choices. Further simulations are then run until the solution is finalized.

5.4.2 Relation Between Flowrate (Q) and Head Losses (H_L) in a Pipe under Pressure

Using the Darcy-Weisbach (2.3) and Hazen-Williams (2.4) equations, we shall show the relation between head losses, H_L, that occur in a pressured pipe, and the flowrate, Q, flowing through this pipe. As flowrate is the product of the cross-sectional area of the flow and the water velocity, and since, in this case, $R_h = d/4$:

$$Q = Av = \left(\frac{\pi d^2}{4}\right) \times 0{,}849\ C_{HW} \left(\frac{d}{4}\right)^{0,63} s^{0,54}$$

As the slope of the energy grade line, *s,* equals the ratio of the head loss (H_L) to the length of the pipe, it yields:

$$Q = K'\left(H_L\right)^{0,54}$$

or

$$H_L = K\,Q^{1,852} = K\,Q^m \tag{5.1}$$

where

$$K = \frac{\beta L}{d^{4,871} C_{HW}{}^{1,852}} \tag{5.2}$$

Table 5.1 shows the different values of the coefficient β as a function of flowrate, Q, pipe diameter, d, pipe length, L, and head losses, H_L, in SI and imperial units.

Table 5.1 Values of β as a function of flowrate, diameter, length, and head losses

	H_L (ft)		H_L (m)			
Q	Gal CA/min	Gal US/min	m^3/s	L/s	L/s	m^3/s
d	in	in	m	mm	cm	mm
L	ft	ft	m	m	m	m
β	14,693	10,467	10,679	$1{,}218 \times 10^{10}$	163 874	$4{,}38 \times 10^{15}$

5.4.3 Equivalent Pipe Method

To reduce the number of computations or amount of computer memory needed to study a network, actual pipes are often replaced with imaginary equivalent pipes in these calculations. Similarly, the computation of pressures often involves the replacement of equipment (elbows, tees, valves, etc.) with equivalent pipes. The equipment installed on pipes is replaced with fictitious pipes of same diameters, but of lengths resulting in the same actual head losses.

In this section, we demonstrate how to simplify a network by using the equivalent pipe method. It consists of replacing pipes installed in parallel or in series, and head losses due to pieces of equipment, with equivalent pipes. We also show how to analyze and design a distribution network using iterative methods, such as the Hardy-Cross method.

Pipes in series. A pipe equivalent to two pipes installed in series (fig. 5.2) carries the same flowrate and creates the same head losses as the two pipes. For pipes in series, the head losses in the equivalent pipe are described by the following equation:

$$(H_L)_{equ} = (H_L)_1 + (H_L)_2 = \Sigma (H_L)_i$$

However $H_L = KQ^m$ (equation 5.1)

therefore

$$(H_L)_{equ} = K_1 Q^{m_1} + K_2 Q^{m_2} = \Sigma\ K_i Q^{m_i}$$

Moreover, we know that for the Hazen-Williams equation:

$$m_{equ} = m_1 = m_2 = m = 1{,}852$$

Therefore

$$\begin{aligned}(H_L)_{equ} &= K_1 Q^m + K_2 Q^m = K_1 Q^{1{,}852} + K_2 Q^{1{,}852} \\ &= Q^m \left(\Sigma\ K_i\right) \\ &= Q^{1{,}852} \left(\Sigma\ K_i\right)\end{aligned}$$

In the above equations:

$(H_L)_{equ}$ = head losses in the equivalent pipe
$(H_L)_i$ = head losses in pipe i
Q_i = flowrate in pipe i
m = 1,852
K_i = proportionality constant of pipe i (equation 5.2)
K_{equ} = proportionality constant of the equivalent pipe

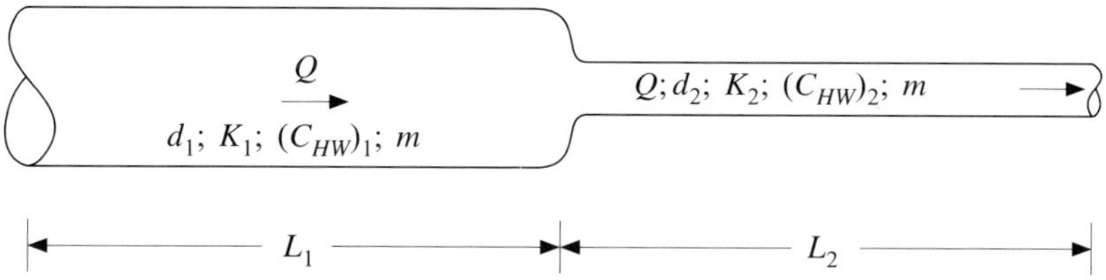

Figure 5.2 Two pipes in series

However, we have seen (equation 5.2) that for a given pressurized pipe:

$$K_i = \frac{\beta L_i}{d_i^{4,871}(C_{HW})_i^{1,852}}$$

where β = 10,679
L_i = length of pipe i (m)
d_i = inner diameter of pipe i (m)
$(C_{HW})_i$ = the Hazen-Williams roughness coefficient of pipe i

As a result:

$$\begin{aligned}(H_L)_{equ} &= Q^{1,852}\ \Sigma \frac{\beta L_i}{d_i^{4,871}(C_{HW})_i^{1,852}} \\ &= Q^{1,852}\,(K)_{equ} \\ &= Q^{1,852} \frac{\beta L_{equ}}{(d_{equ})^{4,871}(C_{HW})_{equ}^{1,852}} \qquad (5.3)\end{aligned}$$

where Q = flowrates in pipes in series (m^3/s)

The above computations are also valid when more than two pipes are installed in series.

Pipes in parallel. A pipe equivalent to two pipes installed in parallel (fig. 5.3), carries a flowrate equal to the sum of the flowrates of the parallel pipes and creates head losses equal to the head losses in each pipe. (At equilibrium, head losses in each branch of a loop – see Law of conservation of energy below – are equal.) For pipes in parallel, the head losses in the equivalent pipe are expressed as follows:

$$(H_L)_{equ} = (H_L)_1 = (H_L)_2 = H_L$$

Moreover, we can state:

$$Q_{equ} = Q_1 + Q_2 = \Sigma Q_i$$

However, for each pipe i:

$$(H_L)_i = K_i Q_i^{m_i} = K_i (Q_i)^m$$

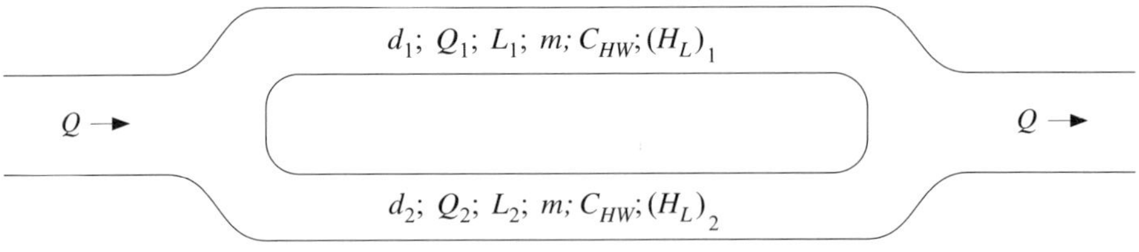

Figure 5.3 Two pipes in parallel

Moreover, we know that:

$$m_{equ} = m_1 = m_2 = m_i = m = 1{,}852$$

Therefore:

$$Q_i = \left[\frac{(H_L)_i}{K_i}\right]^{1/m} = \left[\frac{H_L}{K_i}\right]^{1/m} = \left[\frac{H_L}{K_i}\right]^{1/1{,}852}$$

Equation $Q_{equ} = Q_1 + Q_2$ then becomes:

$$\left[\frac{H_L}{K_{equ}}\right]^{1/m} = \left[\frac{H_L}{K_1}\right]^{1/m} + \left[\frac{H_L}{K_2}\right]^{1/m} = \sum\left[\frac{H_L}{K_i}\right]^{1/m}$$

where m has always the value of 1,852.

Finally, we conclude:

$$\left[\frac{1}{K_{equ}}\right]^{1/m} = \left[\frac{1}{K_1}\right]^{1/m} + \left[\frac{1}{K_2}\right]^{1/m} = \sum\left[\frac{1}{K_i}\right]^{1/m} = \sum\left[\frac{1}{K_i}\right]^{1/1{,}852} \tag{5.4}$$

and:

$$H_L = (H_L)_{equ} = Q^{1{,}852}\,(K)_{equ}$$
$$= Q^{1{,}852}\,\frac{\beta\, L_{equ}}{(d_{equ})^{4{,}871}\,(C_{HW})_{equ}^{1{,}852}}$$

The above reasoning and computations are also valid for more than two pipes installed in parallel.

Head losses due to equipment. To evaluate the behaviour of a network accurately, we may have to estimate the head losses caused by various pieces of equipment. In this method, each piece of equipment is replaced by an imaginary pipe normally of the same diameter as the pipe on which it is installed, that creates the same head losses. The length of the equivalent pipe is calculated with the following relation:

$$L_{equ} = 0{,}0077 K_L Q^{0{,}148} C_{HW}^{1{,}852} d^{0{,}87} \tag{5.5}$$

where L_{equ} = length of equivalent pipe of a piece of equipment (m)
K_L = coefficient related to the piece of equipment
Q = flowrate in pipe on which the equipment is installed (m^3/s)
C_{HW} = the Hazen-Williams roughness coefficient of the pipe on which the equipment is installed
d = inner diameter of the pipe on which the equipment is installed (m)

EXAMPLE 5.1

Calculate the diameter of the equivalent pipe, 150 m long, that replaces two 150 m long pipes installed in parallel. Assume that:

$$\begin{aligned} d_1 &= 205 \text{ mm} \\ d_2 &= 150 \text{ mm} \\ (C_{HW})_1 &= (C_{HW})_2 = C_{HW} = 120 \end{aligned}$$

Solution

We know that:

$$\frac{1}{m} = \frac{1}{1{,}852} = 0{,}54$$

$$K_i = \frac{\beta L_i}{d_i^{4{,}871}(C_{HW})_i^{1{,}852}}$$

where $\beta = 10{,}679$ if L_i and d_i are expressed in metres. Therefore, for the 205 mm pipe, it yields:

$$K_{205} = \frac{10{,}679 \times 150}{0{,}205^{4{,}871} \times 120^{1{,}852}} = 573{,}67$$

For the 150 mm pipe, it yields:

$$K_{150} = \frac{10{,}679 \times 150}{0{,}150^{4{,}871} \times 120^{1{,}852}} = 2329{,}37$$

For pipes in parallel, equation 5.4 can be used:

$$\begin{aligned} \left[\frac{1}{K_{equ}}\right]^{0{,}54} &= \left[\frac{1}{573{,}67}\right]^{0{,}54} + \left[\frac{1}{2329{,}37}\right]^{0{,}54} \\ &= 0{,}047578 \text{ m} \end{aligned}$$

so that

$$K_{equ} = 281{,}48 \text{ m}$$

Using equation 5.2, we find that the diameter of the equivalent pipe is:

$$\begin{aligned} d_{equ} &= \left[\frac{\beta\, L_{equ}}{K_{equ}\; C_{HW}^{1{,}852}}\right]^{1/4{,}871} = \left[\frac{10{,}679 \times 150}{281{,}48 \times 120^{1{,}852}}\right]^{1/4{,}871} = 0{,}0008^{1/4{,}871} \\ &= 0{,}232 \text{ m } (232 \text{ mm}) \end{aligned}$$

Suggestions for computation of equivalent pipes. When choosing equivalent pipes, we generally choose a diameter already in use in the network (whether pipes are in series or in parallel) or a diameter available on the market. To simplify computations, it might be useful to assign constant values to d, L, and C_{HW}. For equivalent pipes in a small water distribution network, we often suggest the following values: $d = 255$ mm; $L = 350$ m; $C_{HW} = 120$.

Law of continuity. For a given node (fig. 5.4), the algebraic sum of the flowrates (including disaggregated flows) is zero. We can therefore say:

$$\Sigma Q_i = 0$$

where Q_i = flowrate in pipe i (see also equation 5.6)

Conventionally, we usually assume that flowrates flowing toward a node are positive, and those flowing from a node are negative.

Law of conservation of energy. In a loop, the algebraic sum of the head losses is zero. Therefore, we can say:

$$\Sigma (H_L)_i = 0$$

where $(H_L)i$ = head losses in pipe i

Conventionally, we generally assume that in a loop, the flowrates flowing clockwise are positive and vice-versa. Similarly, head losses created where flowrates are flowing clockwise are also positive, and so forth.

The law of conservation energy states that for a given node there can only be one pressure.

Equations used to represent flows in a network. For a given condition (e.g., the maximum daily flowrate), a continuity equation can be written for each node, representing the sum of the flowrates at this node. For a given network, the number of such independent and linear equations is equal to the number of nodes, N, minus one; the nth equation is not independent as it is a function of the other $(N - 1)$ equations. For the node in figure 5.4, the continuity equation is as follows:

$$Q_1 + Q_2 + Q_3 - Q_4 = \text{disaggregated flow (constant value)} \tag{5.6}$$

This type of equation is linear, as all the variables Q (flowrates) are raised to the power of one.

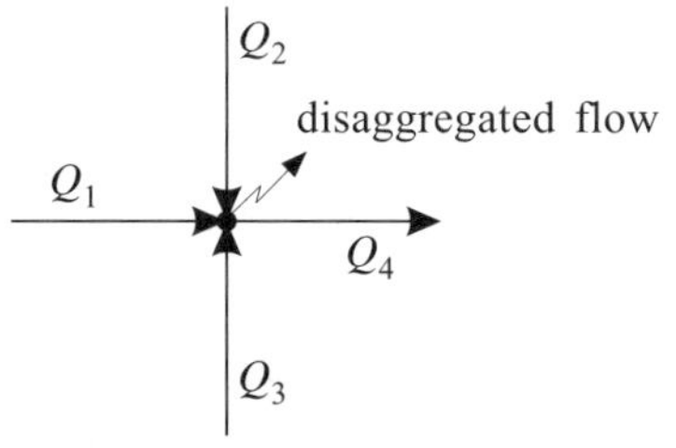

Figure 5.4 Flowrates at one node

For each loop, there is a non-linear equation that represents the conservation of energy (or head losses). For a given network, there are as many independent and non-linear equations as there are loops, *B*. For each loop, the algebraic sum of the head losses is zero; in other words:

$$\Sigma (H_L)_i = 0$$

Therefore (see equation 5.1)

$$\Sigma K_i Q_i^{m_i} = \Sigma K_i Q_i^{m} = 0 \tag{5.7}$$

This type of equation is non-linear, as all variables Q_i (flowrates) are raised to powers other than one, i.e., 1,852.

To summarize, for a network composed of *N* nodes and of *B* loops, we can identify:

- $N - 1$ linear equations (equations related to nodes)
- *B* non-linear equations (equations related to loops)

The total yields *D* equations, where *D* is equal to the number of pipes in the network. Solving the *D* equations enables us to determine the flowrate in each of the pipe sections when the system is in equilibrium. First, however, the *B* non-linear equations must be transformed into linear equations. The mathematical procedure for this is presented in section 5.5.

5.5 ANALYSIS OF WATER-DISTRIBUTION NETWORKS USING THE WOOD-CHARLES LINEARIZATION METHOD

This analytic method has three parts.

1. The first step is to transform the *B* non-linear equations into linear equations, as suggested by Wood and Charles (1972). To this end head loss of each pipe *i*, $(H_L)_i$ $\left[(H_L)_i = K_i Q_i^{m_i} = K_i Q_i^{m}\right]$ is represented as follows:

$$(H_L)_i = K_i (Q_i)_0^{m-1} Q_i = K_i' Q_i \tag{5.8}$$

   where K_i = proportionality constant of pipe *i* (equation 5.2)
   $(Q_i)_0$ = estimated flowrate in pipe *i*
   Q_i = flowrate to be computed (variable) in pipe *i* from estimated flowrate

   This last equation is linear; that is, the exponent of Q_i is 1,0.

2. Solving this system of equations formed with pipes and the system of equations for nodes (see law of continuity) allows us to calculate the flowrates Q_i in every pipe of the network.

3. Finally, the flowrates Q_i are compared to the initial flowrates $(Q_i)_0$. If the difference $(Q_i)_0 - Q_i$ is higher than a predetermined value, within the limits of accuracy, a second iteration can be performed assuming that $(Q_i)_0 = Q_i$.

One of the advantages of this method is that the initial flowrates in the network do not have to be evaluated for the first iteration; therefore, we assume that $K_i = K_i'$, which simply means $(Q_i)_0 = 1$. After two iterations, the authors of this method recommend that the following equation be used to give a new value to $(Q_i)_0$:

$$(Q_i)_0 = \frac{Q_{i-1} + Q_{i-2}}{2} \tag{5.9}$$

where Q_{i-1} and Q_{i-2} represent the values of the flowrate in the preceding two iterations.

Using this relation, convergence toward the final solution is faster since oscillations of the flowrate Q_i around the actual value are eliminated or at least reduced.

5.6 ANALYSIS OF WATER-DISTRIBUTION NETWORKS USING THE NEWTON-RAPHSON METHOD

The Newton-Raphson method is a numerical procedure to find the value of the variable x of a function $F(x) = 0$ from a first estimate of its value. The following well-known relation is used:

$$x_{p+1} = x_p - \frac{F(x_p)}{F'(x_p)} \tag{5.10}$$

where x_{p+1} = new value (calculated) of x from x_p
x_p = estimated value of x or previous value of x_p
$F(x_p)$ = value of $F(x)$ for $x = x_p$
$F'(x_p)$ = slope of the function $F(x)$ at $x = x_p$

For a given pipe in a loop, equation 5.10 becomes:

$$Q_i = (Q_i)_0 - \frac{F(Q_i)_0}{F'(Q_i)_0}$$

We can estimate the speed at which the Newton-Raphson method converges toward the final solution by expanding the function $F(x) = 0$. The Taylor series expansion is used; after assigning a first value to x, i.e. x_p, it yields:

$$0 = F(x)$$

$$0 = F(x) = F(x_p) + (x - x_p)\, F'(x_p) + (x - x_p)^2\, \frac{F''(\xi)}{2} + \ldots \tag{5.11}$$

where $\xi = \dfrac{x + x_p}{2} \quad (x_p < \xi < x)$

It then yields:

$$x = x_{p+1} - \left(x - x_p\right)^2 \frac{F''(\xi)}{2F'\left(x_p\right)}$$

The convergence toward the final solution is quadratic. In other words, for a given iteration, the error reduction in the evaluation of x is proportional to the square of the previous error. Therefore, for an initial error of 20% (0,2), the errors of the following iterations will be: $(0{,}2)^2$, therefore 0,04 (4%), then $(0{,}04)^2$, therefore 0,0016 (0,16%), and so on (Jeppson, 1977: 115).

To use the Newton-Raphson method, the initial flowrates or initial head losses must be evaluated. (WATCAD and the newest AQUAGEO software, which calculate the equilibrium of a distribution network, use the Newton-Raphson method.)

5.7 ANALYSIS OF WATER-DISTRIBUTION NETWORKS USING THE HARDY-CROSS METHOD

The Hardy-Cross method is probably the oldest and most widely used analysis procedure for water-distribution networks. This iterative method was developed before the advent of computers and is still used today, even though computer technology is easily accessible.

The Hardy-Cross method enables you to compute either the flowrates or the head losses (depending on the formulation of the problem and the chosen equations). After each iteration the previous values of the flowrates (or head losses) are modified.

5.7.1 Flow Correction in a Loop After One Iteration

Let us consider a loop in which the flowrates have been estimated for each pipe. If Q_i is the initial estimated flowrate in one pipe and ΔQ is the value of the flowrate correction, the corrected flowrate, after the first iteration, is:

$$Q = Q_i + \Delta Q$$

This flowrate Q is either the new flowrate Q_i of the following iteration, or the final flowrate Q (end of the Hardy-Cross method). For ΔQ, this modification, positive or negative, will correct the flowrates of every pipe in the loop.

For the following iteration, the head loss in the pipe section is:

$$H_L = K\left(Q_i + \Delta Q\right)^m$$

thus

$$H_L = KQ_i^m\left[1 + \frac{\Delta Q}{Q_i}\right]^m$$

Here H_L and K are characteristic of the pipe under study. For the sake of simplicity we do not write subscript i, i.e. $(H_L)_i$ and K_i.

Or, after developing the binomial:

$$H_L = KQ_i^m\left[1 + m\frac{\Delta Q}{Q_i} + \frac{m(m-1)}{2!} \times \frac{\Delta Q^2}{Q_i^2} + \text{terms of negligible values}\right]$$

If the value of ΔQ is low, as it should be after many iterations, the previous equation becomes:

$$H_L = KQ_i^m\left[1 + m\frac{\Delta Q}{Q_i}\right]$$

Then, the algebraic sum of the head losses of all pipes in the loop is computed. This sum yields:

$$\Sigma(H_L)_i = 0 = \Sigma K_i Q_i^m + m\Delta Q\, \Sigma K_i Q_i^{m-1}$$

We assume that $\Sigma(H_L)_i = 0$, since the system should tend toward this point, as seen above (section 5.4.3; equation 5.7).

From the relation above, the algebraic value of the correction ΔQ for a loop is obtained:

$$\Delta Q = \frac{-\Sigma K_i Q_i^m}{m\,\Sigma\left(\frac{K_i Q_i^m}{Q_i}\right)} = \frac{-\Sigma(H_L)_i}{m\,\Sigma\left(\frac{(H_L)_i}{Q_i}\right)} \qquad (5.12)$$

where $m = 1{,}852$ (Hazen-Williams equation)

$$K = \frac{\beta L}{d^{4{,}871}\ C_{HW}^{1{,}852}} \text{ (equation 5.2)}$$

The value of K varies depending on the pipe studied.

Conventionally, flowrates circulating clockwise are positive, and vice-versa. Also, for a given pipe, the sign of the head losses is the same as the corresponding flowrate. Therefore, if ΔQ is positive, the positive flowrates are increased by ΔQ, and negative flowrates are reduced by ΔQ.

Many iterations are required before accurate results are obtained–in other words, until ΔQ is small enough, depending on the needs of the engineer and the nature of the problem. If the initial estimate of the flowrates is realistic, the number of necessary iterations is small and there is rapid convergence. However, if the initial estimate is far off, the iterations may not lead to a solution and the procedure may have to be retried with new initial estimates.

After this stage, it is important to note that the water distribution in a loop, or anywhere in the network, is independent of the height of the nodes and the pipes. However, this height influences the pressure exerted at different locations in the loop.

Step j of section 5.7.2 illustrates how to calculate the pressure exerted at the nodes after the network has reached equilibrium.

5.7.2 Step-by-Step Procedure Using the Hardy-Cross Method

Here is a summary of the step-by-step procedure.

a) Identification of the loops (B) and nodes (N) and verification that the $B + N - 1 = D$ (section 5.4.3) relation is respected.

b) Considering the disaggregated flows (consumption) at each node, estimation is made of initial flowrates, Q_i, in each pipe: at each node, the algebraic sum of the flowrates must be zero.

c) Computation of the head losses, $(H_L)_i$, in each pipe using the $(H_L)_i = K_i Q_i^{1,852}$ (equation 5.1) relation.

d) Computation of the algebraic sum of the head losses for the pipes of the first loop: except in special cases, this sum is not zero, as the initial evaluation of the flowrates is rarely exact; as a result: $\Sigma(H_L)_i \neq 0$.

e) Computation of the value of $\Sigma(H_L)_i/Q_i$ for the first loop: algebraic signs are ignored, as for any pipe, $(H_L)_i$ and Q_i have the same sign.

f) Computation of the correction ΔQ to the flowrates for the pipes of the first loop, using the following relation:

$$\Delta Q = \frac{-\Sigma(H_L)_i}{m\,\Sigma\left(\frac{(H_L)_i}{Q_i}\right)} \quad \text{(the value of } \Delta Q \text{ is either positive or negative) (équat. 5.12)}$$

g) Algebraic correction of the flowrates for the pipes of the first loop, using the following relation:

$$Q_i = Q_i + \Delta Q$$

h) Return to steps (d) to (g) for all the other loops of the network.

i) Return to steps (d) to (h) until desired accuracy is obtained: in general, operations end when the correction values ΔQ are smaller than a predetermined value.

j) From known pressures for certain nodes, for which the elevation is known, (e.g., pressures exiting at the treatment plant outlet, required pressures for consumers, etc.), the pressure for every node can be calculated using the Bernoulli equation (equation 2.5) by solving progressively node by node; for example, relating to nodes A and B located at the end of the same pipe, it yields:

$$\left(\frac{p_A}{\rho_e g} + \frac{v_A^2}{2g} + h_A\right) - \left(\frac{p_B}{\rho_e g} + \frac{v_B^2}{2g} + h_B\right) = H_L$$

If p_A and h_A are known at node A, if h_B is the height of node B linked to node A by the pipe A-B, and if the pipe diameter is constant between nodes A and B, then $v_A = v_B$ and the previous equation becomes:

$$\frac{p_A}{\rho_e g} + h_A = \frac{p_B}{\rho_e g} + h_B + H_L$$

Finally, as the head losses H_L can be calculated (because we know the flowrate and the pipe characteristics between nodes A and B), the pressure $p_B/\rho_e g$ at node B can be calculated.

5.7.3 Limitations of the Hardy-Cross Method

Computers have been used for a long time to analyze, modify, and design distribution networks. However, the use of computer technology can create certain problems; the following problems are inherent to the Hardy-Cross method:

a) Slow convergence to the final solution.

b) In some cases, problems are impossible to solve. This situation motivated engineers to find analytical and computational techniques that were better suited to computers. Some software uses the Newton-Raphson method or the Wood-Charles linearization method, as both easily allow integration of elements such as pumping stations, pressure-regulating valves, elevated reservoirs, etc.

5.8 PARALLEL WATER-DISTRIBUTION NETWORKS

As the consumption water volume represents only a small percentage of the total volume of water used by the community, some authors advocate the use of two water-distribution networks: one for drinking water, and one for water of lesser quality, intended for other purposes than consumption.

This view has been long rejected, as it entails important expenses. Nowadays, with the constant and foreseeable rise in water-treatment costs, this idea is once again being studied. The reason for this renewed interest arises from the treatment of raw water into drinking water. The need to eliminate ever-growing numbers of micro-pollutants (pesticides, toxic minerals, overabundance of minerals, etc.) and viruses will probably, in the near future, make dual water-distribution networks more financially attractive.

EXAMPLE 5.2

Using the Hardy-Cross method, calculate the flowrate circulating in each of the pipe of the following network, made of ductile cast iron. See figure 5.5.

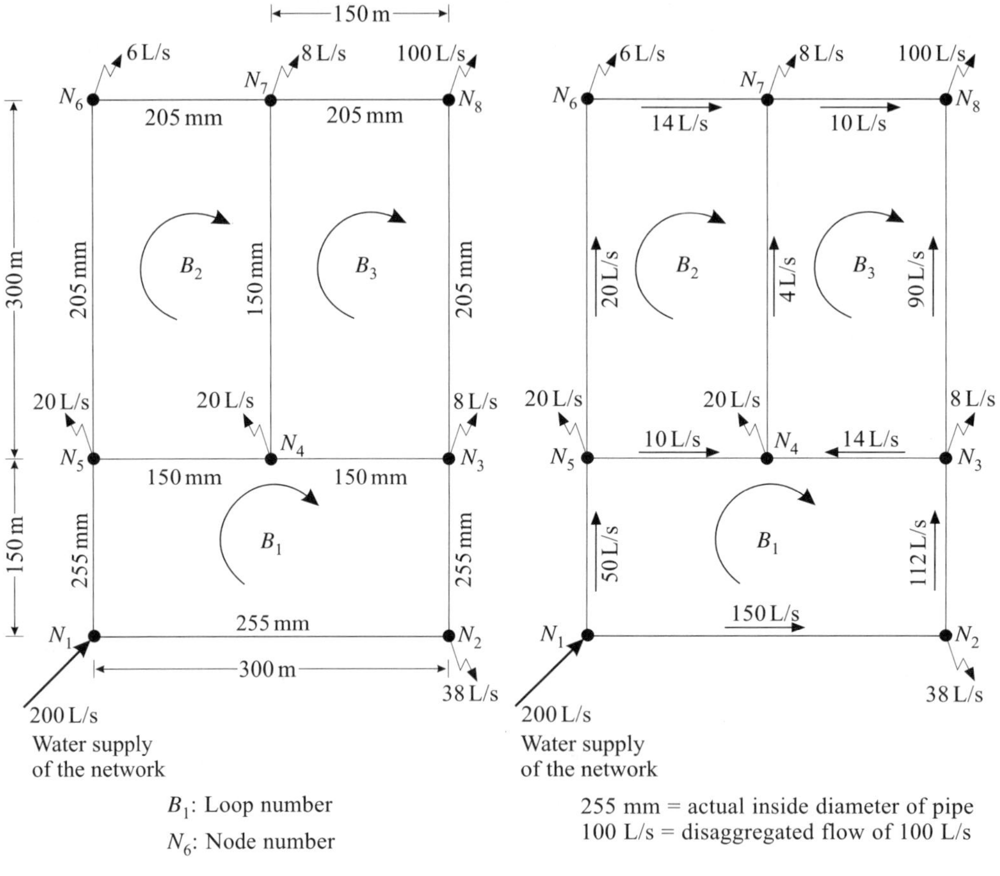

Figure 5.5 Balancing of a water-distribution network in ductile cast iron (example 5.2): a) network nodes, loops, diameters, and disaggregated flows; b) initial estimate of flowrates and direction of flows

Table 5.2 shows the pipe characteristics of the network, made of ductile cast iron, described in example 5.2 (fig. 5.5), and table 5.3 shows the first four steps in analysis of this network.

Table 5.2 (ex. 5.2) Pipe characteristics of the network and initial flowrates

Pipe	L (m)	d_{act} (mm)	C_{HW}	Initial Flowrate (Q_i) (L/s)
$N_1 - N_2$	300	255	120	150
$N_1 - N_5$	150	255	120	50
$N_2 - N_3$	150	255	120	112
$N_3 - N_4$	150	150	120	14
$N_4 - N_5$	150	150	120	10
$N_5 - N_6$	300	205	120	20
$N_6 - N_7$	150	205	120	14
$N_7 - N_4$	300	205	120	4
$N_7 - N_8$	150	205	120	10
$N_8 - N_3$	300	205	120	90

Solution

In the first iteration, the three loops of the network are studied in turn. The following tables present the results of all computations described in section 5.7. First, the computations for loop 1 were completed, then for loops 2 and 3. Note that the corrected flowrates of a given loop are used as estimated flowrates in pipes common to subsequent loops. For example, in pipe N_4-N_5 of the first loop, the estimated flowrate of 0,010 m^3/s was corrected to yield a flowrate of 0,043 m^3/s. This new corrected flowrate is used as the first approximation for the same pipe when studying the second loop, since pipe N_4-N_5 is common to both loops.

Note (section 5.7.2) that the balancing of the network depends neither on its altitude nor on altitude differences between various parts of the network. However, if the water pressure is known at the water supply source (N_1), and if the altitude of different nodes is also known, we can deduce the pressure exerted at every node in the network. Therefore, the pressure at node N_2 is equal to the pressure at node N_1, minus the head losses in pipe N_1-N_2, plus or minus the difference in elevation between the two nodes (depending on whether node N_2 is lower or higher than node N_1).

Table 5.3 (ex. 5.2) Four steps in the network analysis: a) first iteration

Loop	Pipe	d_{act} (mm)	C_{HW}	L (m)	K*	Q_i (m³/s)	$(H_L)_i$** (m)	$\frac{(H_L)_i}{Q_i}$	$-\Sigma(H_L)_i$ (m)	$m\Sigma\frac{(H_L)_i}{Q_i}$	ΔQ*** (m³/s)	$Q_i+\Delta Q$ (m³/s)
1	N_1-N_2	255	120	300	351,36	-0,150	-10,468	69,787				-0,117
	N_2-N_3	255	120	150	175,68	-0,112	-3,047	27,205				-0,079
	N_3-N_4	150	120	150	2329,37	-0,014	-0,859	61,357				0,019
	N_4-N_5	150	120	150	2329,37	0,010	0,461	46,100				0,043
	N_5-N_1	255	120	150	175,68	0,050	0,684	13,680	13,229	403,974	0,033	0,083
								$\Sigma\frac{(H_L)_i}{Q_i}$ = 218,129				
2	N_5-N_6	205	120	300	1017,31	0,020	0,726	36,300				0,035
	N_6-N_7	205	120	150	508,66	0,014	0,188	13,429				0,029
	N_7-N_4	205	120	300	1017,31	-0,004	-0,037	9,250				0,011
	N_4-N_5	150	120	150	2329,37	-0,043	-6,862	159,581	5,985	404,773	0,015	-0,028
								$\Sigma\frac{(H_L)_i}{Q_i}$ = 218,560				
3	N_4-N_7	205	120	300	1017,31	-0,011	-0,240	21,818				0,019
	N_7-N_8	205	120	150	508,66	0,010	0,101	10,100				0,040
	N_8-N_3	205	120	300	1017,31	-0,090	-11,768	130,756				-0,060
	N_3-N_4	150	120	150	2329,37	-0,019	-1,512	79,579	13,419	448,653	0,030	0,011
								$\Sigma\frac{(H_L)_i}{Q_i}$ = 242,253				

* $K = \dfrac{\beta L}{d^{4,871} C_{HW}^{1,852}}$ where $\beta = 10{,}679$ if L and d are in m and if Q is in m³/s.

** $(H_L)_i = K_i (Q_i)^m = K_i (Q_i)^{1,852}$

*** $\Delta Q = \dfrac{-\Sigma (H_L)_i}{m\Sigma\left(\dfrac{(H_L)_i}{Q_i}\right)}$

Table 5.3 (ex. 5.2) b) second iteration

Loop	Pipe	d_{act} (mm)	C_{HW}	L (m)	K	Q_i (m³/s)	$(H_L)_i$ (m)	$\frac{(H_L)_i}{Q_i}$	$-\Sigma(H_L)_i$ (m)	$m\Sigma\frac{(H_L)_i}{Q_i}$	ΔQ (m³/s)	$Q_i+\Delta Q$ (m³/s)
1	N_1-N_2	255	120	300	351,36	-0,117	-6,607	56,470				-0,109
	N_2-N_3	255	120	150	175,68	-0,079	-1,596	20,203				-0,071
	N_3-N_4	150	120	150	2329,37	-0,011	-0,549	49,910				-0,003
	N_4-N_5	150	120	150	2329,37	0,028	3,100	110,714				0,036
	N_5-N_1	255	120	150	175,68	0,083	1,749	21,072	3,948	478,499	0,008	0,091
								$\Sigma\frac{(H_L)_i}{Q_i}$ = 258,369				
2	N_5-N_6	205	120	300	1017,31	0,035	2,047	58,486				0,041
	N_6-N_7	205	120	150	503,66	0,029	0,722	24,910				0,035
	N_7-N_4	205	120	300	1017,31	-0,019	-0,660	34,737				-0,013
	N_4-N_5	150	120	150	2329,37	-0,036	-4,938	137,167	2,829	469,112	0,006	-0,030
								$\Sigma\frac{(H_L)_i}{Q_i}$ = 253,300				
3	N_4-N_7	205	120	300	1017,31	0,013	-0,327	25,154				0,025
	N_7-N_8	205	120	150	508,66	0,040	1,311	32,775				0,053
	N_8-N_3	205	120	300	1017,31	-0,060	-5,554	92,567				-0,047
	N_3-N_4	150	120	150	2329,37	0,003	0,050	16,667	3,866	309,59	0,013	0,016
								$\Sigma\frac{(H_L)_i}{Q_i}$ = 167,163				

Table 5.3 (ex. 5.2) c) third iteration

Loop	Pipe	d_{act} (mm)	C_{HW}	L (m)	K	Q_i (m³/s)	$(H_L)_i$ (m)	$\frac{(H_L)_i}{Q_i}$	$-\Sigma (H_L)_i$ (m)	$m\Sigma \frac{(H_L)_i}{Q_i}$	ΔQ (m³/s)	$Q_i + \Delta Q$ (m³/s)
1	$N_1 - N_2$	255	120	300	351,36	-0,109	-5,795	53,165				-0,104
	$N_2 - N_3$	255	120	150	175,68	-0,071	-1,310	18,451				-0,066
	$N_3 - N_4$	150	120	150	2329,37	-0,016	-1,100	68,750				-0,011
	$N_4 - N_5$	150	120	150	2329,37	0,030	3,523	117,433				0,035
	$N_5 - N_1$	255	120	150	175,68	0,091	2,074	22,791	2,608	519,653	0,005	0,096
								$\Sigma \frac{(H_L)_i}{Q_i} = 280{,}590$				
2	$N_5 - N_6$	205	120	300	1017,31	0,041	2,744	66,927				0,045
	$N_6 - N_7$	205	120	150	508,66	0,035	1,023	29,229				0,039
	$N_7 - N_4$	205	120	300	1017,31	-0,025	-1,098	43,920				-0,021
	$N_4 - N_5$	150	120	150	2329,37	-0,035	-4,687	133,914	2,018	507,430	0,004	-0,031
								$\Sigma \frac{(H_L)_i}{Q_i} = 273{,}990$				
3	$N_4 - N_7$	205	120	300	1017,31	0,021	0,795	37,857				0,024
	$N_7 - N_8$	205	120	150	508,66	0,053	2,207	41,642				0,056
	$N_8 - N_3$	205	120	300	1017,31	-0,047	-3,533	75,170				-0,044
	$N_3 - N_4$	150	120	150	2329,37	0,011	-0,549	49,910	1,080	378,880	0,003	0,014
								$\Sigma \frac{(H_L)_i}{Q_i} = 204{,}579$				

Table 5.3 (ex. 5.2) d) fourth iteration

Loop	Pipe	d_{act} (mm)	C_{HW}	L (m)	K	Q_i (m³/s)	$(H_L)_i$ (m)	$\frac{(H_L)_i}{Q_i}$	$-\Sigma (H_L)_i$ (m)	$m\Sigma \frac{(H_L)_i}{Q_i}$	ΔQ (m³/s)	$Q_i + \Delta Q$ (m³/s)
1	$N_1 - N_2$	255	120	300	351,36	-0,104	-5,313	51,087				-0,101
	$N_2 - N_3$	255	120	150	175,68	-0,066	-1,144	17,333				-0,063
	$N_3 - N_4$	150	120	150	2329,37	-0,014	-0,859	61,357				-0,011
	$N_4 - N_5$	150	120	150	2329,37	0,031	3,743	120,742				0,034
	$N_5 - N_1$	255	120	150	175,68	0,096	2,290	23,854	1,283	508,139	0,003	0,099
								$\Sigma \frac{(H_L)_i}{Q_i} = 274,373$				
2	$N_5 - N_6$	205	120	300	1017,31	0,045	3,260	72,444				0,047
	$N_6 - N_7$	205	120	150	508,66	0,039	1,251	32,077				0,041
	$N_7 - N_4$	205	120	300	1017,31	-0,024	-1,018	42,417				-0,022
	$N_4 - N_5$	150	120	150	2329,37	-0,034	-4,442	130,647	0,949	514,087	0,002	-0,032
								$\Sigma \frac{(H_L)_i}{Q_i} = 277,585$				
3	$N_4 - N_7$	205	120	300	1017,31	0,022	0,866	39,364				0,020
	$N_7 - N_8$	205	120	150	508,66	0,056	2,244	43,643				0,054
	$N_8 - N_3$	205	120	300	1017,31	-0,044	-3,127	71,068				-0,046
	$N_3 - N_4$	150	120	150	2329,37	0,011	0,549	49,910	-0,732	377,780	-0,002	0,009
								$\Sigma \frac{(H_L)_i}{Q_i} = 203,985$				

CONCLUSION

In this chapter, all aspects of analysis and design of a water-distribution network have been presented. We learned that the entire network must be simplified using the equivalent pipe method: a) to reduce the number of loops; b) to take into account the influence of pieces of equipment (valves, etc.). A network can also be analyzed using one of the methods described below.

Hardy-Cross method. This method is based on iterations and was invented before the use of computers. Therefore, it is solved by manual computations or with a calculator. The initial flowrates in the pipes must be estimated.

Newton-Raphson method. This method is based on iterations and allows a rapid convergence toward the final solution. Here, again, the initial flowrates in the pipes must be estimated. The WATCAD and the newest AQUAGEO[1] models use this method. There are other water-distribution models available on the market. EPANET is free and available on the net.

Methods in which equations are linearized. These methods are very interesting, as they do not require an initial estimate for the flowrates.

PROBLEMS

5.1 Find the length of the equivalent pipe, 305 mm in diameter, of a network made of ductile cast iron, shown in figure 5.6. The Hazen-Williams coefficient, for all pipes, is 120. Also, find the flowrate in each branch of the network.

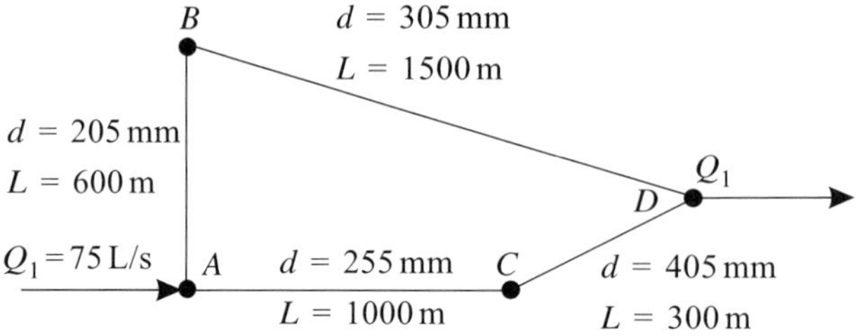

Figure 5.6 (problem 5.1)

5.2 Calculate the head losses between points A and C of a network made of ductile cast iron, shown in figure 5.7. The Hazen-Williams coefficient, for all pipes, is 120. Also, find the flowrate in each branch of the network.

1. CIVIL DESIGN, 61 St-Charles St. West, Office 50, Longueuil (Qc) Canada, J4H 1C5; Phone : (450) 674-0657; Fax: (450) 674-0665.

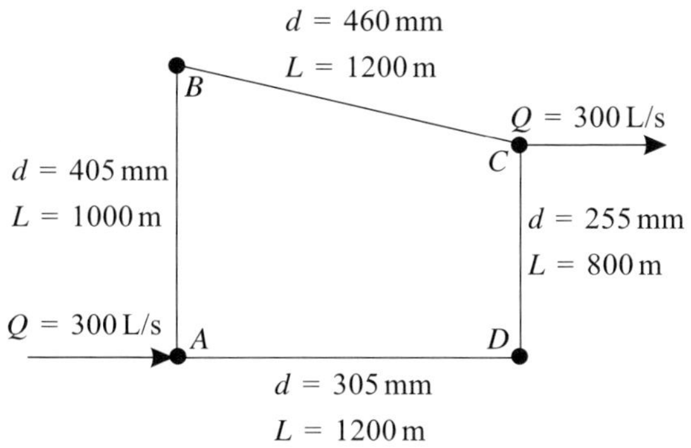

Figure 5.7 (problem 5.2)

5.3 Find the length of the equivalent pipe of a network made of ductile cast iron (C_{HW} = 120), as shown in figure 5.8. The diameter of the equivalent pipe, d_{equ}, is 205 mm; its Hazen-Williams coefficient, $(C_{HW})_{equ}$, is 120; and the flowrate, Q_1, is 75 L/s. How is the flowrate distributed between the branches of the network and what are the head losses between nodes *A* and *D*?

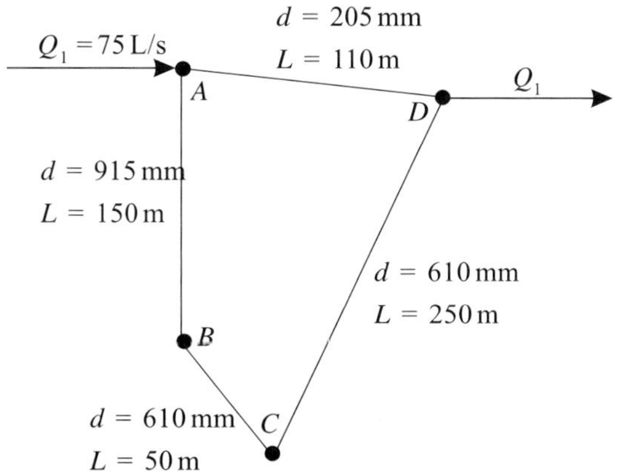

Figure 5.8 (problem 5.3)

Table 5.4 shows the characteristics of the pipes of the network, shown in Figure 5.8.

Table 5.4 (problem 5.3) Characteristics of the pipes

Pipe	**d* (mm)**	C_{HW}	**L (m)**
A-D	205	100	110
A-B	915	120	150
B-C	610	125	50
C-D	610	130	250

* Inside diameter.

5.4 Using the Hardy-Cross method, find the flowrates in each pipe section of a network made of ductile cast iron, as shown in figure 5.9. All pipes have a Hazen-Williams coefficient of 100. If, for the first approximation, the flowrate in pipe *A-B* is estimated at 200 L/s, calculate the flowrate in each pipe after 5 approximations. What is the pressure exerted on node *F*, if the pressure on node *A* is equivalent to 46 m of water and the network is in a horizontal plane?

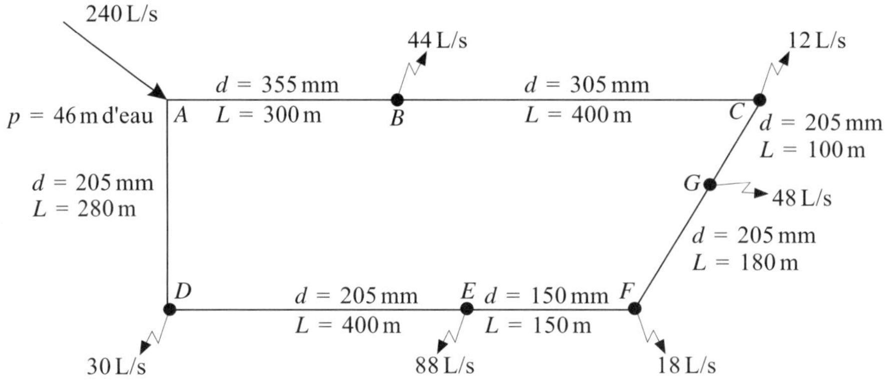

Figure 5.9 (problem 5.4)

5.5 Calculate the flowrate in each pipe of a network made of ductile cast iron, as shown in figure 5.10, and determine the direction of the flow in each section. All pipes are identical, and their characteristics are as follows: L = 100 m; d = 255 mm; and C_{HW} = 100 (pipes 1-2 and 1-3 are 50 m long).

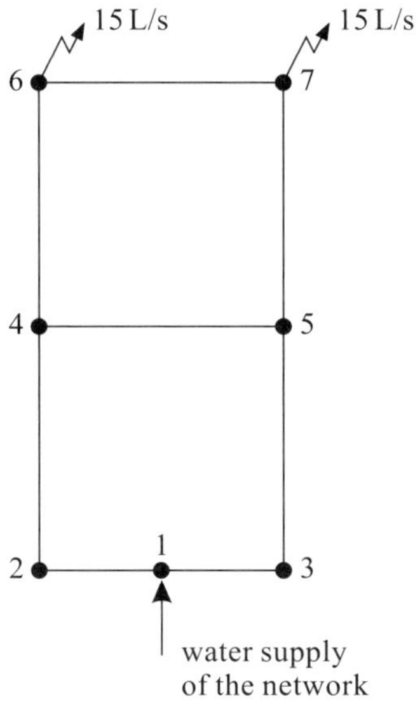

Figure 5.10 (problem 5.5)

5.6 What would the answers to problem 5.5 be if the network were moved around an axis passing through nodes 3 and 6, so that nodes 2 and 7 would be located at 0,5 m under and 0,5 m over nodes 3 and 6?

5.7 Calculate or find all the missing values: the disaggregated flows S_2, S_4, S_6, S_7, S_8, and S_9, the pressures p_2 and p_3, the flowrates in all pipes, and the length of pipe N_1–N_3 in a network made of ductile cast iron, as shown in figure 5.11, assuming that:

- the water is supplied to the network through node 1 with a flowrate of 160 L/s and a pressure of 550 kPa;
- the Hazen-Williams coefficient, C_{HW}, for every pipe is 130;
- all pipes have a length of 420 m, except for pipe N_1–N_2, which has a length of 210 m, and pipe N_1–N_3, for which the length is unknown.

Solve this problem using equations 5.1 and 5.2. Do not use Hardy-Cross Method.

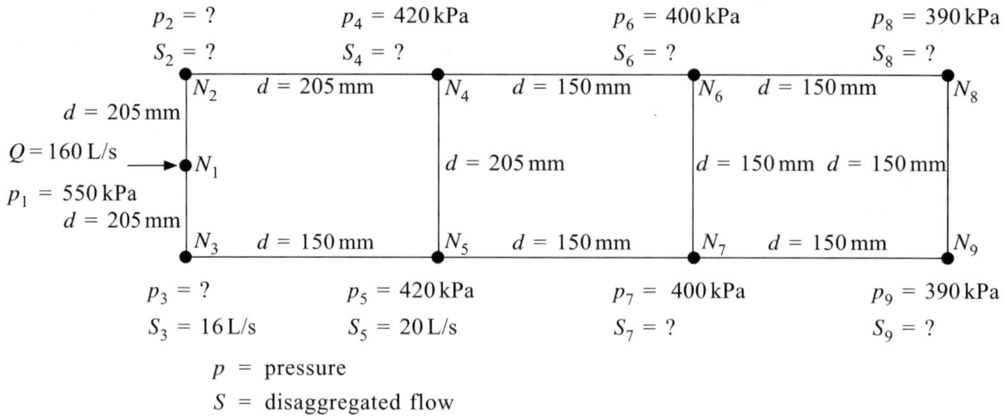

Figure 5.11 (problem 5.7)

5.8 Using the Bernoulli equation, show that the head losses $(H_L)_1$ and $(H_L)_2$ are equal in the case of two pipes installed in parallel (fig. 5.3).

Chapter 6

VOLUMES AND FLOWS OF SANITARY SEWAGE

INTRODUCTION

In the past, surface runoff was evacuated in ditches. When volumes and flows of domestic sewage were small, especially before drinking water distribution networks were in place, domestic wastewater was evacuated on the land surrounding houses. As the density of population grew, ditches were replaced by pipes buried under the streets. Moreover, because properties grew smaller, wastewater flows increased, and public health requirements grew, city dwellers soon started evacuating their domestic wastewater in the pipes meant to rid cities of urban runoff. Since ultimately sewers evacuate sewage in watercourses, surface water became polluted, sometimes heavily, especially in urban areas.

To reduce water pollution while keeping in mind the imperative to build sewage treatment plants, management of urban waters was re-evaluated. The study concluded that treatment was necessary for domestic wastewater but not for runoff.

From a hydraulics point of view, it meant that, again, there was a tendency to separate runoff and domestic wastewater. This is why in urban areas, we find:

- districts equipped with one sewer pipe network (combined sewers) to collect and evacuate domestic wastewater as well as runoff;
- other, more recent, districts equipped with two systems: one for the evacuation of runoff (storm sewer system), and one set up in parallel to evacuate domestic wastewater (sanitary sewer system – also mistakenly called the domestic or separate system, section 6.1.1).

To design these sewer systems, we must determine the variations in the flowrates of the different types of wastewater to be transported. The present chapter will demonstrate how to determine the values of the average, minimum, and maximum domestic flowrates. From these values, we determine the pipe sizes of the sanitary sewer system, as well as the sizes of intercepting sewers, and pumping stations. Finally, the requirements of pumps and other equipment can be determined.

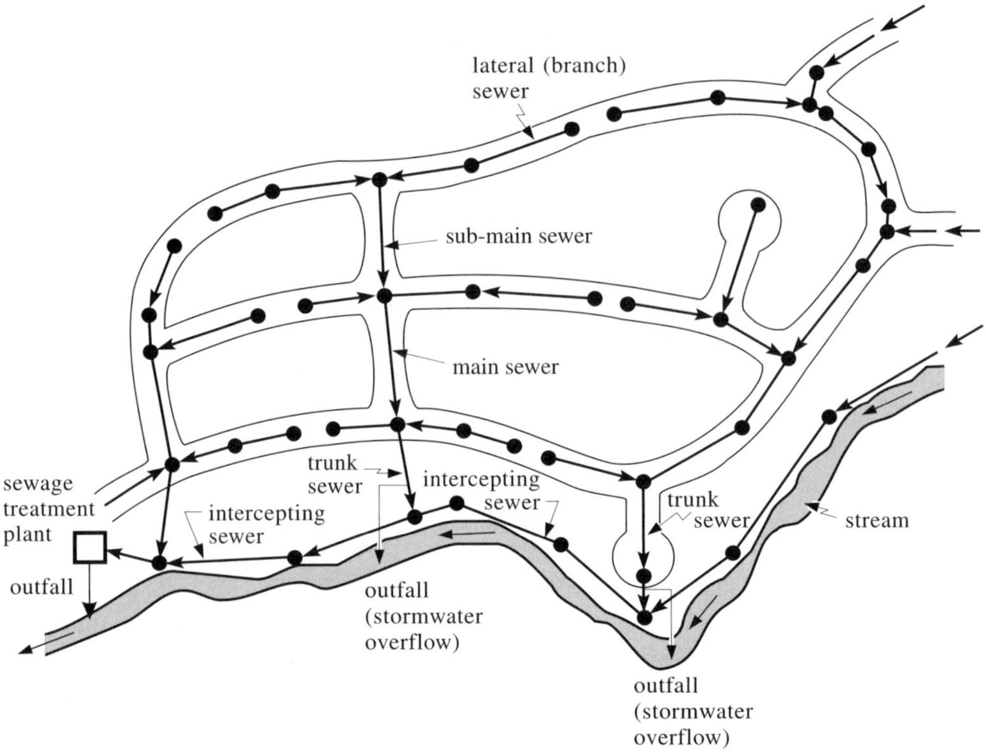

Figure 6.1 Structure of a Sewer System (Sewer System Layout).

Figure 6.1 shows a sewer system with lateral (branch) sewers, main and sub-main sewers, trunk sewers, an intercepting sewer, outfall sewers, and a treatment plant; the black dots represent manholes and arrows indicate the direction of flow (stormwater overflows should only occur in combined sewers).

6.1 DEFINITIONS

The various sewer systems can be classified by the wastewater they transport. Therefore, we classify them as: sanitary sewer systems, combined sewer systems, and pseudo-sanitary sewer systems.

The following sections describe and explain these systems and the different types of pipes with which they are built.

6.1.1 Sanitary Sewer System

Mistakenly called the domestic sewer system, a sanitary sewer system transports domestic wastewater, which includes water used for domestic consumption, water from commercial districts and industrial establishments, (generally, excluding water used as cooling agent) and water considered as parasitic (section 6.2.6).

6.1.2 Storm Sewer System or Storm Water System

A storm sewer system is designed to evacuate runoff resulting mostly from rainfall and snow melt. Usually, runoff is discharged into a receiving watercourse without undergoing treatment. As we will see in section 6.1.3, for a given area or district, the maximum runoff flowrates are much more important than the maximum domestic wastewater flowrates.

6.1.3 Combined Sewer System

A combined sewer system evacuates all the wastewater in a given region. The wastewater can be of domestic, storm, industrial or commercial origin. However, as maximum rainwater flowrates are much greater than maximum domestic flowrates, the design of a combined sewer does not take into account the maximum domestic flowrates. It may or may not consider the industrial flowrates, depending on the nature and importance of the industrial presence in the area. Nowadays, the construction of combined sewers is prohibited in Quebec and in most regions of America since such systems make water treatment difficult and costly, mainly because of the large flow variations and the strong dilution of domestic wastewater when it rains or during snow melt.

6.1.4 Pseudo-Sanitary Sewer System

The pseudo-sanitary sewer system collects domestic wastewater and some rainfall. The runoff evacuated comes from foundation drain pipes (pipes intended for the evacuation of water surrounding the foundations of buildings), flat roof drains, basements and entrances to garages located below ground level. Houses linked to this type of sewer system have only one service connection, which evacuates both domestic wastewater and authorised rainwater. The purpose of the pseudo-sanitary sewer system is to avoid backflow problems in basements through basement drain pipes, foundation drain pipes and foundations cracks. Houses located close to lakes and streams are mostly affected when the water level is high (during spring thaw or after heavy rains) and if their storm service pipes are linked to the storm sewer systems. Since rainwater is not treated, storm outfalls can be installed in the vicinity of these houses. When water level is high, water from lakes and streams backs up in outfalls, putting storm sewers and storm service pipes under pressure. On the other hand, sanitary and pseudo-sanitary sewers evacuate wastewaters far away towards the wastewater treatment plant. Therefore, these sewers do not run the risk of pressurization because of the high water level in watercourses since they are not linked directly to these watercourses. However, most pseudo-sanitary sewers were installed before the decision was made to treat domestic wastewater, and since there are now other means to prevent backflow in basements, the construction of pseudo-sanitary sewer systems is generally prohibited.

6.1.5 Foundation Drain Pipes

Foundation drain pipes are perforated pipes installed around the foundations of houses and other constructions. They maintain the water table at an acceptable depth so that basements are not flooded. Foundation drain pipes are part of the storm sewer system, the pseudo-sanitary sewer system, or the combined sewer system.

6.1.6 Service Connection or Service Pipe

The service connections are the pipes linking the sewer system to houses, stores, businesses and industries. Most often, the section of the pipe located on the private property (houses, stores, businesses and industries) belongs to the owners, and the other section belongs to the city.

6.1.7 Lateral (or Branch) Sewer, Submain, Main Sewer, and Trunk Sewer

The lateral or branch sewer serves only one or a small number of streets. The submain collects sewage from two or more lateral sewers and discharges into a main. Large mains are often called trunks. As for the sewer systems, the main or trunk sewers can be categorized as sanitary, storm, combined and pseudo-sanitary. At all times, the main and trunk sewers must be able to transport the totality of flows evacuated by the connecting pipes according to standardized design (see chapter 11): in other words there is no designed overflow on these pipes except for exceptional event.

6.1.8 Intercepting Sewer

The intercepting sewer receives the wastewater from the main or trunk sewers and sends it to the wastewater treatment plant. As all its wastewater is directed towards the treatment plant, the intercepting sewer is not equipped with any evacuation or overflow system to evacuate non-treated wastewater into receiving watercourses. When wastewater cannot be treated, for example in combined sewer systems, the excess wastewater is evacuated to a receiving watercourse, through an outfall (stormwater overflow), prior to its interception. Therefore, the intercepting sewer is not necessarily designed to carry, at all times, all the wastewater flows brought by the pipes (as planned by the designer). Sometimes, to increase the percentage of treated wastewater, reservoirs are used – often called detention/retention reservoirs – to store the wastewater overflow (during rainy periods) for later treatment during low flow periods. At this end, manholes may be equipped with vortex flowrate controllers to create head losses and slow down the wastewater flow. Wastewater is thus backed up into reservoirs when flows exceed the capacity of the intercepting sewers. To reduce the frequency of overflows into the watercourse, engineers often design the interceptors to accept a fraction of rainwater in addition to the maximum sanitary sewage flowrate (domestic infiltration flowrates): they often use a flowrate of 0,5 m^3/person·d.

6.1.9 Outfall

The outfall pipe carries the raw or treated wastewater towards the receiving watercourse, to be discharged. Treated wastewater from the treatment plant is thus discharged into the receiving watercourse through an outfall. When wastewater is not treated, the last section of a main or trunk sewer, located downstream from a watershed, is usually an outfall. Moreover, when wastewater treatment is in place but part of the wastewater cannot be treated, this untreated sewage is also evacuated through an outfall. These overflows occur during snow melt or heavy rainfalls, due to a high infiltration flowrate and, in general, when the water table is high and close to ground level. This also happens during rainy weather, when an intercepting sewer collects wastewater from a combined sewer system: we refer then to stormwater overflows.

6.2 WASTEWATER FLOWS IN SANITARY SEWERS

To evaluate the wastewater flowrates a sanitary sewer system must carry, we successively evaluate the flowrates of domestic, industrial (except water used as a cooling agent), commercial, institutional, and community wastewater, as well as parasitic flowrates, i.e. infiltration and inflow.

6.2.1 Domestic Wastewater Flowrates

Most domestic wastewater is the water left over from consumptive uses. This wastewater flow is the water returning to the sewer. The water used for lawn sprinkling, fire-fighting, street cleaning, filling up swimming pools, and to cool industrial equipment is eventually collected by the storm sewer system. However, part of this water is lost through evaporation and ground infiltration. The volume of domestic wastewater discharged in the sanitary sewers represents, in general, 60 to 80% of the total drinking water volume distributed. In dry regions, where the water table is located below the sewer system, a large portion of wastewater can leak from flawed joints or breaks in the system, infiltrate the ground, and contaminate the underground water. In general, in Quebec and other wet areas, the volume of wastewater returning to the sewer is greater than the volume of distributed water. This phenomenon is mainly due to infiltration (when the water table is high), to illegal connections of storm sewers, storm service pipes or water inlets, or to water losses from the water distribution network.

In Quebec's residential areas, wastewater returning to the sewer is, on average, equal to a wastewater flowrate of 200 to 225 L/(person·d). However, when considering wastewater from community based buildings (churches, police stations, small businesses, community centres, service-stations, etc.), the average domestic wastewater flowrate is equal to 320 L/(person·d). This flowrate is an average annual flowrate, evaluated with the following equation:

$$\frac{\text{Total volume of wastewater evacuated during one year}}{365\ \text{days} \times\ \text{number of users}} \tag{6.1}$$

In general, total domestic wastewater flowrates are expressed in m^3/d when designing a wastewater treatment plant, and in m^3/s when designing a sewer system.

Variations in Domestic Wastewater Flowrate. The domestic wastewater flowrate – mostly water for consumptive use returning to the sewer after use (parasitic waters are added to these waters) – varies with seasons, temperature, days of the week, hours of the day, etc.

During the day, variations in the flowrate generally follow the curve represented in Figure 6.2. We note that between 2 a.m. and 6 a.m., the flowrate is minimum, in other words, when the drinking water consumption is also minimum. It is during this period and after at least two days without rain, that the infiltration flowrate can be measured. Nowadays, people do not tend to do laundry on the same day (before, the laundry was traditionally done on Mondays) and the variations in domestic wastewater flowrate during the week (but not week-ends) are often minimal.

In some regions, variations in flowrates are important depending on the seasons or the days of the week; this mostly occurs in cities where temporary residents engage in seasonal activities (skiing, summer sports, cultural events, etc.).

Peak Factors. Usually, to determine minimum and maximum flowrates, peak factors (FP) are used. These are obtained by dividing a given flowrate by the average flowrate. To obtain the hourly peak factor, for example, the maximum hourly flowrate of a given year is divided by the average annual hourly flowrate for the same year. The daily peak factor is obtained in the same way by substituting the appropriate variables.

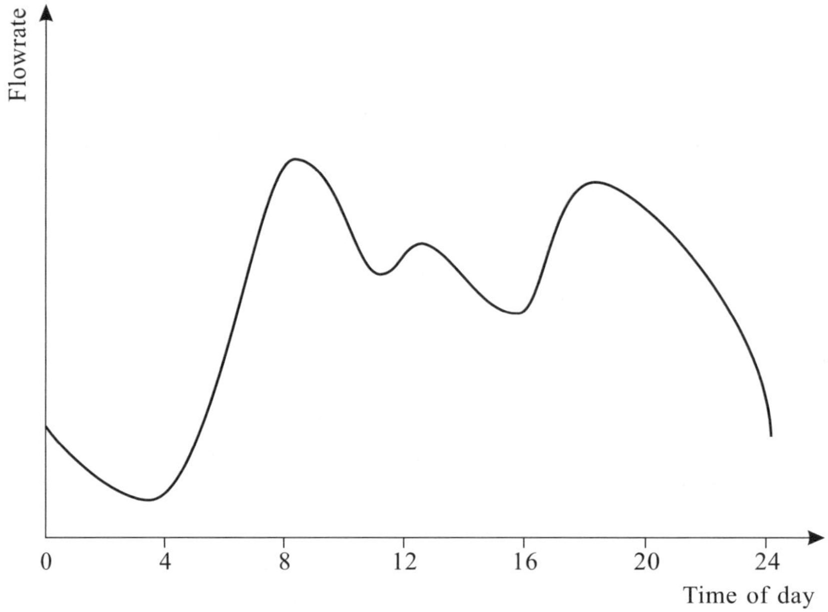

Figure 6.2 Variation of the Domestic Wastewater Flowrate During One Day in a Residential Area.

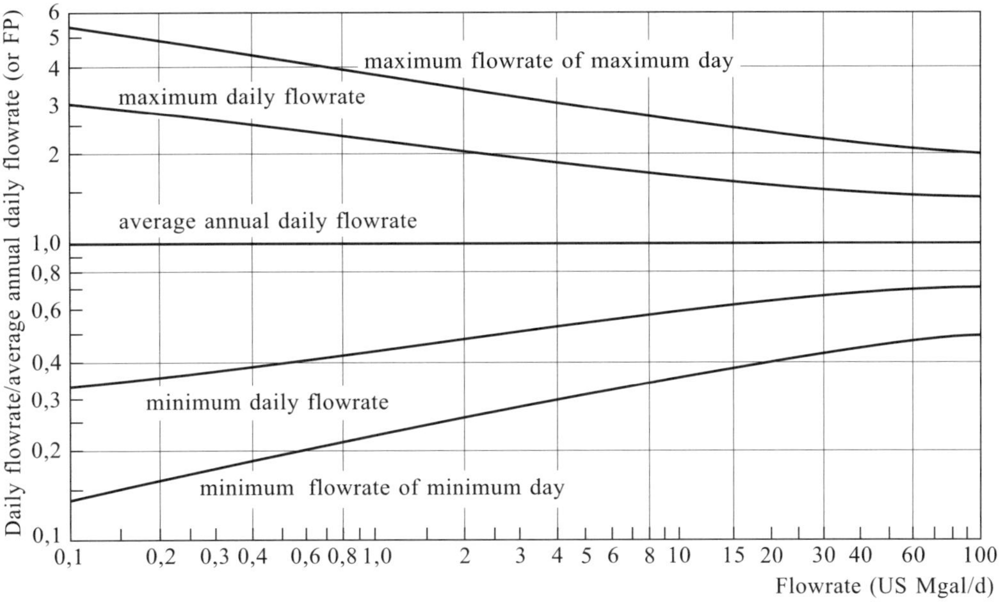

Figure 6.3 Minimum and Maximum Peak Factors of the Sanitary Wastewater Flowrates (Domestic Flowrates and Parasitic Flowrates), Factors suggested in New England (US) for the Design of Sanitary Sewer Systems; 1 US Mgal /d = 3785 m^3/d.

From: ASCE and WPCF, Gravity Sanitary Sewer Design and Construction, ASCE Manuals and Reports on Engineering Practice, no 60, WPCF, Manual of Practice, no FD-5, 1982, p. 40.

Peak factors vary according to regions and characteristics of the users. Thus, in small cities where activities are limited (dormitory towns, for example), peak factors are higher than in big cities, where activities are diversified. When a sewer system is already in place, these peak factors are preferably determined by measuring the flowrates. However, when there are no existing sewer systems or when the construction of a new sewer system must be planned, we rely on peak factors suggested by government agencies or by specialists in the field. In Figure 6.3, the values of the maximum peak factors for the sanitary wastewater flowrates (domestic and parasitic flowrates; section 6.2.6) decrease when the average daily flowrate increases. When new systems must be designed, it is advisable to use the peak factors from the upper curve (maximum flowrate of the maximum day). These peak factors can also be calculated from the equation presented in Table 6.1.

The curves shown in Figure 6.3 and the equations of Table 6.1 were determined from studies conducted in 18 Massachusetts cities (US).

Table 6.1 Equations to Compute Maximum Peak Factors FP_{max} of Sanitary Wastewater Flowrates (Domestic and Parasitic Flowrates), Factors Suggested in New England (US) for the Design of New Sanitary Sewer Systems

Units of the average annual sanitary wastewater flowrate	FP_{max}
m^3/s	$2{,}25Q^{-0{,}165}$
m^3/d	$14{,}68Q^{-0{,}165}$
ft^3/s	$4{,}05Q^{-0{,}165}$
CA MGal/d	$3{,}89Q^{-0{,}165}$
US MGal/d	$3{,}77Q^{-0{,}165}$

Table 6.2 Peak Factors Suggested by the Quebec Ministry of Environment to Calculate Maximum and Minimum Domestic Wastewater Flowrates

Average annual domestic wastewater flowrate (m^3/s)	FP_{max}	FP_{min}
$Q < 0{,}004$	4,0	0,25
$0{,}004 \leq Q \leq 0{,}4$	$1{,}742Q^{-0{,}156}$	$Q^{0{,}1506}/1{,}742$
$Q \geq 0{,}4$	2,0	0,50

From: Quebec Government, Ministry of Environment, Guideline 004, p. 6.1.5, October 25th 1989, (in French).

Nowadays in Quebec, the Ministry of Environment suggests the use of peak factors shown in Table 6.2, to calculate domestic wastewater flowrate.

When designing a sewer system, peak factors must be chosen carefully after conferring with local authorities (ministries and municipalities), because peak factors have a great influence on sizes and costs of construction.

Note, however, that the values of the maximum peak factors, shown on Figures 6.3 and 6.4 and on Tables 6.1 and 6.2, are suggested values for the design of sewer systems. They are not necessarily hourly peak factor values, although they are relatively close.

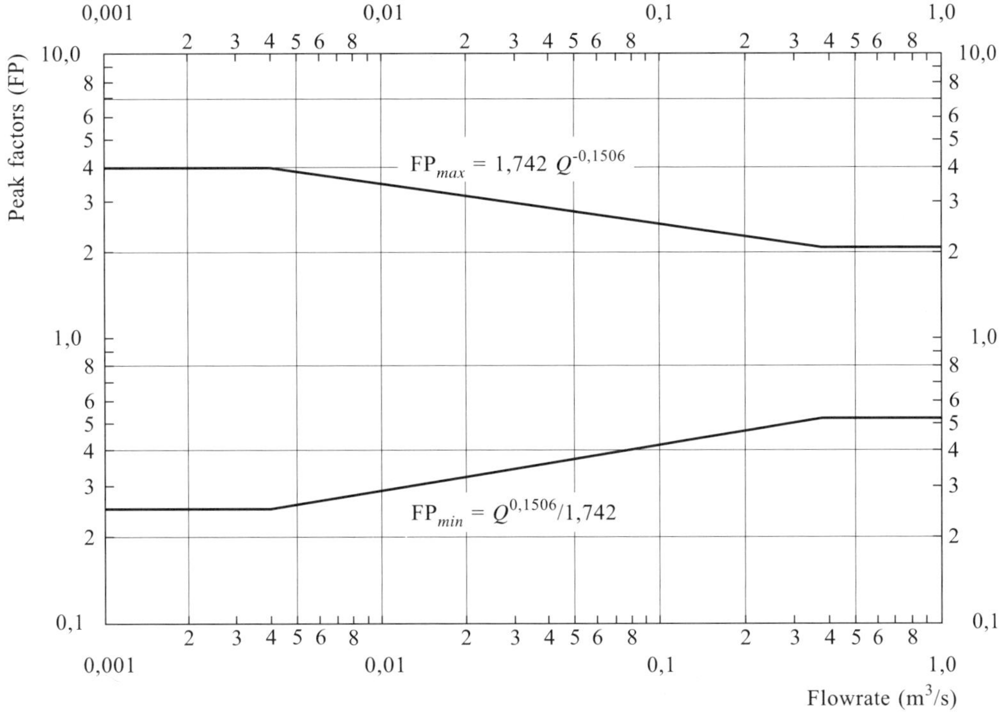

Figure 6.4 Minimum and Maximum Peak Factors Suggested by the Quebec Ministry of Environment to Compute Domestic Wastewater Flowrates.

From: Quebec Government, Ministry of Environment, Guideline 004, p. 6.1.5, October 25th 1989, (in French).

In practice, the variations of the domestic wastewater flowrate are not as significant as the flowrate variations of the water for consumption. The cause is related to the following facts:

- the great storage capacity of the pipes tends to reduce peak flows;
- in general, the time for wastewater to reach a given measurement point is not the same for all sub-basins discharging to this point of measure: the maximum flow at this point is smaller than the sum of the peak flows for each of these sub-basins.

6.2.2 Industrial Wastewater Flowrates

In general, existing industries are equipped with water meters. However, the proportion of used water evacuated to the sewers must be evaluated accurately because the water returning to the sewer might be very different from that returned to the sewer in residential areas. Thus, the quantity and quality of wastewater evacuated through the sewer by the main industries of a municipality must be carefully measured and analysed. If flowrates are impossible to measure, average values like the ones shown in Table 6.3 can be used. In general, water used as cooling agent is evacuated in the storm sewer system.

Table 6.3 Industrial Wastewater*

Type of industry	Average wastewater flowrate $[m^3/(ha\cdot d)]$
Industries for which the manufacturing process does not require water (clothing, transportation material, etc.)	10
Industries for which the manufacturing process requires small amounts of water (textile, wood, etc.)	25
Industries requiring large amounts of water (canning, pulp and paper, steel mill, etc.)	> 50
The surfaces of the infrastructures (road, parking lots, etc.) are included in the total surface area. Water used as a cooling agent is evacuated in the storm sewer system.	

6.2.3 Wastewater Flowrates from Commercial areas

The commercial wastewater flowrate in residential areas* is estimated to 75 $m^3/(ha\cdot d)$. To determine the total surface area, we add the different surfaces used for commercial purposes. If the areas of parking lots and other related spaces are added to this total, we obtain a wastewater flowrate of about 35 $m^3/(ha\cdot d)$. (Significant variations exist between the actual flowrate values and these average values.)

6.2.4 Wastewater Flowrates from Institutional Facilities

The institutional wastewater flowrate in residential areas** is estimated to be 25 $m^3/(ha\cdot d)$. Here again, it is important to note that this flowrate value is an average, and can vary significantly.

6.2.5 Wastewater Flowrates from Areas Dedicated to the Community

In a municipality with areas dedicated to the community (industrial parks, commercial areas, and institutional areas), it is important to study and evaluate consumption flowrates and wastewater flowrates. Results of these studies and measurements are very helpful when designing sewer systems, pumping stations, and water treatment plants. In general, zones dedicated as green areas and land occupied by major roads and highways are not considered in the computation of the sanitary wastewater flowrate.

---

* Quebec Government, Ministry of Environment, Guideline 004, p. 6.1.2 (Table 6.3) and p. 6.1.3, October 25th 1989, (in French).

** Idem, p. 6.1.3.

6.2.6 Parasitic Flowrate

First, let us define the following terms: "parasitic water", "infiltration", and "inflow".

Parasitic Water. This is the water, other than domestic, commercial, and industrial wastewater, which seeps into a *sewer system*: it gets there by *infiltration* or *inflow*. In general, we consider that the infiltration flowrate (Q_{inf}) is the constant portion of the parasitic flowrate and that the inflow (captured flow) (Q_i) is its variable portion.

Infiltration. This is parasitic water originating from the ground and constantly seeping into the sewer system through network defects (poor quality joints and cracks in the pipes, manholes or service connections).

Inflow. Inflow is parasitic water intermittently seeping into the sewer system. It mostly occurs during wet-weather and is caused by various illegal practices, such as foundation drain connections, street inlets or storm sewers connecting to the sanitary sewer system. Inflow can also seep into the sewer system through the lifting holes of manhole covers when they lie in a depression of the ground. Diversion into the sewers of streams and ditches that flow during dry-weather is also prohibited. However, the flow of ditches during wet-weather is part of the storm water flow and should be evacuated by the storm sewer system.

Ideally, sanitary sewer systems should only transport wastewater originating from users. In Quebec, watertight systems were only built after 1978, since the introduction of watertightness tests for all newly built sanitary sewer systems. Thus, inflow and infiltration flow problems are often significant in sewer systems built prior to 1978. Table 6.4 lists, according to source, different kinds of inflow into various types of sewer systems.

Table 6.4 Sources of Inflow (to Be Eliminated) in Various Sewer Systems (Yes = Flow)

Water sources	**Type of sewer system**			
	pseudo-sanitary	**sanitary**	**combined**	**storm***
Ditches flowing during dry-weather (flowrate during dry-weather)	yes	yes	yes	yes/no
Foundation drain pipe	no	yes	no	no
Street inlets	yes	yes	no	no
Cooling systems	yes	yes	no	no

* Ditches may be integrated to the storm sewer network.

Transporting (including the cost of energy at pumping stations) and treating this water increases energy expenditure. Water treatment is also affected because substances found in low concentrations are harder to eliminate. For these reasons it is important to reduce the flow of parasitic water.

Infiltration Flowrate. Groundwater infiltration is subject to fluctuations lasting for several days or weeks. However, in Quebec and in many other regions of America, infiltration is considered to be continuous and constant. For this reason, the infiltration flowrate constitutes a constant portion of the parasitic flowrate.

The infiltration flowrate is measured, especially when the water table level is high, during dry-weather (after at least two days without rain), at night between 2 a.m. and 6 a.m. (when wastewater flowrates of various origins are minimal). In the spring, during the snow melt, infiltration flowrates often reache record values: this exceptional situation requires special attention. As seen previously, parasitic water (including infiltration) increases the costs of transportation and treatment of wastewater. Therefore it is economical to minimize parasitic water by improving the watertightness of the sewer system.

In a new sewer system, infiltration flowrate should equal zero. Unfortunately, as construction materials and workmanship are not ideal, a certain amount of infiltration must be tolerated. This infiltration is much less than seen in the past.

Infiltration Flowrate in a New Sewer System (in Quebec). In new sewer systems, we generally consider the infiltration flowrate to be a function of the diameter and length of the pipes. In Quebec, for various computations, we use an infiltration flowrate equal to 225 L/(cm·km·d) (225 litres per day, per diameter of the pipe in centimetres, and per length of the pipe in kilometres). This value is also used for service connections; in fact, because service connections make up an important part of the system, they must be considered when evaluating the infiltration flowrate.

Infiltration Flowrate for an Existing Sewer System which has been Rehabilitated (in Quebec). For sewer systems built prior to 1978, the year watertight sanitary sewer systems became mandatry, infiltration flowrate is generally very high. Infiltration alone can dilute wastewater sufficiently to make water treatment economically impossible. They must then be repaired to reduce the infiltration flow. The value of the infiltration flowrate used to estimate the total flows in the sewers is then the value to be measured after restoring the sewer system. When repairing a sanitary sewer system, the goal is to minimise the flow of infiltration. However, it is difficult to reduce infiltration below 20 L/(cm·km·d). It is thus forbidden to use a value smaller than the previous one in the computation of infiltration flowrate.

Table 6.5 shows the values of infiltration flowrates measured in Quebec sewer networks.

In general, peak factors are not used for infiltration. However, if on-site measurements show large variations in the flowrate, a peak factor calculated from these infiltration measurements (maximum infiltration/average infiltration), can be used in the computations of peak flows.

Table 6.5 Infiltration Flowrates Observed in the Sewer Networks of the Province of Quebec, before Rehabilitation

Infiltration flowrate	**Type of sewer network**			
	sanitary		**pseudo-sanitary or combined**	
	[m³/(ha·d)]	**[L/cm·km·d)]**	**[m³/(ha·d)]**	**[L/cm·km·d)]**
Low	6	1500	8	2000
Average	12	3000	16	4000
High	24	6000	36	9000

From: Government of Quebec, Ministry of Environment, Guideline 004, p. 6.1.4, October 25th 1989, (in French).

Inflow Flowrate in Sanitary and Pseudo-Sanitary Sewer Systems. During and after precipitation, some runoff is captured (mainly through various illegal connections) and can cause significant flows lasting for one to several days. Most of the time, it causes intercepting sewers or pumping stations to exceed their capacities, and sometimes, raw wastewater to be discharged into streams through structures designed for that purpose, such as weirs, pumping stations or outfalls (stormwater overflows). Moreover, this parasitic water may cause a backflow in the sewer system. There are two categories of inflow: direct inflow, which occurs immediately after precipitation; and delayed inflow, which occurs during the two days following precipitation (e.g. from foundation drain connections). After this period, parasitic water is considered to originate only from infiltration flows.

Inflow flowrates are measured and their sources are identified by instantaneous or continuous measurements in wet weather, and by site inspections in dry and wet weather, using various techniques such as the tracer test and the smoke test. A tracer (dyes, for example) is dissolved in the sewage and is tracked through the system to make sure that the sewage is evacuated at the proper location. In dry weather, smoke is blown into sewer pipes. By observing where the smoke emerges, we can identify openings in the pipes and identify illegal water sources. However, the absence of smoke does not guarantee the absence of illegal connections, as water, for example, may block the passage of smoke. The corrective measures undertaken to reduce inflow consist mainly of reducing the number of illegal storm sewer connections to the sanitary sewer system. To measure the instantaneous flowrates, we apply the following procedure: measurements at the end of the rain, then measurements 12, 24, 48 and 72 hours after the end of the rain.

In a new sanitary sewer system, inflow flowrate should theoritically equal zero. However, a flowrate reaching 25 L/(person·d) is acceptable when the system is new. After 15 to 20 years, we expect the flowrate to reach 50 L/(person·d). In existing systems, the aim is to reduce inflow as much as possible.

In combined and storm sewer systems, inflow is not considered, mainly because these systems must evacuate all rainwater. In practice, however, some draining ditch connections and stream connections introduce unwanted and unauthorized water which is also categorised as inflow.

Parasitic Water Flowrate for Areas Without Sewer Systems. When planning a sewer system, infiltration or inflow flowrates cannot be evaluated *in situ*. In this case, the parasitic water flowrate is estimated at 2,8 and even to 5,6 $m^3/(ha{\cdot}d)$. When the population is known, the use of 60 and 50 L/(person·d) is suggested for infiltration and inflow flowrates. Table 6.6 shows suggested values for parasitic water flowrates in sanitary sewer systems in Quebec.

Table 6.6 Parasitic Water Flowrates in Sanitary Sewer Systems in Quebec

Type of flowrate	Sanitary sewer systems			
	old	**new***	**rehabilitated**	**planned**
Infiltration	measured	225 L/(cm·km·d)**	measured	60 L/(person·d)
Inflow	measured	25 L/(person·d) [50 L/(person·d) when system is older]	measured	50 L/(person·d)
Total (parasitic)	total	total	total	110 L/(person·d) or 2,8 to 5,6 $m^3/(ha{\cdot}d)$***

* Quebec Government, Ministry of Environment, Guideline 004, p. 6.1.4 to 6.1.6, October 25th 1989, (in French).

** cm = diameter of pipes in centimetres; km = length of pipes in kilometres.

*** We use one formula or the other depending on wether the population is known.

6.2.7 Minimum Flowrates

Minimum flowrates must be calculated for the main following reasons:

a) It is important to know the flow conditions in the pipes when flowrates are at minimum. If the flow velocity and the duration of wastewater in the pipes are known, we can evaluate the quantity of various types of deposit, the risk of bad odor production, and toxic or explosive gas release.

b) In addition to maximum flowrates, minimum flowrates must also be known in order to compute part sizes and volumes of pumping stations, and to choose the type of pump required. Here, again, wastewater remaining in pumping stations too long may produce bad odors and release toxic or explosive gases.

c) Engineers responsible for the design of sewage treatment plants must also know the minimum flowrates in order to build an efficient treatment plant, even during periods of low flowrates.

6.3 TOTAL MAXIMUM FLOWRATES

To design sewer systems, pumping stations, various equipment, and treatment plants, the engineer must compute the total maximum flowrates. The total maximum flowrates are made up of domestic flowrates, infiltration flowrates, and inflow flowrates. The engineer must also take into account commercial and industrial flowrates, when present.

6.3.1 Sanitary Sewage

The peak factors (FP) used in Quebec only apply to domestic wastewater flowrates, thus the maximum sanitary flowrates are obtained using the following relation:

$$(Q_{san})_{max} = (Q_{dom})_{avg} \times \text{FP} + Q_{inf} + Q_i \tag{6.2}$$

where Q_{inf} = infiltration flowrate
Q_i = inflow flowrate

As seen in Figure 6.3, the maximum values of the peak factors fall when the average daily flowrate increases. However, as is usual in Quebec, maximum peak factors below 2,0 should never be used to calculate the maximum flowrate. When using the curves shown in Figure 6.3 (these curves refer to the sanitary flowrate and not to the domestic flowrate only), we base our decision on local engineers' practices to determine with which of the two maximum flowrate curves, the peak factors will be chosen. As a safety measure, we suggest to use the upper curve (the one representing the maximum flowrate during the day of the year when the daily flowrate is maximum), at least for flowrates below 3785 m^3/d (1 US Mgal/d).

In existing industrial areas, we can use a peak factor specific to each establishment. This peak factor does not necessarily decrease when the average flowrate increases.

6.3.2 Infiltration Flowrates

As seen previously, no peak factors are applied to the infiltration water flowrate, except when on-site measurements show significant variations. In this case, we use the peak factor obtained from these observations to compute the maximum flowrate.

6.3.3 Inflow Flowrates

In a sanitary sewer system, the variations of the inflow flowrates must be carefully monitored. For a new pipe, inflow must not exceed 25 L /(person·d), but 50 L/(person·d) is acceptable for older pipes (15 to 20 years).

In a pseudo-sanitary sewer system, the authorized runoff flowrate is added to the infiltration flowrate. (Note that pseudo-sanitary sewer systems are no longer built in Quebec.)

In a combined sewer system, there are no unauthorised connections, as this type of system is designed to collect all runoff. In practice, however, there are some connections to draining ditches and streams which contribute to the inflow in combined sewers. These connections must be eliminated because the water they carry is usually not polluted, it dilutes wastewater that must be treated. Finally, it should be evacuated through the surface draining system. (This water should naturally flow towards lakes and streams without the use of pipes.)

EXAMPLE 6.1

A region in Quebec has a sanitary sewer system and a storm sewer system. For the sanitary sewer system, we have the following data:

- Actual population: 7500 persons
- Population in 10 years: 12 500 persons
- Ultimate population (i.e. in 30 years): 25 575 persons
- Total surface area of the region: 165 ha
- Actual measure of parasitic flowrate: 15 000 m^3/d
- Actual measure of infiltration flowrate: 2000 m^3/d
- Actual measure of flowrate for water used as cooling agent: 12 000 m^3/d
- For design purposes, we tolerate:
  · a maximum infiltration flowrate: 60 L/(person·d)
  · a maximum inflow flowrate: 50 L/(person·d)

A new sanitary trunk sewer will be built at the end of this year for this region. Calculate the maximum and minimum flowrates carried by this trunk sewer this year, in 10 years and in 30 years.

Repairs to the existing system will be made before the construction of the new trunk sewer.

Solution

According to the information presented in this chapter, we know that:

- the average domestic wastewater flowrate is 320 L/(person·d) (section 6.2.1);
- the water used as a cooling agent must be excluded from the sanitary sewer system (section 6.1.1 and section 6.2).

Computation of the Domestic Flowrates and the Maximum and Minimum Sanitary Flowrates

For the purpose of this example, we will compute the initial flowrates (first year of the trunk sewer).

Average domestic flowrate

$$\left(Q_{dom}\right)_{avg} = (7500 \times 320) \div 1000 = 2400 \text{ m}^3/\text{d} \left(0{,}028 \text{ m}^3/\text{s}\right)$$

Maximum peak factor in Quebec (for domestic flowrate)

$$\begin{aligned} \text{FP} &= 1{,}742\, Q^{-0{,}1506} = 1{,}742 \times 0{,}028^{-0{,}1506} \\ &= 2{,}99 \end{aligned}$$

Minimum peak factor in Quebec (for domestic flowrate)

$$\begin{aligned} \text{FP} &= Q^{0,1506}/1,742 = 0,028^{0,1506}/1,742 \\ &= 0,335 \end{aligned}$$

Maximum domestic flowrate

$$(Q_{dom})_{max} = 2400 \times 2,99 = 7172 \text{ m}^3/\text{d}$$

Minimum domestic flowrate

$$(Q_{dom})_{min} = 2400 \times 0,335 = 803 \text{ m}^3/\text{d}$$

Infiltration flowrate

$$Q_{inf} = (7500 \times 60) \div 1000 = 450 \text{ m}^3/\text{d}$$

Inflow flowrate

$$Q_i = (7500 \times 50) \div 1000 = 375 \text{ m}^3/\text{d}$$

Maximum sanitary flowrate

$$\begin{aligned} (Q_{san})_{max} &= (Q_{dom})_{max} + Q_{inf} + Q_i \\ &= 7172 + 450 + 375 \\ &= 7997 \text{ m}^3/\text{d} \end{aligned}$$

Minimum sanitary flowrate

The minimum sanitary flowrate occurs when it does not rain; in other words, when inflow flowrate is zero. However, the infiltration flowrate is constant.

$$\begin{aligned} (Q_{san})_{min} &= (Q_{dom})_{min} + Q_{inf} \\ &= 803 + 450 \\ &= 1253 \text{ m}^3/\text{d} \end{aligned}$$

The remaining results of computations are presented in the following Tables.

Domestic Wastewater Flowrates

	$(Q_{dom})_{avg}$		FP (Q_{dom} for Quebec)		Q_{dom} (m^3/d)	
	(m^3/d)	**(m^3/s)**	**max.**	**min.**	**max.**	**min.**
Present	2 400	0,028	2,99	0,335	7 172	803
In 10 years	4 000	0,046	2,77	0,361	11 068	1 446
In 30 years	8 184	0,095	2,48	0,403	20 331	3 294

Sanitary Wastewater Flowrates

	(Q_{dom}) (m³/d)		Q_{inf} 60 L/(pers.·d) (m³/d)	Q_i 50 L/(pers.·d) (m³/d)	Q_{san} (m³/d)	
	max.	min.			max.	min.
Present	7 172	803	450	375	7 997	1 253
In 10 years	11 068	1 446	750	625	12 443	2 196
In 30 years	20 331	3 294	1 535	1 279	23 144	4 829

The authorized infiltration and inflow flowrates are lower than the actual flowrates measured; these actual flowrates must then be reduced by rehabilitation of the network (mainly sewer pipes and service connections).

EXAMPLE 6.2

A region in Quebec is equipped with a combined sewer system. To design the intercepting sewer that will serve this community, we rely on the following data:

- Actual population: 5250 persons
- Population in 10 years: 10 300 persons
- Ultimate population (i.e. in 30 years): 18 275 persons
- Total surface area of the region (present and future): 215 ha
- Actual measure of the infiltration flowrate: 850 m³/d
- Actual measure of the inflow flowrate originating from a ditch that flows in dry-weather: 3000 m³/d
- For design purposes, we tolerate:
  - a maximum infiltration flowrate at the intercepting sewer: 60 L/(person·d)
  - a maximum inflow (rainwater) flowrate at the intercepting sewer: 100 L/(person·d)*

The intercepting sewer will be built by the end of this year. Calculate the maximum and minimum flowrates carried by this sewer this year, in 10 years, and in 30 years.

Repairs to the combined sewer network will be made before the construction of the new interceptor.

---

* Since it may rain anytime, it may rain when wastewater flowrate is maximum (domestic and infiltration flowrates). Thus, to decrease the number of storm overflows into the watercourse, it is usual to design an interceptor receiving wastewater from a combined sewer network so that it can accept a fraction of the rainwater. Sometimes engineers use 0,5 m³/person·d.

Solution

The intercepting sewer only carries wastewater sent to the wastewater treatment plant. Its flowrate corresponds to the maximum sanitary flowrate. The excess (often called stormwater overflow) is discharged into the watercourse or stored in reservoirs to be treated later, when the treatment plant can handle this volume of wastewater.

According to the information presented in this chapter, we know that:

- the average domestic wastewater flowrate is equal to 320 L/(person·d) (section 6.2.1);
- the inflow originating from a ditch which flows during dry-weather must not be captured.

Computation of Domestic and Sanitary Maximum Flowrates. For the purpose of this example, we will compute the actual flowrates.

Average Domestic Flowrate

$$\left(Q_{dom}\right)_{avg} = (5250 \times 320) \div 1000 = 1680 \text{ m}^3/\text{d} \left(0{,}0194 \text{ m}^3/\text{s}\right)$$

Maximum Peak Factor in Quebec (for domestic flowrate)

$$\begin{aligned} \text{FP} &= 1{,}742\, Q^{-0{,}1506} = 1{,}742 \times 0{,}0194^{-0{,}1506} \\ &= 3{,}15 \end{aligned}$$

Minimum Peak Factor in Quebec (for domestic flowrate)

$$\begin{aligned} \text{FP} &= Q^{0{,}1506}/1{,}742 = 0{,}0194^{0{,}1506}/1{,}742 \\ &= 0{,}317 \end{aligned}$$

Maximum Domestic Flowrate

$$\left(Q_{dom}\right)_{max} = 1680 \times 3{,}15 = 5297 \text{ m}^3/\text{d}$$

Minimum Domestic Flowrate

$$\left(Q_{dom}\right)_{min} = 1680 \times 0{,}317 = 533 \text{ m}^3/\text{d}$$

Infiltration Flowrate

$$Q_{inf} = (5250 \times 60) \div 1000 = 315 \text{ m}^3/\text{d}$$

Inflow Flowrate

$$Q_i = (5250 \times 100) \div 1000 = 525 \text{ m}^3/\text{d}$$

Maximum Sanitary Flowrate

$$\begin{aligned} \left(Q_{san}\right)_{max} &= \left(Q_{dom}\right)_{max} + Q_{inf} + Q_i \\ &= 5297 + 315 + 525 \\ &= 6137 \text{ m}^3/\text{d} \end{aligned}$$

Minimum Sanitary Flowrate

The minimum sanitary flowrate occurs when it does not rain; in other words, when the inflow is zero. However, the infiltration flowrate is constant.

$$\begin{aligned}(Q_{san})_{min} &= (Q_{dom})_{min} + Q_{inf} \\ &= 533 + 315 \\ &= 848 \text{ m}^3/\text{d}\end{aligned}$$

The remaining results of computations are presented in the following Tables.

Domestic Wastewater Flowrates

	$(Q_{dom})_{avg}$		FP (Q_{dom} for Quebec)		Q_{dom} (m^3/d)	
	(m^3/d)	(m^3/s)	max.	min.	max.	min.
Present	1 680	0,019	3,15	0,317	5 297	533
In 10 years	3 296	0,038	2,85	0,351	9 390	1 157
In 30 years	5 848	0,068	2,61	0,383	15 282	2 238

Sanitary Wastewater Flowrates (flows that the intercepting sewer must transport to the treatment plant)

	(Q_{dom}) (m^3/d)		Q_{inf} 60 L/(pers.·d) (m^3/d)	Q_i 100 L/(pers.·d) (m^3/d)	Q_{san} (m^3/d)	
	max.	min.			max.	min.
Present	5 297	533	315	525	6 137	848
In 10 years	9 390	1 157	618	1 030	11 038	1 775
In 30 years	15 282	2 238	1 096	1 828	18 206	3 334

We notice significant variations of the sanitary flowrates between the beginning and the end of the intercepting sewer's design period (e.g. from 6137 to 18 206 m^3/d for the maximum flowrates).

CONCLUSION

Sanitary sewer systems must be designed to evacuate the maximum sanitary wastewater flowrates (maximum domestic wastewater flowrate, infiltration, and inflow). Their hydraulic capacity must be adequate for a long period, in general

30 years. The large variation in flowrates between the beginning and the end of the design period requires careful planning from the design engineer, who must ensure that the various requirements are respected at all times.

The average, maximum, and minimum sanitary flowrates are computed in the following manner:

Average Flowrate

Q_{avg} = population × Q_{avg} per person + Q_{inf} + Q_i + other flowrates (from industries, stores, etc.)

Maximum Flowrate

Q_{max} = population × Q_{avg} per person × FP_{max} + Q_{inf} + Q_i + other flowrates (from industries, stores, etc.)

Minimum Flowrate

Q_{min} = population × Q_{avg} per person × FP_{min} + Q_{inf} + other flowrates (from industries, stores, etc.)

N.B. There is no inflow in minimum flowrate.

PROBLEMS

6.1 In a suburban residential area of Quebec, a new sanitary sewer system must ultimately be used by 1200 inhabitants, spread over an area of 100 ha. Calculate, in m^3/d, the average domestic flowrate and the average, minimum, and maximum sanitary flowrates that the trunk sewer serving this community must carry.

6.2 In the United States, a sanitary trunk sewer must be built able to carry an average sanitary flowrate of 9464 m^3/d.Determine the value of the maximum peak factor that should be applied to this average flowrate if the population is 25 000 inhabitants. Evaluate the maximum peak factor applied to the domestic wastewater flowrate if the trunk sewer was built in the Province of Quebec. In each case, indicate the sources from which you determined the values of the peak factors.

6.3 Calculate, in m^3/d, the actual and ultimate maximum sanitary flowrates of a new sewer system located in a Quebec residential area of 100 ha. In this area, there are no stores, businesses, or institutions. The 12 000 inhabitants who presently live in this area occupy 80 of the 100 hectares.

Chapter 7

RUNOFF IN URBAN AREAS

INTRODUCTION

Progressive urbanization produces a considerable increase in surface imperviousness (due to paved streets, roofs, driveways, etc.). This results in less rain absorbed by the ground surface and more water running on the surface, called runoff.

For a given quantity of rainwater on an urban watershed, runoff flows more easily and maximum runoff is much greater (up to 100 times greater) after urbanization has taken place. The engineer in charge of water management must minimize problems and damage caused by runoff (and by wastewater from combined sewers, including runoff).

In this chapter, we first discuss methods for evaluating runoff volumes in urban areas during rainfalls. Then we determine the maximum runoff for watersheds of different shapes and characteristics, and for different types of rainfall. Finally, we present techniques for reducing runoff flows.

7.1 DEFINITIONS

Catchment (or watershed). Drainage area discharging its water toward a single output: the discharge.

Rainfall hyetograph. Histogram representing the height of rainfall as a function of time at a given site or on a watershed.

Runoff hydrograph. Graphic representation of the runoff variation as a function of time at the discharge of a watershed following rainfall.

Urban sub-basin. Catchment for which the outflow is discharged in a street inlet.

7.2 RAINFALL AND RUNOFF IN URBAN AREAS

For a watershed located in an urban area or elsewhere, the runoff and its volume are determined by:

– rainfall characteristics (intensity, duration, and space-time distribution);

- characteristics of the watershed surface (nature, permeability, slope, and landscaping);
- hydraulic equations regarding water flowing freely on the surface.

Table 7.1 displays sequentially the relationship between a rainfall (and its characteristics), a watershed (and its characteristics), the sewer network and the characteristics of the runoff hydrograph at the discharge of a watershed.

Table 7.1 Components of the relationship between a rainfall and a runoff hydrograph at the discharge of a watershed

Function	Characteristics
Source: rain	Atmospheric - space and time distribution of the rainfall or synthetic hyetogram
Transformation: watershed	Physical - surfaces so-called impermeable, directly drained so-called impermeable, indirectly drained non-drained (i.e. not connected to sewer systems) permeable, drained - water initially retained: by permeable surfaces in settlements of so-called impermeable surfaces - conditions prior to rainfall - degree of water infiltration in the ground during rainfall
Transportation: sewer system	Hydraulic - characteristics of the storm-sewer or combined-sewer system - propagation of runoff hydrograph in sewer systems - storage of water (in pipes or detention basins)
Results: runoff hydrograph	Runoff hydrograph at the discharge of the watershed

Adapted from: Patry, Gilles and Raymond, Louise. "Simulation du ruissellement en milieu urbain. Le modèle Illudas : présentation du modèle", *Eau du Québec*, vol. 11, no. 4, pp. 280-290 (1978).

7.2.1 Rainfall

Rain falling on a watershed plays three important roles. First, it dampens the surface of the watershed and the equipment installed on site. Second, it infiltrates the ground to reach the water table. Finally, should the occasion arise, it runs off to a collecting location (ditch, street inlet, or river).

7.2.2 Runoff

In urban areas, surface-runoff management is not usually meant to protect human lives, as it often is in flood areas; rather, it is aimed mostly at avoiding damage to property. Thus, since the turn of the century many large North American cities have developed infrastructure to rid their lands of frequent basement floods.

Over the years, a large number of lawsuits related to damage caused by urban runoff have led to the development of legal precedents, which have affected civil engineering practices or guidelines of infrastructure design. Municipalities following these guidelines cannot be held responsible for damage caused by exceptional rainfall (such as the deluge of July 14, 1987 in Montreal). However, these guidelines vary depending by region and, especially, by country, because they are established according to historical court rulings and engineering practices. Thus, in well-to-do municipalities where houses are more luxurious, management may take precautions against rare events that may not be taken in other municipalities. Finally, new techniques can be used to minimize flooding, such as surface water retention and the use of vortex regulators or hydro-brakes.

Many computational methods can determine runoff in urban areas, depending on whether the maximum runoff alone or the runoff variation as a function of time (hydrograph) is considered. In North America, the "rational" computation method calculates maximum runoff and is still quite popular. Because of easy access to affordable and efficient personal computers with extended memory, engineers have developed computer models to compute and plot the characteristics of runoff hydrographs for various types of rainfall, watersheds, and collection infrastructures. These tools allow for efficient computation of storage reservoir volumes, storage capacity in pipes, duration of flow in pipes, and so on.

7.3 RATIONAL METHOD

The rational method was developed before computers were invented. It is used to quickly determine the maximum runoff of uniform rainfall on small-surface watersheds (less than 5 km^2 according to Viessman and Hammer, 1993, p. 232; up to 25 km^2 according to the Ministry of Transportation of Quebec in its manual on culverts – manuel de conception des ponceaux – 1993); therefore, it is the perfect method for small urban catchments.

7.3.1 Origin and Diffusion

It seems that the rational method was used for the first time in 1851 in Ireland. It appeared in North America in 1889, when Kuichling (E. Kuichling, *The Relation Between the Rainfall and the Discharge of Sewers in Populous Districts,* Transactions – ASCE, vol. 20, 1889, pp. 1-56) used it to study the behaviour of several urban runoff-collection systems. It was introduced to England in 1906, where it was called the Lloyd-Davies method (D.E. Lloyd-Davies, *The Elimination of Storm-Water from Sewerage Systems*, Institute of Civil Engineers, vol. 64, 1906). In 1975, a survey of 37 Canadian municipalities revealed that 97% of them used the rational method to design storm-sewer systems.

7.3.2 Development of the Rational Equation

To explain the rational equation, let us describe the runoff hydrograph (fig. 7.1b) at the watershed's discharge shown in figure 7.1a. Several **isochrones** are plotted. These curves connect every point that has the same runoff time to the watershed's outlet.

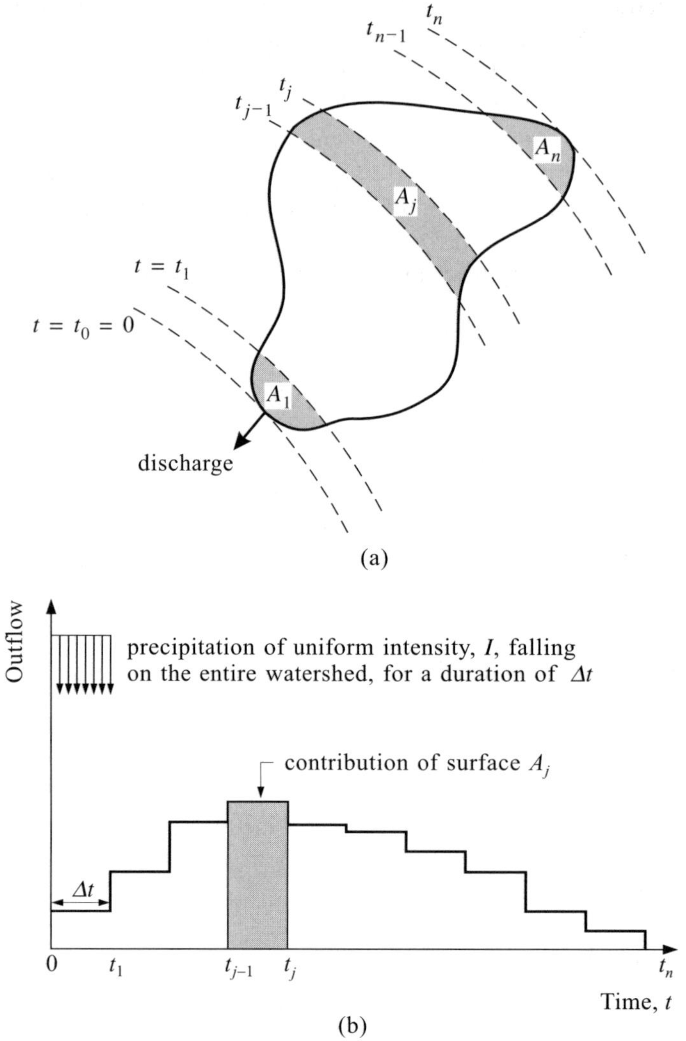

Figure 7.1 a) Watershed and isochrones; b) runoff hydrograph of a uniform rainfall falling on the entire watershed over a period of time, Δt

Adapted from: P. Lemieux, *La méthode rationnelle dans le drainage urbain* (Premier colloque sur le drainage urbain, Association québécoise des techniques de l'eau, 1977), pp. 85–107.

The plotted lines in fig. 7.1a determine surface areas A_1, A_2, ... A_n (surface A_j is determined by isochrones t_{j-1} and t_j). The variable t represents the time for the runoff to flow from isochrone to the discharge. It also represents for the present demonstration the time from the beginning of the rainfall. For the purpose of our demonstration, the choice of isochrones is such that:

$$
\begin{aligned}
t_1 &= t_1 - t_0 = \Delta t \\
t_2 &= t_2 - t_0 = 2\Delta t \\
t_3 &= t_3 - t_0 = 3\Delta t \\
t_j &= t_j - t_0 = j\Delta t \\
t_n &= t_n - t_0 = n\Delta t
\end{aligned}
$$

In this example, we will study the three cases shown below.

Case 1. Let us assume that on the entire area of the watershed, rain falls uniformly with an intensity I and a duration Δt; let us also assume that R_j is the fraction of the fallen rain over surface area A_j that runs off (runoff coefficient, section 7.9). Therefore, the flow $Q(t)$ is determined at the discharge of the watershed by the following method:

$$
\begin{aligned}
&\text{at } t_0 = 0, && Q(t_0) = Q(0) = 0 \\
&\text{at } t_1 = \Delta t, && Q(t_1) = A_1 I R_1 \\
&\text{at } t_2 = 2\Delta t, && Q(t_2) = A_2 I R_2 \\
&\text{at } t_j = j\Delta t, && Q(t_j) = A_j I R_j \\
&\text{at } t_n = n\Delta t, && Q(t_n) = A_n I R_n
\end{aligned}
$$

where $A_1 I$ = volume of the fallen rain on surface area A_I from the beginning of the rainfall to time $t_1 = \Delta t$

$A_1 I R_1$ = runoff volume at the discharge of the watershed, originating from area A_1, at time t_1.

Figure 7.1b represents the runoff hydrograph – that is, the variation in flow $Q(t)$ at the discharge of the watershed (shown in figure 7.1a) as a function of time. This variation follows precipitation of duration t_1 ($t_1 = \Delta t$) with uniform intensity, I, falling on the entire watershed.

Case 2. For a precipitation of the same intensity, I, but of twice the duration, $2\Delta t$, and falling on the entire watershed, we find the runoff hydrograph at the discharge of the watershed by adding to the flows of the previous runoff hydrograph the flows of an identical hydrograph happening Δt later. The new outflow $Q(t)$ (runoff hydrograph) can then be computed as follows:

$$
\begin{aligned}
&\text{at } t_0 = 0, && Q(t_0) = Q(0) = 0 \\
&\text{at } t_1 = \Delta t, && Q(t_1) = A_1 I R_1 \\
&\text{at } t_2 = 2\Delta t, && Q(t_2) = A_1 I R_1 + A_2 I R_2 \\
&\text{at } t_3 = 3\Delta t, && Q(t_3) = A_2 I R_2 + A_3 I R_3 \\
&\text{at } t_j = j\Delta t, && Q(t_j) = A_{j-1} I R_{j-1} + A_j I R_j \\
&\text{at } t_n = n\Delta t, && Q(t_n) = A_{n-1} I R_{n-1} + A_n I R_n \\
&\text{at } t_{n+1} = (n+1)\Delta t, && Q(t_{n+1}) = A_n I R_n \\
&\text{at } t_{n+2} = (n+2)\Delta t, && Q(t_{n+2}) = 0
\end{aligned}
$$

Case 3. For precipitation of the same intensity, I, but for a duration equal to $n\Delta T$, falling on the entire watershed, the outflow, $Q(t)$ (runoff hydrograph), is computed as follows:

$$
\begin{array}{llll}
\text{at } t_0 = 0, & Q(t_0) = Q(0) = 0 \\
\text{at } t_1 = \Delta t, & Q(t_1) = A_1 I R_1 \\
\text{at } t_j = j\Delta t, & Q(t_j) = \sum_{k=1}^{j} A_k I R_k \\
\vdots & \\
\text{at } t_n = n\Delta t, & Q(t_n) = \sum_{k=1}^{n} A_k I R_k \\
\text{at } t_{n+1} = (n+1)\Delta t, & Q(t_{n+1}) = \sum_{k=2}^{n} A_k I R_k \\
\vdots & \\
\text{at } t_{2n-1} = (2n-1)\Delta t, & Q(t_{2n-1}) = A_1 I R_1 \\
\text{at } t_{2n} = 2n\Delta t, & Q(t_{2n}) = 0
\end{array}
$$

The three cases described above, as well as many others that could easily be designed, emphasize the following facts, **for uniform precipitation falling on the entire watershed, with a constant intensity *I***:

a) If the duration of the rainfall is t_n (t_n being the time necessary for the water droplet most hydraulically distant from the discharge to run off to the outlet; in other words, the longest time for a water droplet that falls on the watershed to reach the discharge; t_n is usually referred to as the time of concentration of the watershed, t_c), the entire surface area of the watershed contributes to the outflow $Q(t_n)$ at the discharge at the end of the rainfall; in a nutshell, the n areas (fig. 7.1a) then contribute to the runoff at the discharge.

b) If the precipitation duration is longer than t_n, the outflow value remains equal to $Q(t_n)$ for a time equal to the difference between the duration of the rainfall and t_n.

c) When the duration of the rainfall is less than t_n, the maximum runoff occurs at the end of the rainfall and is smaller than the runoff for a precipitation of t_n duration.

d) The maximum runoff flow is always reached at the latest by the end of the rainfall.

Note: The surface on which a rain falls is always assumed to be a surface with a time of concentration equal to the duration of the rainfall. The importance of this statement will become clear when discussing Intensity-Duration-Frequency curves (section 7.7).

7.3.3 Mathematical Expression of the Rational Equation

At the discharge of a watershed such as the one shown in figure 7.1a, the maximum runoff due to precipitation of uniform intensity I, falling on the entire surface of the watershed, and of duration t_n equal to the time of concentration t_c of the watershed, is obtained from the following equation:

$$Q(t_n) = Q(t_c) = \sum_{k=1}^{n} A_k I R_k = I \sum_{k=1}^{n} A_k R_k \qquad (7.1)$$

In general, the value of R for the entire watershed is known or is determined from *in situ* data taken from each sub-surface A_k. As we will see in art. 7.9, the variable R is the runoff coefficient – that is, the fraction of the rainwater volume that runs off.

Moreover, as $\sum_{k=1}^{n} A_k$ represents the surface area of the watershed, equation 7.1 becomes:

$$Q(t_n) = Q(t_c) = AIR \tag{7.2}$$

This equation is the Rational equation, usually expressed as follows:

$$Q = KAIR \tag{7.3}$$

where K is a conversion factor.

The variables in equation 7.3 are defined in table 7.2 with their corresponding units.

Table 7.2 Variables and units in the Rational equation $Q = KAIR$

Variable	Units		
Q = maximum runoff	m^3/s	m^3/s	ft^3/s
A = surface area of watershed	ha	km^2	acre
I = rain intensity	mm/h	mm/h	in/h
R = runoff coefficient of watershed	–	–	–
K = conversion factor	$2{,}75 \times 10^{-3}$	0,275	1,0

7.3.4 Theoretical Basis of the Rational Equation

The rational equation and its use are based on the following assumptions:

- Rainfall intensity is constant throughout the duration of precipitation.
- Rainfall intensity is uniform over the entire watershed.
- Rain falls on the entire watershed throughout the duration of precipitation.
  The runoff coefficient has a constant value throughout the duration of the precipitation, and this value is the same regardless of rain intensity; in practice, however, the runoff coefficient tends to increase during the rainfall (art. 7.9.2).
- For a given frequency of occurrence of rainfall (art. 7.5), the maximum runoff occurs when the duration of precipitation is equal to the time of concentration of the watershed.
- For a given frequency of rainfall (art. 7.5), the surface on which rain falls is the surface with a time of concentration equal to the duration of the rainfall.
- So-called impermeable surfaces (streets, driveways, roofs [section 7.1.4], etc.) are uniformly distributed in the watershed.
- The return period or frequency of the maximum runoff is the same as the one used to compute rainfall. This means that for a rainfall occurring, on average, once every ten years, we expect that the calculated maximum runoff will occur, on average, once every ten years.
- The shape of the watershed is rectangular; thus, at any time after the rain begins to fall uniformly on the entire watershed, the percentage of the watershed's surface

contributing to the runoff at the discharge is approximately equal to the period from the beginning of the rainfall expressed as a percentage of the time of concentration – this last percentage must not exceed 100% (thus $\% A = \% t_c$).

The **normal maximum runoff** is the flow computed with the rational equation and based on the previous assumptions.

Despite these simplifying assumptions, even today the rational equation enables us to easily obtain maximum runoff values accurate enough to design most storm-sewer systems serving small drainage areas (less than 5 km^2 but even as large as 25 km as we have seen at the beginning of art. 7.3).

Computerized methods have the main advantage of calculating all flows, including maximum runoff, for a wide range of conditions (rainfalls of variable intensity in time and space, variable runoff coefficients, etc.).

It should also be mentioned that when the rational method is used for the design of storm-sewer systems (chapter 9), **we always consider the water-flow velocity in the pipes to be that of a pipe in full flow**. This assumption is valid considering the flow-velocity characteristics in sewers flowing partly full (chapter 2).

7.4 CATCHMENT (WATERSHED)

Every storm-sewer pipe is used to evacuate rainwater from a given area. So that the pipe has an adequate hydraulic capacity without being too large, every element that might influence the flow to be evacuated must be determined with maximum accuracy. For each watershed or sub-basin (if the watershed is spread over a large area), we must determine:

- the current and future layout of the area (to evaluate runoff coefficients);
- the current and future permeability of the ground;
- the slopes of the surfaces of the watershed and sub-basins;
- the routes used currently and in the future by the runoff.

7.5 RETURN PERIOD (FREQUENCY OF OCCURRENCE)

The return period is the average time between two rainfalls of the same intensity and duration: it is the inverse of the frequency of occurrence. Thus, a rainfall with a return period of 10 years occurs on average once every 10 years. For storm-water design purposes the greater the chosen return period, the greater the maximum runoff (computed with the rational equation), which results in better protection against flooding – at, however, a higher cost. The choice of a return period depends on the engineering rules and conventions in the studied regions; it may also be determined by the value of the properties under threat or by the importance of damage caused by backed-up sewer systems.

In Quebec, the design of combined sewer systems (chapter 8) is generally calculated over a return period of 10 years. For storm-sewer systems (chapter 8), a return period of 5 years is usually chosen. Note, however, that the cost of a runoff collection system is not directly proportional to the return period chosen in the computations. Thus, the cost of a system built for a return period of 10 years may only be 10% above the cost of a system built for a return period of 5 years. The average cost of the additional transported water volume is lower than the average cost of the total transported water volume: this is what economists call economies of scale. Nowadays, for the design of equalizing basins, we often use return periods of up to 50 or even 100 years.

7.6 RAINFALL INTENSITY

The use of the rational equation requires the choice of an appropriate rain intensity, *I*. This intensity is obtained by the intensity-duration-frequency curves (IDF curves) plotted for the region where the watershed is located. To make an appropriate choice, we must evaluate the time of concentration of the watershed, t_c, from which the maximum runoff is calculated (art. 7.3.4), and determine a workable return period; this last decision relies mostly on engineering practices.

7.7 INTENSITY-DURATION-FREQUENCY CURVES (IDF CURVES)

IDF curves graphically represent the variation in average rain intensity as a function of its duration, for short rainfalls (< 3 h and even < 1 h) and for various return periods. Figure 7.2 shows the IDF curves for the Montreal and Quebec City regions; the most frequently used return periods have always been 25 years, 15 years, 10 years, 5 years, 2 years, 1 year, 6 months, and 3 months. Nowadays, as we have said, return periods of 50 and 100 years are used for the design of equalization basins.

Note: Since rain falls on a surface with a time of concentration equal to the duration of the rainfall, for a given return period, the longer the duration of the rainfall (the greater the size of the surface area, *A*, on which it falls), the lower its intensity, and vice versa.

Typical equations describing the IDF curves are as follows:

$$I = \frac{CN^x}{(a + t)^n} \tag{7.4a}$$

$$I = \frac{M}{t + B} \tag{7.4b}$$

$$I = \frac{M}{t^n + B} \tag{7.4c}$$

where I = average rainfall (mm/h)
N = return period of the precipitation (year)
t = duration of rainfall (min)
a, M, C, B, x, n = constants for each return period and geographic region

For the Montreal region, we generally use the equations suggested by Mitci (fig. 7.2a and table 7.3).

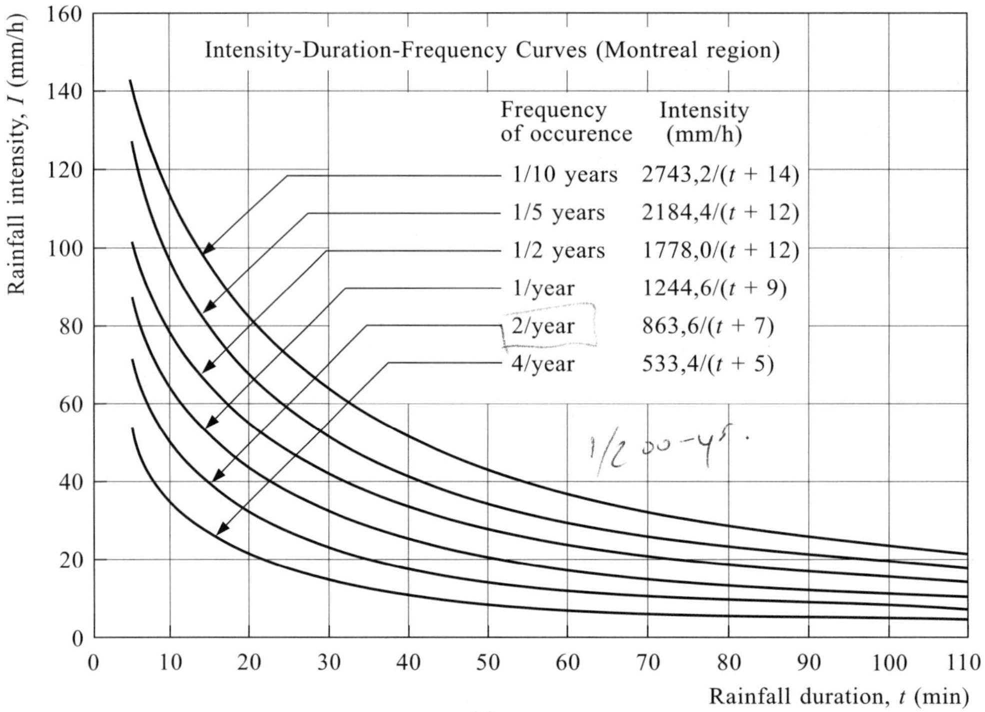

(a)

Figure 7.2 Intensity-duration-frequency curves for the regions of a) Montreal; b) Quebec City; c) Dorval (Montreal Airport)

From: a) and b) C. Mitci, "Nouvelles courbes de pluie et orage type pour la région de Montréal," *Eau du Québec*, vol. 7, no. 1 (1974): 37–44; c) Rivard, G. *Gestion des eaux pluviales en milieu urbain*, Alias, Sainte-Dorothée, Quebec, Canada (1998) (in french).

Table 7.3a Values of rain intensity for different return periods: Montreal (Mitci) (t is in min)

Return period (years)	Rainfall intensity, *I*	
	(mm/h)	(in/h)
0,25	$533,4/(t + 5)$	$21/(t + 5)$
0,50	$863,6/(t + 7)$	$34/(t + 7)$
1	$1244,6/(t + 9)$	$49/(t + 9)$
2	$1778,0/(t + 12)$	$70/(t + 12)$
5	$2184,4/(t + 12)$	$86/(t + 12)$
10	$2743,2/(t + 14)$	$108/(t + 14)$

t = minutes

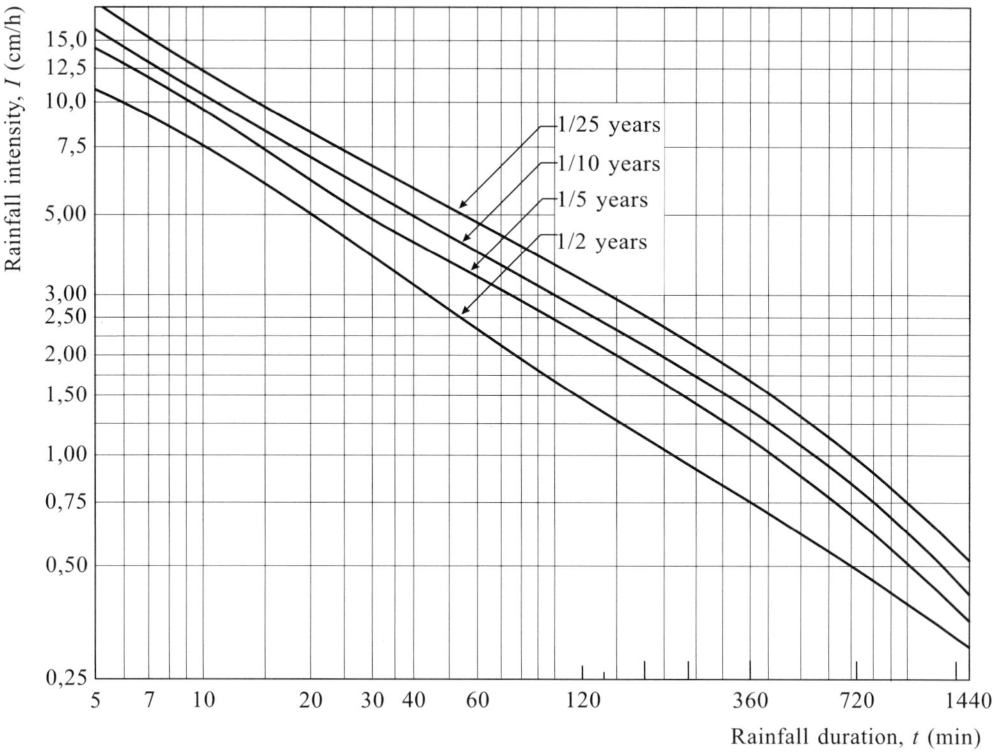

(b)

Figure 7.2 continued

From: TUBECON, *Manuel technique, canalisations et éléments préfabiqués en béton* (1985) p. 4.41.

Table 7.3b Values of rain intensity for different return periods : Quebec Airport and Montreal Dorval Airport (Rivard 1998) – 5 min $\leq t \leq$ 120 min

Return period (years)	Rainfall intensity, *I* (mm/h)	
	Quebec Airport (26 years of data)	Montreal Dorval Airport (48 years of data)
2	$399{,}945/(t + 3{,}012)^{0{,}712}$	$657{,}204/(t + 5{,}273)^{0{,}806}$
5	$416{,}819/(t + 2{,}259)^{0{,}645}$	$1121{,}542/(t + 7{,}507)^{0{,}856}$
10	$457{,}077/(t + 2{,}253)^{0{,}626}$	$1562{,}794/(t + 9{,}094)^{0{,}892}$
25	$477{,}037/(t + 1{,}711)^{0{,}594}$	$2134{,}165/(t + 10{,}501)^{0{,}921}$
50	$479{,}021/(t + 1{,}500)^{0{,}571}$	$2440{,}834/(t + 10{,}781)^{0{,}927}$
100	$540{,}535/(t + 1{,}529)^{0{,}572}$	$2838{,}476/(t + 11{,}344)^{0{,}937}$

t = minutes

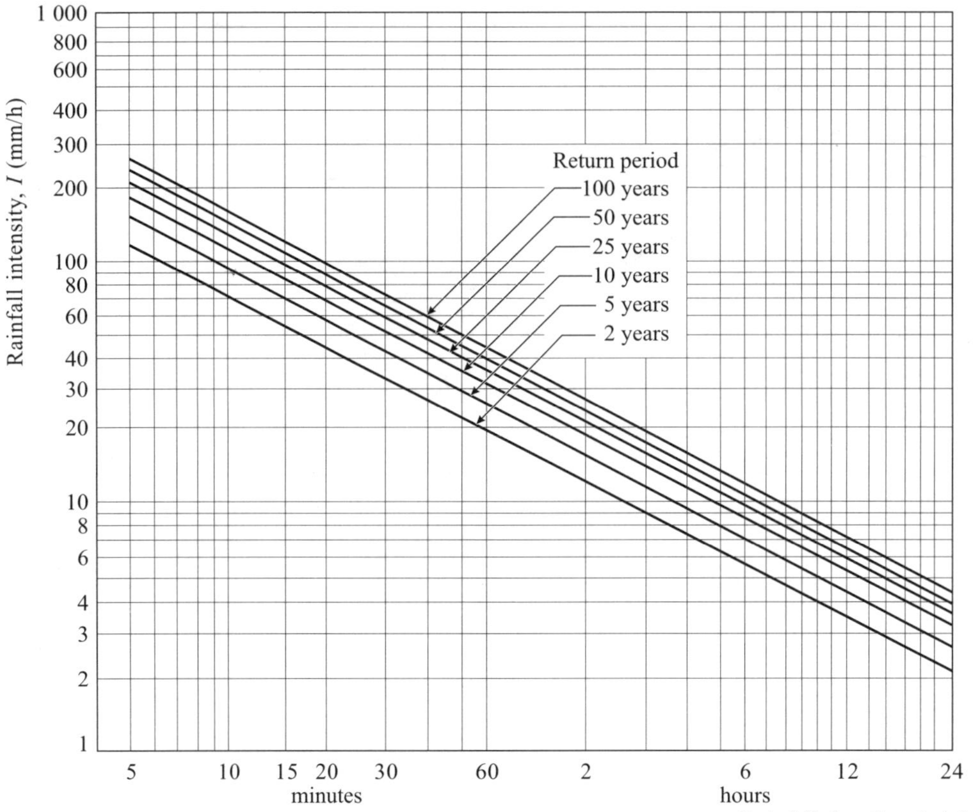

(c)

Figure 7.2c Intensity-Duration-Frequency curves – Dorval airport (Montreal region) Adapted from: Rivard G., *Gestion des eaux pluviales en milieu urbain*, Alias, Sainte-Dorothée, Quebec, Canada (1998).

7.8 TIME OF CONCENTRATION

The time of concentration of a watershed is the longest time that a water droplet will take to run off this watershed to its discharge.

In urban areas, the smallest watershed – **the urban sub-basin** – has a small surface (typically including several lots with a building on each one, and a street section). Its discharge is a street inlet. The **inlet time**, t_e, is the time of concentration of a sub-basin of this type.

A larger urban watershed is composed of several urban sub-basins; its time of concentration is normally equal to the sum of the inlet time, t_e, of the first urban sub-basin (the one located upstream, hydraulically speaking) and the time of flow, t_f, of the water in the sewer pipes from the urban sub-basin to the discharge of the watershed. The time of concentration, t_c, of the urban watershed is therefore typically:

$$t_c = t_e + t_f \tag{7.5}$$

where t_f, **the time of flow in pipes, is computed assuming the sewers are flowing full**.

Figure 7.3 shows an urban watershed.

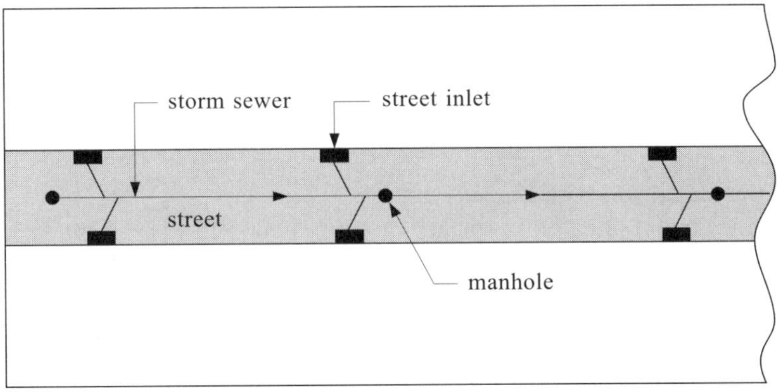

Figure 7.3 Urban watershed

7.8.1 Inlet Time of Urban Sub-basins

The inlet time, t_e, of an urban sub-basin is the longest time the water takes to run off this watershed to reach the street inlet. The value of the inlet time is a function of:

- the average slope of the land's surface to the street inlet;
- the distance the water must travel on the surface to reach the street inlet;
- the type of surface on which the water must run off.

The inlet time of urban sub-basins generally varies from 5 to 30 minutes, depending on their size. In highly populated areas where the proportion of so-called impermeable surfaces is high, inlet times are much shorter. In residential areas, however, where there are single-family dwellings and lots are relatively flat, inlet times vary from 10 to 15 minutes. To evaluate inlet times, several models are proposed: the Kerby model, the Kirpich model, the Schaake model, and the Federal Aviation Agency model.

Kerby Model (W.S. Kerby, "Time of Concentration for Overland Flow", *Civil Engineering*, vol. 29, no. 3 [1959]: 174). This model evaluates the time necessary for a thin layer of water to run off on different types of surfaces. This model seems to give the best results for urban sub-basins. In Kerby's equation, t_e is in minutes and is expressed as follows:

$$t_e = \left(\frac{2{,}187 Ln}{s^{0{,}5}} \right)^{0{,}467} \qquad (7.6)$$

where L = maximum distance the water travels on the surface (m) (this distance cannot normally exceed 365 m – when L exceeds 365 m, water flows faster than the model predicts)

n = Manning's coefficient of roughness of the surface (table 7.4 shows the values of n for different surfaces)
s = average slope of the route travelled by water (m/m)

Table 7.4 Manning's coefficient of roughness, n, for different surfaces

Surface	n
Flat so-called impermeable surface	0,02
Flat grass surface	0,10
Compacted soil with flat surface	0,10
Poorly maintained grass surface or cultivated field	0,20
Pasture lands	0,40
Deciduous forest	0,60
Coniferous forest	0,60
Deciduous forest and tall grass	0,80

Note: When water runs over various surfaces before reaching the discharge (usually the street inlet), the inlet time, t_e, must be calculated for each type of surface. The inlet time (t_e) of the sub-basin is then the sum of all t_e values so calculated. This applies for all models.

Kirpich Model (K.N. Mutreja, *Applied Hydrology* [New Delhi: McGraw-Hill, 1986]). Kirpich proposed the following equation, which is used mostly for rural areas (where t_e is in minutes):

$$t_e = \frac{0{,}0195 L^{0{,}77} F}{s^{0{,}385}} \qquad (7.7)$$

where L = maximum distance travelled by water on the surface (m) (this distance is normally between 30 and 3050 m)
s = average slope of the route travelled by water (m/m)
F = factor related to the surface; enables this equation to be used for regions other than rural areas (table 7.5 gives different values of F)

Table 7.5 Different values of factor F (Kirpich equation)

Surface	F
Cleared area with flat surface (rural watershed)	1,0
Grass surface	2,0
Concrete or asphalt surface	0,4
Grass borders well maintained	1,0
Concrete channel	0,2

The Kirpich equation can be solved easily using the nomograph shown in figure 7.4. On this nomograph, Δh is the total difference in height between the beginning and the

0.71

end of the route travelled by water. The value of inlet time, t_e, which is read on the nomograph, must be multiplied by F to obtain the inlet time of a watershed that is not located in a rural area.

Schaake Model. Schaake developed the following equation for urban sub-basins with streets and street inlets along curbs or sidewalks (where t_e is in minutes):

$$t_e = \frac{1{,}8L^{0{,}24}}{s^{0{,}16}A_{imp}^{\;0{,}26}} \qquad (7.8)$$

where L = maximum distance travelled by water along the curb to the street inlet (m)

s = average slope of the route travelled by water (m/m)

A_{imp} = fraction of the total surface of the sub-basin made of so-called impermeable surfaces (roofs, paved streets, sidewalks, etc.)

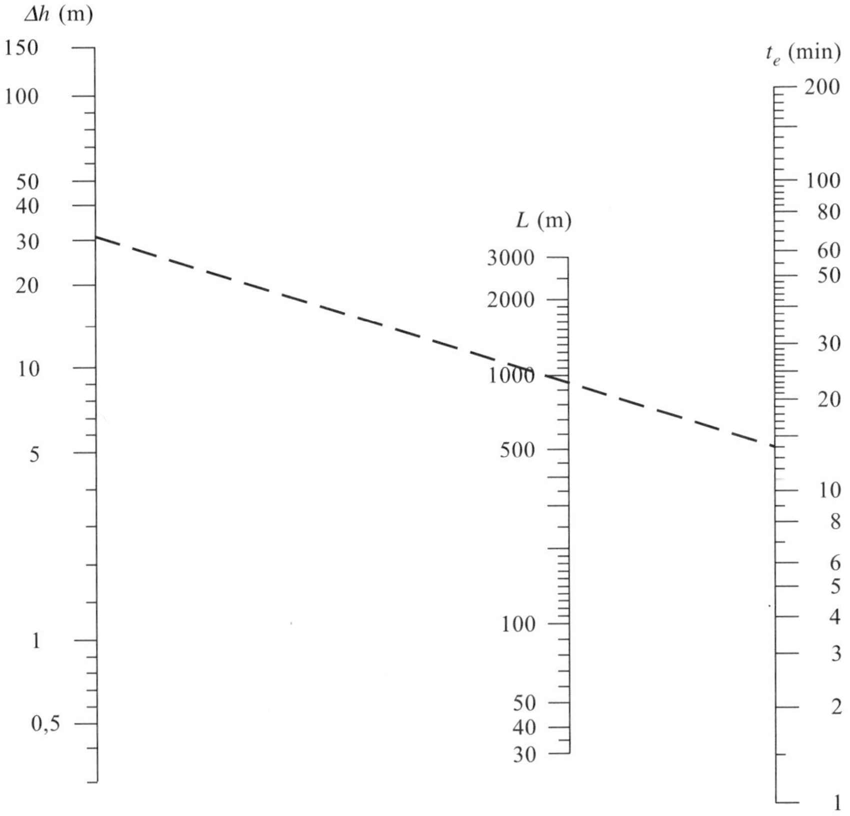

Figure 7.4 Nomograph for finding the inlet time of small watersheds (Kirpich model); Δh = difference in height of the route travelled by water. The result obtained with the nomograph must be multiplied by the appropr[illegible] F value.

Federal Aviation Agency model (N. Ashford and P.H. Wright, *Airport Engineering* [New York: John Wiley & Sons, 1979]). This American agency proposed the following equation, developed for draining airports (where t_e is in minutes):

$$t_e = \frac{3{,}26(1{,}1 - R)L^{0{,}5}}{s^{0{,}33}} \tag{7.9}$$

where R = runoff coefficient (section 7.9)
L = maximum distance travelled by water on the surface (m)
s = average slope of the route travelled by water (%)

7.8.2 Time of Flow in Pipes Flowing Full

The time of flow in a pipe flowing full, t_f, is calculated using the Manning equation:

$$\left(v = \frac{1}{n} R_h^{2/3} s^{1/2} \right) \tag{2.2}$$

One then gets the following equation:

$$t_f = \frac{Ld^2}{76{,}3944Q_p} \tag{7.10}$$

where t_f = time of flow (min)
L = length of the pipe (m)
d = inner diameter of the pipe (m)
Q_p = flowrate in the pipe flowing full (m^3/s)

7.9 RUNOFF COEFFICIENT

The runoff coefficient represents the fraction of rainwater that runs off on a given surface. Its value depends on:

- the characteristics of the ground surface;
- the average slope of the ground surface toward the discharge;
- rain intensity;
- the amount of rainwater stored in ground settlements;
- atmospheric conditions preceeding rainfall (such as periods of drought or dampness prior to precipitation).

The runoff coefficient can be considered either a constant characteristic of the ground surface or variable, depending on various situations.

7.9.1 Constant Runoff Coefficient

Usually, the runoff coefficient is considered constant during a rainfall, regardless of the characteristics of the rainfall. Table 7.6 presents values of the runoff coefficients of various surfaces, and table 7.7 presents runoff coefficients of different types of district.

Table 7.6 Runoff coefficients of different surfaces

Surface	Runoff coefficient, *R*
Concrete or asphalt pavement	0,70 - 0,95
Brick pavement	0,70 - 0,85
Roof	0,75 - 0,95
Grass area, sandy soil	
- flat (slope < 2%)	0,05 - 0,10
- average slope (2 to 7%)	0,10 - 0,15
- steep slope (> 7%)	0,15 - 0,20
Grass area, dense soil	
- flat (slope < 2%)	0,13 - 0,17
- average slope (2 to 7%)	0,18 - 0,22
- steep slope (> 7%)	0,25 - 0,35
Gravel driveway	0,15 - 0,30

Table 7.7 Runoff coefficients of different districts

District	Runoff coefficient, *R*
Commercial	
- downtown	0,70 - 0,95
- suburb	0,50 - 0,70
Residential	
- single-family dwellings	0,30 - 0,50
- multiple-family dwellings, detached	0,40 - 0,60
- multiple-family dwellings, row houses	0,60 - 0,75
- houses in suburbs with little development	0,10 - 0,25
- houses in suburbs	0,25 - 0,40
- apartment buildings	0,50 - 0,70
Industrial	
- average density	0,50 - 0,80
- high density	0,60 - 0,90
Parks, cemeteries, prairies	0,10 - 0,25
Play fields	0,25 - 0,35
Wastelands	0,10 - 0,30

Adapted from (tables 7.6 and 7.7): ASCE and WPCF, *Design and Construction of Sanitary and Storm Sewers* (1970), p. 51.

In general practice, the value of the runoff coefficient of a watershed or a given area is determined by evaluating the various surfaces located in a sample are representative of this watershed or area. The runoff coefficient is then evaluated the following equation:

$$R = \frac{\sum_{i=1}^{n} A_i R_i}{\sum_{i=1}^{n} A_i} \tag{7.11}$$

where A_i = surface of type i (ha, m^2, etc.)
R_i = runoff coefficient of the surface A_i
ΣA_i = A, total surface (ha, m^2, etc.)

EXAMPLE 7.1

Compute the runoff coefficient of an urban residential area of 1 420 000 m^2 (142 ha) with surfaces in the following proportions: roofs, 25%; paved streets, 14%; concrete sidewalks, 5%; gravel driveways, 7%; grass areas with small slope, 49%.

Solution

This problem is solved using equation 7.11, and the values of the variables are presented in table 7.8.

Table 7.8 Values of the variables in equation 7.11 (example 7.1)

Type of surface	Area, A_i (ha)	Runoff coefficient, R_i	Surface A_iR_i (ha)
Roofs	35,5	0,90	31,95
Paved streets	19,9	0,85	16,92
Concrete sidewalks	7,1	0,90	6,39
Gravel driveways	9,9	0,25	2,48
Grass areas with small slope	69,6	0,15	10,44
Total	142		68,18

Runoff coefficient of the district:

$$R = \frac{\sum A_i R_i}{\sum A_i} = \frac{68{,}18 \text{ ha}}{142 \text{ ha}}$$
$$= 0{,}48$$

7.9.2 Time-Variable Runoff Coefficient

The runoff coefficient of a watershed is not really constant. It varies during rainfall: it is lower at the beginning because some water is absorbed by the ground and some fills ground settlements. This is why some authors recommend the use of runoff coefficients with values that increase over time. However, since the Rational equation (equation 7.3) makes use of average rain intensity without taking into account variations in rain intensity in time and space, the use of time-variable runoff coefficients is delicate; it requires great care and at least some knowledge of the rainfall hyetograph. The use of variable runoff coefficients is only really possible through computerized models.

Figures 7.5, 7.6, and 7.7 present the curves of the runoff coefficient variation as a function of time (plotted by the City of Montreal) for various absorptivity percentages of the watershed's permeable surfaces. When sewer systems are designed, the runoff coefficients drawn from the curves are used only when their values exceed the corresponding constant runoff coefficients – that is, if runoff is greater with variable runoff coefficients. Most of the time, absorptivity is considered average.

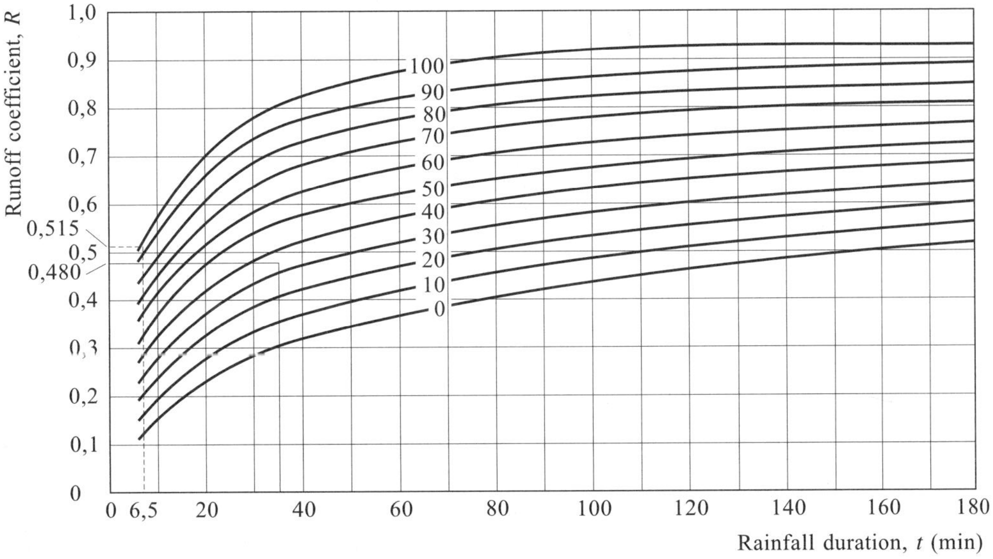

Figure 7.5 Curves of the variation in runoff coefficient, R. The numbers 0 to 100 identifying each curve correspond to percentages of so-called impermeable surfaces (concrete, asphalt, etc.) in the districts. **Absorptivity of the permeable surfaces: low**

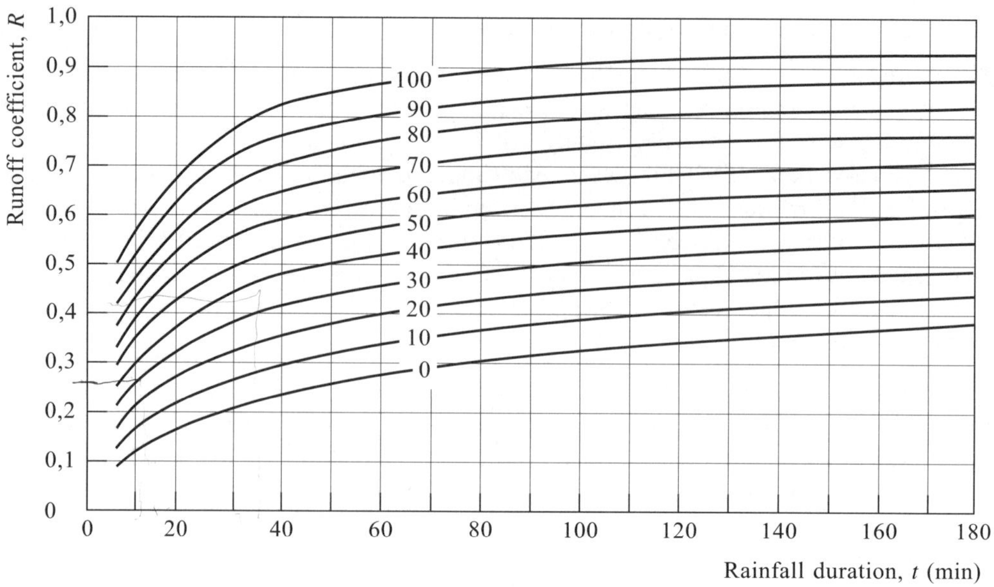

Figure 7.6 Curves of the variation in runoff coefficient, *R*. The numbers 0 to 100 on each curve correspond to percentages of so-called impermeable surfaces (concrete, asphalt, etc.) in the districts. **Absorptivity of the permeable surfaces: average**

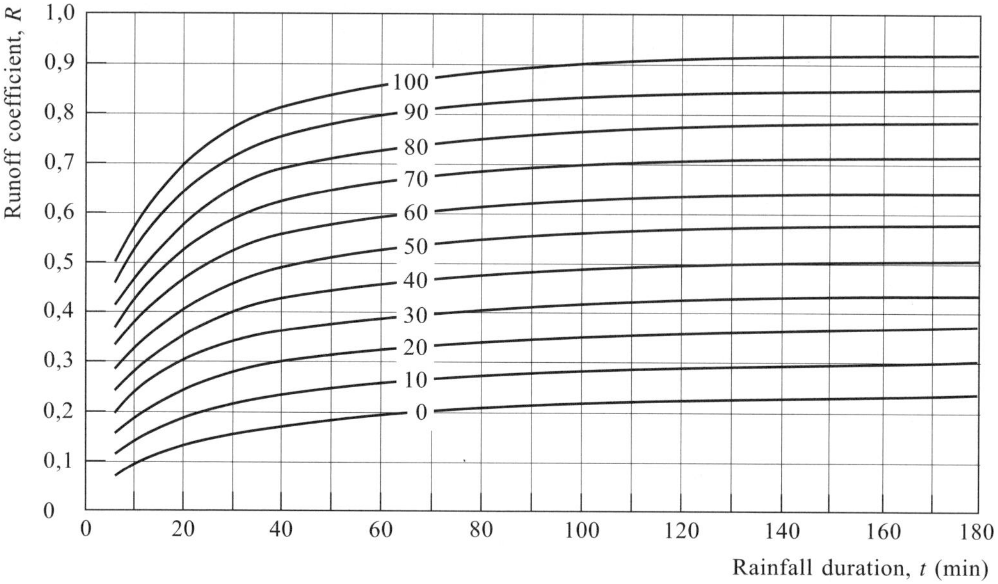

Figure 7.7 Curves of the variation in runoff coefficient, *R*. The numbers 0 to 100 on each curve correspond to percentages of so-called impermeable surfaces (concrete, asphalt, etc.) in the districts. **Absorptivity of the permeable surfaces: high**.

EXAMPLE 7.2

In a dense industrial zone, the inlet time of a 5 000 m^2 (0,5 ha) urban sub-basin is estimated at 6,5 minutes. The absorptivity of the permeable ground is low, and the percentage of so-called impermeable surfaces is 100. Calculate the maximum runoff at the street inlet, for a rainfall with a return period of 10 years.

Solution

Rational equation 7.3 is used to solve this problem, and the values of the variables are presented in table 7.9.

Table 7.9 Values of the variables of Rational equation 7.3 (example 7.2)

Variable	Runoff coefficient, R	
	constant	variable
A	5000 m^2 (0,5 ha)	5000 m^2 (0,5 ha)
t_e	6,5 min	6,5 min
R	0,90*	0,515
$I = \dfrac{2743{,}2}{t_e + 14}$	133,8 mm/h	133,8 mm/h
$Q = 2{,}75 \times 10^{-3} AIR$	0,166 m^3/s (166 L/s)	0,095 m^3/s (95 L/s)

* A value of 0,90 is obtained for the runoff coefficient, R, in table 7.7.

Using figure 7.5, we find a value of 0,515 for the variable runoff coefficient, R, for t_e of 6,5 and for 100% of so-called impermeable surfaces. As a safety measure, it is better to choose a runoff of 0,166 m^3/s – the one calculated with a constant runoff coefficient.

EXAMPLE 7.3

In a residential area of 250 000 m^2 (25 ha), the percentage of so-called impermeable surfaces is 35. Assuming the time of concentration of this watershed is 35 min and the absorptivity of the permeable ground is low, determine the runoff for a rainfall with a return period of 5 years.

Solution

This problem is solved using Rational equation 7.3, and the values of the variables are presented in table 7.10.

Table 7.10 Values of the variables of Rational equation 7.3 (example 7.3)

Variable	Runoff coefficient, R	
	constant	variable
A	250 000 m^2 (25 ha)	250 000 m^2 (25 ha)
t_c	35 min	35 min
R	0,35	0,48
$I = \dfrac{2184{,}4}{t_c + 12}$	46,5 mm/h	46,5 mm/h
$Q = 2{,}75 \times 10^{-3} AIR$	1,12 m^3/s (1120 L/s)	1,54 m^3/s (1540 L/s)

In this case, it is better to choose 1,54 m^3/s – the runoff obtained with the variable runoff coefficient.

7.10 RELATIONSHIP BETWEEN THE SHAPE OF THE WATERSHED AND THE RUNOFF

The shape of the watershed has great influence on the maximum runoff and on the time this flow appears, as we will see in this section. Therefore, when the shape of the watershed is neither square nor rectangular, the rational equation must be used with care.

7.10.1 Watershed Shaped Like an Equilateral Triangle

For a watershed shaped like an equilateral triangle, with its discharge located in the middle of one side (fig. 7.8a), the relationship between the duration of the rainfall (expressed as the percentage of the time of concentration) and the fraction of the surface of the watershed contributing to the runoff (expressed as the percentage of the watershed's total surface) gives results similar to those presented in table 7.11. When the rainfall lasts longer than the watershed's time of concentration, t_c, the entire surface contributes to the runoff from the time of precipitation $t = t_c$ to the end of precipitation (i.e. for $t_c \leq t \leq$ end of precipitation).

(a) (b)

Figure 7.8 Watersheds shaped as: a) equilateral triangle; b) circle segment; c) rectangle; d) square.

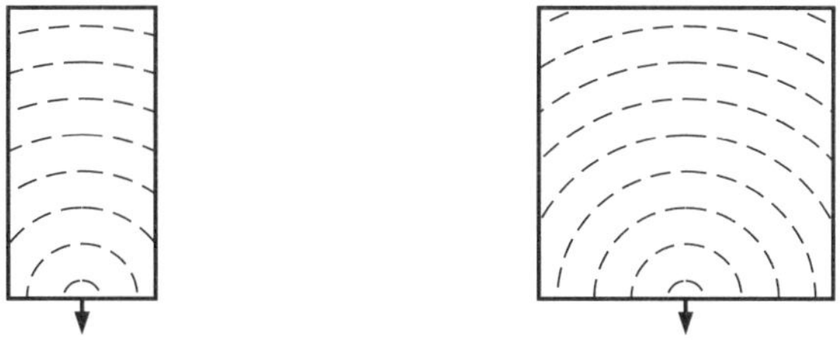

(c) (d)

Figure 7.8 continued

Table 7.11 Variation in the percentage of the watershed's surface contributing to runoff during a rainfall, as a function of time: cases of watersheds shaped as an equilateral triangle, circle segment, rectangle, and square, each with time of concentration of 10 minutes

Rainfall duration (min)	**Equilateral triangle**		**Circular sector**		**Rectangle**		**Square**	
	$\%t_c$	$\%A$	$\%t_c$	$\%A$	$\%t_c$	$\%A$	$\%t_c$	$\%A$
0	0	0	0	0	0	0	0	0
2	20	25	20	4	20	18	20	8
4	40	63	40	16	40	39	40	30
6	60	83	60	36	60	59,5	60	58,5
8	80	96	80	63	80	80	80	83
10	100	100	100	100	100	100	100	100
12	100	120	120	100	120	100	120	100

In this case, as a high proportion of the watershed's surface is located close to its discharge, runoff increases quickly after the beginning of the rainfall. The runoff is therefore maximum for a rainfall of less duration than the time of concentration of the watershed: for the same return period (frequency of occurrence), a shorter rainfall has greater intensity but falls on only part of the watershed. As a result, by the end of such a rainfall, the product AI and consequently the flow $Q = KAIR$ are higher than the corresponding values for a rainfall as long as the time of concentration, t_c, of the watershed, assuming again that the runoff coefficient is constant and uniform over the entire surface (see example 7.4a and b).

EXAMPLE 7.4

A 0,4 ha watershed in the Montreal area is shaped like an equilateral triangle and its outlet is at the centre of one side. It has a 19 min time of concentration (t_c) and its runoff coefficient is 0,85.

a) What is the maximum flow at the outlet of this watershed following a precipitation that lasts 0,60 t_c (i.e., 0,60 × 19 min = 11,4 min). Precipitations of 5-year frequency of occurrence must be considered.

A 5-year frequency of occurrence precipitation with a 11,4 min duration falls on a surface with a concentration time, t_c, of 11,4 min for 11,4 min. Its maximum intensity is:

$$I = \frac{2184{,}4}{t + 12} = \frac{2184{,}4}{11{,}4 + 12} = 93{,}35 \text{ mm/h}$$

After 0,60 t_c of precipitation, the entire surface on which the rain (with a duration of 0,60 t_c) falls contributes to the runoff at the outlet. In this case, a watershed in the form of an equilateral triangle, the proportion of the entire watershed that contributes to the runoff at the end of precipitation is 83% (table 7.11).

Thus, the maximum flow computed with the Rational equation is:

$$\begin{aligned} Q_{max} &= KAIR \\ &= 2{,}75 \times 10^{-3} \times 0{,}83A \times 93{,}35 \text{ mm/h} \times 0{,}85 \\ &= 2{,}75 \times 10^{-3} \times (0{,}83 \times 0{,}4) \text{ ha} \times 93{,}35 \text{ mm/h} \times 0{,}85 \\ &= 0{,}072 \text{ m}^3\text{/s} \end{aligned}$$

b) What is the maximum runoff at the outlet of a watershed following precipitation that lasts 1,0 t_c – that is, 19 min? Precipitations of 5-year frequency of occurrence must be considered.

A precipitation of 19 min duration falls on a surface with a time of concentration t_c of 19 min for 19 min. Its maximum intensity is:

$$I = \frac{2184{,}4}{t + 12} = \frac{2184{,}4}{19 + 12} = \frac{2184{,}4}{31} = 76{,}47 \text{ mm/h}$$

After 19 min, the entire surface of the watershed contributes to the runoff. Then:

$$\begin{aligned} Q_{max} &= KAIR \\ &= 2{,}75 \times 10^{-3} \times 0{,}4 \text{ ha} \times 76{,}47 \text{ mm/h} \times 0{,}85 \\ &= 0{,}0715 \text{ m}^3\text{/s} \end{aligned}$$

This flow is the maximum normal runoff.

c) What is the maximum runoff at the outlet of the watershed following a precipitation that lasts 1,2 t_c (i.e. 1,2 × 19 min = 22,8 min)? Precipitations of 5-year frequency of occurrence must be considered.

A precipitation with a duration of 1,2 t_c (i.e., 1,2 × 19 min = 22,8 min) falls on a surface with a time of concentration t_c of 22,9 min for 22,9 min. Its maximum intensity is:

$$I = \frac{2184{,}4}{t + 12} = \frac{2184{,}4}{22{,}9 + 12} = \frac{2184{,}4}{34{,}9} = 62{,}59 \text{ mm/h}$$

Note that the maximum intensity of precipitation with a given return period is lower if its duration is longer.

After 19 min and up to the end of precipitation, the entire surface of the watershed contributes to the constant runoff value; then:

$$Q_{max} = 2{,}75 \times 10^{-3} \times 0{,}4 \text{ ha} \times 62{,}59 \text{ mm/h} \times 0{,}85$$
$$= 0{,}0585 \text{ m}^3/\text{s}$$

Conclusion: because of the particular form of the watershed surface, the maximum runoff flow is generated by a rainfall of shorter duration than the time of concentration t_c of the watershed.

7.10.2 Watershed Shaped Like a Circle Segment

For a watershed shaped like a circle segment, with its discharge located at the tip of the sector (fig. 7.8b), the relationship between the duration of the rainfall (expressed as the percentage of the time of concentration) and the surface of the watershed contributing to the runoff (expressed as the percentage of the watershed's surface) gives results similar to those in table 7.11. At the beginning of the rainfall, the outflow increases slowly. When the rainfall lasts longer than the time of concentration of the watershed, the entire surface contributes to the runoff until the end of the precipitation.

7.10.3 Rectangular Watershed

In the case of a rectangular watershed with a length four times its width, its discharge located in the middle of one of the short sides, and the top of the watershed located on the short side opposite the discharge (fig. 7.8c), we obtain the results in table 7.11.

7.10.4 Square Watershed

For a square watershed with its discharge located in the middle of one side (fig. 7.8d), we obtain the results shown in table 7.11.

7.10.5 Comments about the shapes of urban watersheds

As most watersheds in urban areas are either rectangular or square, we generally consider that from the beginning of the rainfall, the watershed's surface percentage contributing to runoff, $\%A$, increases at the same rate as the duration of the rainfall, expressed as a percentage of the time of concentration, $\%t_c$. When the rainfall outlasts the time of concentration, t_c, the entire surface, $100\%A$, contributes to the runoff and the outflow at the discharge is constant until the end of the precipitation. Figure 7.9 illustrates the relationship between the rainfall duration, $\%t_c$, and the watershed's surface, $\%A$, for the four watershed shapes described above.

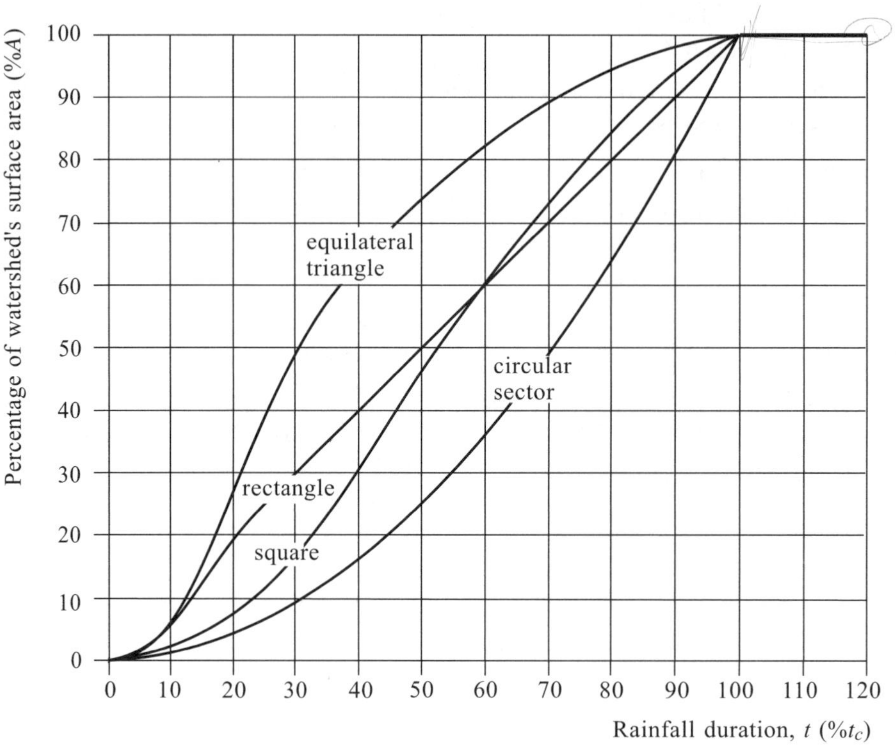

Figure 7.9 Relation between rainfall duration (percentage of the time of concentration, $\%t_c$) and the surface of the watershed contributing to runoff at the discharge (percentage of the total surface of the watershed, $\%A$).

7.11 LIMITATIONS OF THE RATIONAL EQUATION

The rational equation cannot always be used, as it gives only partial information about the runoff at the discharge of a watershed. Here are the limitations of the rational equation:

- When using the rational method to compute the maximum runoff at the discharge, we consider that rain falls uniformly on the entire surface of the watershed for as long as the time of concentration of this watershed. This method yields only a single flow, and little information is obtained about the real shape of the runoff hydrograph.
- The rational method may be used only for small watersheds. Viessman and Hammer (1993) recommended that its use be limited to watersheds smaller than 5 km^2 but according to the Ministry of Transportation of Quebec in its "Manuel des ponceaux" (1993), the rational method may be used for watersheds as large as 25 km^2.

- The rational method only partially accounts for the influence of the sewer system on the runoff hydrograph at the discharge of the watershed. The time of flow, t_f, in the pipes cannot be determined easily and accurately, but this time is an important part of the time of concentration, t_c, of the watershed.
- The variation in the outflow as a function of time (i.e., the runoff hydrograph) is not known with certainty; thus, when using this method, the capacity of the detention basin needed to regulate the runoff (i.e., to lower the crest of the hydrograph) cannot be evaluated accurately.
- The rational method does not take into account conditions prior to precipitation although runoff varies considerably depending on whether the watershed is dry (has not had rain for a long time) or damp (has received rain recently). To address the problem at least partially, we can use a method with a variable runoff coefficient as a function of time.
- The rational method does not consider the amount of water stored in ground settlements or furrows.
- The rational method is limited to the study of precipitation falling uniformly on part or all of the surface of the watershed.
- In order for the maximum runoff of a watershed to correspond to the result of a uniform rainfall for as long as the time of concentration of this watershed, we have to assume the watershed has a regular, more or less rectangular shape. This assumption leads to a proportional relationship (fig. 7.9) between the duration of the rainfall (expressed as a percentage of the time of concentration of the watershed) and the surface of the watershed contributing to runoff (expressed as a percentage of the total surface of the watershed). This flow is called the **normal maximum runoff**. If the shape of the watershed greatly differs from a rectangle or a square, the maximum flow may occur with rainfalls shorter than the watershed's time of concentration and of a higher intensity, considering the same return period.

7.12 DRAINAGE OF PARKING LOTS

The surfaces of parking lots are shaped like inverted cones with water inlets at the bottom. The slope of all sides of the cones in the direction of water inlets must be as small as possible, but never less than 2%.

EXAMPLE 7.5

A concrete rectangular parking lot (Manning's n = 0,013) 30 m by 50 m, in the Montreal area drains into a single water inlet in one of its lower corners. The two corners of the highest small side of the parking lot are at an elevation of 100 m and the two others at 99,5 m. At what level should the water inlet be installed? What in the value of the inlet time? What is the maximum runoff flow at the water inlet knowing that the surface of the parking lot must be shaped like a cone with a slope toward the inlet as small as possible but not less than 2%? Precipitations with a 5 year return period must be chosen and the runoff coefficient of the parking lot is 0,85.

Use Kerby equation. Knowing that if maximum flow at a water inlet must not exceed 30 L/s, will the unique proposed water inlet be sufficient? If not, what would be your suggestion? N.B. When repetitive computations are required, do not make more then three judicious ones.

Solution

Elevation of the water inlet

The water inlet must be at a level such that the slope of the surface towards it (see arrows) is always at least 2%. To this end, three pertinent computations are performed, taking into account flows in the three directions shown by the arrows, i.e.:

- 100 – 2% × 58,31 m = 98,83 m
  (58,31 m = length of the diagonal)
- 99,5 – 2% × 30 m = 98,90 m
- 100 – 2% × 50 m = 99,00 m

The water inlet must be located at 98,83 m.

The slope in the three directions are shown on the figure 7.10.

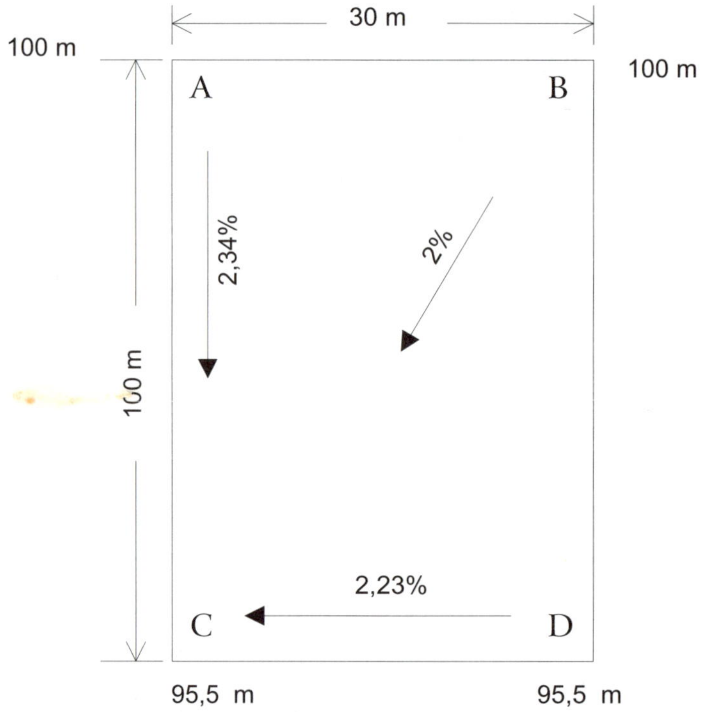

Figure 7.10 (ex. 7.5)

Computation of the entrance time (t_e)

The longest inlet time in directions AC, BC and DC is chosen.

- From B to C:

$$t_e = \left[\frac{2{,}187\ Ln}{s^{0,5}}\right]^{0,487} = \left[\frac{2{,}187 \times 58{,}31 \times 0{,}013}{(0{,}02)^{0,5}}\right]$$

$$= 3{,}16\ \text{min}$$

- From A to C:

$$t_e = 2{,}83\ \text{min}$$

- From D to C:

$$t_e = 2{,}62\ \text{min}$$

Inlet time (t_e) of the parking lot is thus 3,16 min.

Rainfall intensity

$$I = \frac{2184{,}4}{t_e + 12} = \frac{2184{,}4}{3{,}16 + 12} = 144{,}1\ \text{mm/h}$$

Flow

$$\begin{aligned} I &= KAIR \\ &= 2{,}75 \times 10^{-3}\left[(30 \times 50)/10\,000\ \text{ha}\right] \times \left(144{,}1\ \text{mm/h}\right) \times 0{,}85 \\ &= 0{,}05\ \text{m}^3/\text{s}\ \left(50\ \text{L/s}\right) \end{aligned}$$

This flow is greater than 30 L/s, the maximum flow permitted through a water inlet. Two water inlets must be installed. It is left to you to make the appropriate suggestions and computations.

7.13 COMPUTER MODELS FOR SIMULATION OF RUNOFF IN URBAN AREAS

Growing investments related to the construction of infrastructure for runoff collection, the high cost of treating combined sewers' wastewater (which contains runoff and sanitary wastewater, section 6.1.3), and the need to plan and manage combined sewage overflows (wastewater occuring during major rainfalls that cannot be treated, or can be treated only with great difficulty, and must be directly evacuated into the receiving watercourse) have resulted in development of fundamental rules for runoff management in urban areas.

Previously, the goal of runoff management was to minimize problems caused by precipitation. The surface runoff was evacuated as quickly as possible toward water collection infrastructures. In growing urban areas however, this practice was a source of impressive runoff, which forced local authorities to build large, even gigantic, infrastructures. Nowadays, the practice tends toward delaying runoff (water from roofs, for example, is evacuated into grassy areas instead of street inlet), and is even retained in lakes or reservoirs.

The rational method is mainly used to compute maximum flows and does not give accurate information about runoff volume. Therefore this new concept is not a good tool to apply to water management. This is why, as powerful computers became more accessible, engineers and researchers developed computer models to simulate runoff hydrographs – that is, to compute the variation in runoff caused by various episodes of precipitation as a function of time – and to evaluate runoff volumes. Some of these models are now available on personal computers.

All models are not equal: some concentrate only on the quantitative aspects of runoff; others take into account qualitative aspects to quantify pollution discharge; still others attempt to replicate the continuous simulation of runoff and pollution discharge to quantify polluting discharge caused by combined sewage overflow.

However, we can safely say that the more complex the model, the more time, energy, and money it requires from its user. This is why the objectives of a study must be clearly defined before one of the existing models is chosen. In general, it is recommended that the model used be the simplest that meets the needs of the given study. In his book *Gestion des eaux pluviales en milieu urbain*, Rivard thoroughly discusses the computerized simulation of urban drainage and presents most models used in America and, more specifically, in Canada. From this work, we present some characteristics of those models, grouped into families.

The SWMM family models

1. The original version of SWMM (Storm Water Management Model) created by the United States Environmental Protection Agency (EPA) in 1971, is still widely used in America. It is often used to calibrate newer models. With SWMM it is possible to study sewer network working under pressure;
2. OTTSWMM The (Ottawa StormWater Management Model) has been specifically designed to model major and minor urban drainage networks;
3. With PC-SWMM for Windows, created in 1994 by Computer Hydraulics International (CHI), it is possible to present much nicer graphs than with the original SWMM model.

The HYMO family models

1. HYMO (HYdrologic MOdel) worked out in 1972 by the United States Soil Conservation Service (SCS) is rarely used nowadays in its original form;

2. OTTHYMO (OTTawa HYdrologic MOdel) exists in two versions (1984 and 1989). The latter is the most popular in Canada, where engineers appreciate its capacity to simulate drainage of rural watersheds and pipe networks. It can also simulate major and minor urban drainage networks as well as simple flood routing in sewer pipes.

The ILLUDAS family models

1. ILLUDAS (ILLinois Urban Drainage Area Simulation) created in 1974 is rarely employed today;
2. SIRDU (SImulation d'un Réseau de Drainage Urbain) and SIRDU-PC models built up by Béron *et al.* in 1981 and 1989 respectively present many improvements over the original ILLUDAS model. However, they cannot be used to study sewer pipes under pressure, nor major and minor urban drainage networks.

The HEC family models

1. HEC-1 (Hydrologic Engineering Model) models runoff from a precipitation hyetogram in watersheds with complex river networks or in watersheds equipped with drainage network;
2. HEC-2 models back-water curves in rivers and canals; HEC-RAS is an improved version of HEC-2.

The STORM family models

STORM (Storage, Treatment, Overflow, Runoff Model) is used less nowadays.

Much other software is continuously being put on the market, like Sewer CAD and Storm CAD by Haestad Mehtods, USA.

7.14 SYNTHETIC AND HISTORICAL STORMS

With the development of computers, specifically of micro-computers, as well as the general use of computer models in urban drainage, hydrologists developed various theoretical design storms to simulate real rainfalls on watersheds. Nowadays one more often uses real rainfalls, called historical rainfalls.

In all cases, the designer is seeking the best evaluation at least cost of the maximum flows (for the design of sewer pipes, for example) or the maximum runoff volumes (for the design of retention basins, for example).

Synthetic rainfall. According to Rivard (1998) there are two categories of synthetic rainfall:

- Rainfall derived from Intensity-Duration-Frequency curves, to which is given a more or less arbitrary distribution;

– Rainfall to which is given temporal distribution, using analyses from real rainfall events.

In the first category one finds Chicago rainfall (1957), still largely used in its original form in North America, SAE rainfall (Atmospheric Service of Environment Canada, 1980 and 1982), WATT rainfall (1986 and 1993), SCS rainfalls (United States Soil Conservation Service 1986), HUFF rainfall (1987), Pilgrim and Corderey (1975) and Yen and Chow rainfall (1980).

Historical rainfalls. With currently available computer models, one can more easily analyse rainfall characteristics of a region and choose the most critical precipitation.

7.15 LIMITATIONS OF MAXIMUM RUNOFF IN URBAN AREAS

Investments required to channel and evacuate rare peak flows (occurring once every 5 to 10 years or longer) are expensive. For this reason, city management has encouraged engineers to look for new means of reducing these flows. Engineers have two ways to reach this objective:

– developing techniques to keep water on site, even if for only a few minutes;
– using large-sized retention/detention basins on the watershed's site.

7.15.1 Delaying Runoff at Its Source

For management of runoff in urban areas, it is increasingly clear that runoff must be retained to avoid damage caused by major, rapid runoff. It is suggested that evacuation of runoffs should be contained in the site where they originate. This significantly reduces costs related to the construction of infrastructure for water collection.

To retain or delay runoff, the engineer can:

– regulate the evacuation of rainwater fallen on roofs;
– evacuate rainwater fallen from roofs on to grassy surfaces;
– regulate the evacuation of rainwater fallen on parking lots;
– pump water from drain foundations and evacuate it on to grassy surfaces;
– evacuate runoff in to ditches, even in residential areas.

7.15.2 Delaying (or Reducing) Maximum Runoff Using Detention/Retention Basins within the Sewer System

When runoff water cannot entirely or in part be retained or delayed using the methods discussed above, the maximum runoff can nevertheless be reduced significantly by using basins or lakes to retain runoff (fig. 7.11a). A detention basin stores water and reduces the size of peak flows. This lowers the crest of the natural runoff hydrograph after urbanisation (but without a detention basin; fig. 7.11b). The volume and average surface area of the detention basin are a function of:

– the maximum water volume to be retained;
– the maximum acceptable variation in the water surface level in the detention basin;

- savings through construction of an urban runoff collection system downstream from the detention basin;
- other possible uses for the detention basin (boating basin, landscaping design, etc.).

Figure 7.11 shows the influence of the detention basin on the runoff.

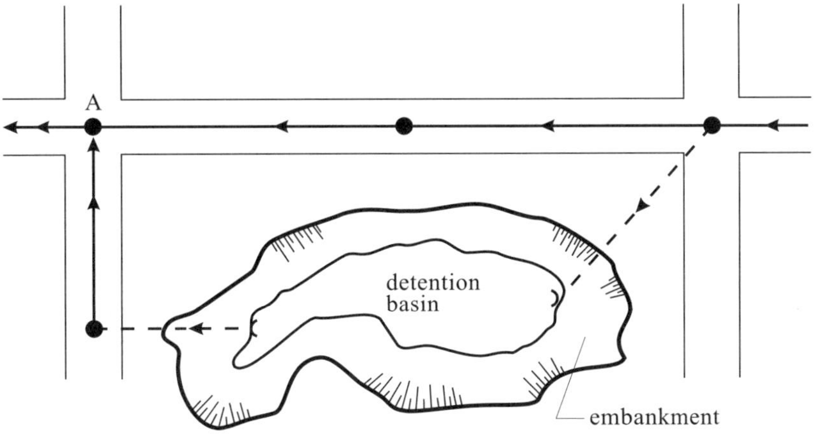

(a)

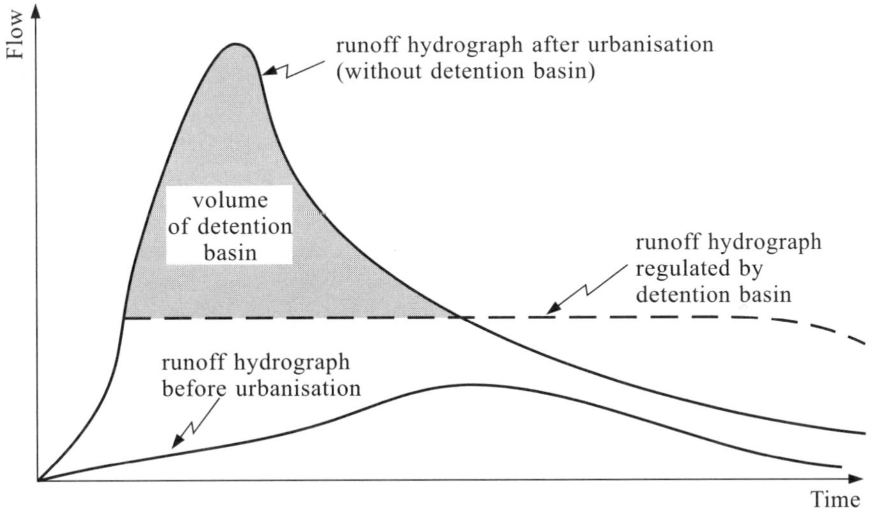

(b)

Figure 7.11 Use of a detention basin to regulate runoff at point A: a) view in plan of basin; b) runoff hydrographs at point A before and after construction of basin

However, water managers must keep in mind that the design of a detention lake or basin requires that the runoff hydrograph be known for the entrance of the planned detention basin. The hydrograph can be determined using the models described in section 7.12.

7.15.3 Delaying Runoff with Vortex Regulator (Hydro-Brake Regulator)

When runoff is high, pipes can be protected by reducing flow. The installation of flow regulators upstream from the pipes creates a hydraulic vortex (turbulence) (fig. 7.12). The head losses created by this mechanism force the excess flow back to detention basins through pipes located upstream. Regulators can be installed in sewer manholes.

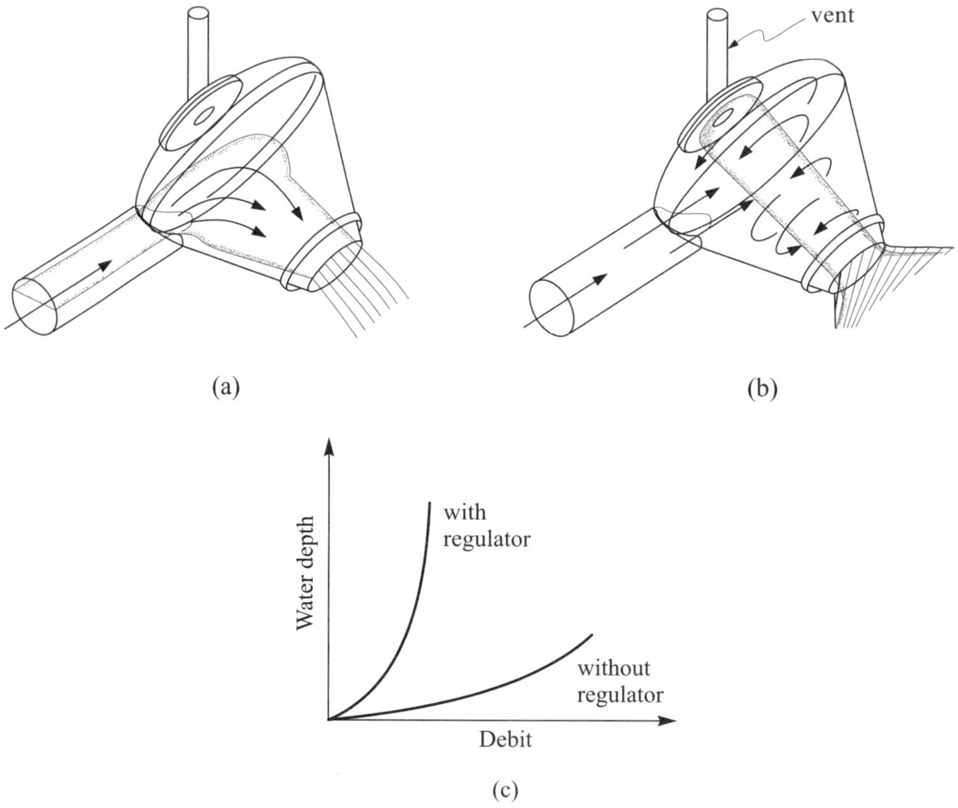

Figure 7.12 Flow regulator: a) without vortex – low flow; b) with vortex – high flow; c) relation between depth and debit

Adapted from: *Régulateur Hydrovex* (catalogue), John Meunier Inc.

7.15.4 Double Drainage Networks (Minor and Major Networks)

Up to the end of the sixties, only standard storm sewer networks were considered to evacuate storm flows. Less frequent but more intensive precipitations than those used for the design of the networks were the source of sewer back-flows and floddings.

More recently, strategies of storm flow management have been modified to take into account the hydraulic capacity of street surfaces, ditches and rivers that comprise the major network.

The street inlet is the heart of the double network. Il must be designed to control the flow within the hydraulic capacity of the minor network. Excess flow is retained on the surface in the major network, until the minor network can receive it.

The hydraulic capacity of the major network being limited, detention basins may have to be constructed, conditions permitting.

Flow control at the entrance of the minor network is guaranteed by one of the following tools:

- the street inlet itself, if its hydraulic capacity is known;
- flow restrictions at the entrance of the pipe that the links street inlet to the sewer network (minor network); the restriction may have the form of a calibrated opening in a plate set in the pipe: a diamond-shaped opening was studied and recommended by research workers at Ottawa University (Canada) as early as the 80's.
- a vortex regulator (art. 7.15.3) that may be installed in manholes.

7.16 STEP-BY-STEP PROCEDURE FOR THE DESIGN OF A STORM SYSTEM

Before submitting to the client the design for a storm-sewer system to drain an area, the engineer must proceed systematically through the following steps:

a) Find useful documents: topographic maps (showing ground elevation and existing structures), geographic maps, soil maps, urban and zoning plans.

b) Identify the boundaries of the area to be drained, draw the boundaries of the various watersheds and sub-basins, measure their surface areas, and identify the locations of streets and sewage pipes.

c) Measure the areas of so-called impermeable surfaces (sidewalks, streets, driveways, roof, etc.) using urban and zoning plans and topographic maps. Two types of roof must be considered: roofs of suburban houses equipped with gutters and down-spouts, and roofs, usually flat, of multiple-family dwellings built in highly populated areas. In the first case, if rainwater is evacuated onto the grass around the house, the roof is indirectly drained; if rainwater is evacuated onto a paved surface, the roofs are directly drained. In the second case, runoff collected from the roofs is transferred to the house connections: these roofs are also directly drained.

d) Identify the intensity-frequency-duration curves (IDF curves) of the given region and choose a precipitation frequency that complies with local practices: in Quebec, the chosen IDF curves are usually related to precipitation with a minimum return period of 5 years (frequency of occurrence 1/5 years).
e) Identify the Manning coefficients related to the surfaces of various watersheds and sub-basins and to pipes to be installed.
f) When using a computer, evaluate the water retention on permeable and impermeable surfaces.
g) Determine the amount of water that infiltrates the ground.
h) When using the rational method, calculate the times of entry related to the various sub-basins using methods such as the Kerby formula.
i) Calculate the flows to be transported and choose the size (diameter) of pipes and their slopes.

CONCLUSION

To determine the sizes of pipes for storm sewers, combined sewers, and intercepting sewers in urban regions, the maximum runoff of watersheds must be known. For small watersheds, the rational equation (equation 7.3) may be used:

$$Q = KAIR$$

The **maximum normal flow** is the flow calculated when all assumptions of the rational equation are met.

Complex desktop computer models now enable us to directly plot the runoff hydrograph at the watersheds' discharge. Among these models, the SWMM, OTTSWMM, and ILLUDAS models are most often used in Quebec. They enable us to compute the volume and surface area of retention basins that reduce maximum flows – that is, to lower the crest of the runoff hydrograph. Nowadays, to reduce maximum flows, we rely on retention of runoff on the surface of urban watersheds – more specifically, on street surfaces. We therefore have a major drainage system (watershed surface) and a minor drainage system (storm-sewer system). For the design of minor systems, the return period used is 5 years; for major systems, the return period is as high as 100 years.

PROBLEMS

7.1 Calculate the runoff coefficient, *R,* of the suburban residential district shown in figure 7.13.

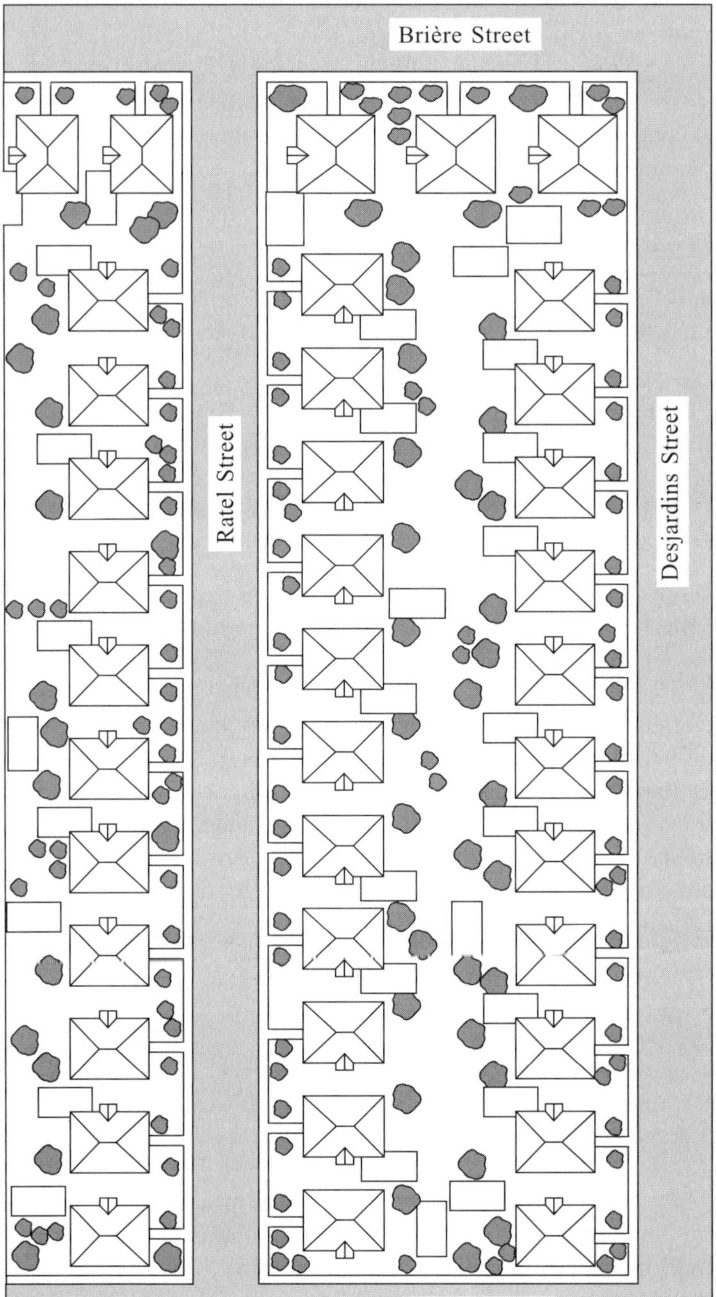

Figure 7.13 Residential area: single-family dwelling (problem 7.1)

7.2 Calculate the runoff coefficient of a commercial district with a time of concentration (or time of entry in this case) of 30 min. The soil in this area is mostly clay and the water table is not deep (absorptivity of the permeable surfaces is low). The different elements of the watershed's surface and the percentages of the total area used by these elements are:

Roofs:	40%	Lawns with strong slope (dense soil):	6%
Paved streets:	18%	Sidewalks:	6%
Parking lots:	25%	Driveways:	5%

7.3 For runoff in urban areas, explain the difference between a constant runoff coefficient and a runoff coefficient varying over time. State when they should be used.

7.4 A rectangular watershed, with a length 4 times its width, has a time of concentration (or time of entry in this case) of 30 min. The runoff flows from the upper short side to the other short side, in the middle of which is the discharge. If the runoff coefficient is constant, calculate the maximum outflow following a precipitation of:

a) 100 mm/h for 10 min, over the entire surface of the watershed;
b) 25 mm/h for 40 min, over the entire surface of the watershed.

Answer in terms of AR m^3/s.

7.5 On a rectangular watershed with a time of concentration (or time of entry in this case) of 40 min, the runoff flows widthwise (parallel to the widths). When it reaches the lowest length, the runoff is collected in chutes carrying it to the discharge located in the middle of the long side. If the water-flow velocity is constant until it reaches the outlet and the runoff coefficient is constant and uniform, calculate the maximum flow caused by each of the following rainfalls:

a) 100 mm/h for 10 min, over the entire surface of the watershed;
b) 25 mm/h for 50 min, over the entire surface of the watershed.

Answer in terms of AR m^3/s.

7.6 A rectangular parking lot measuring 150 by 300 m makes a watershed for which the inlet time, t_e, is 20 min (the sewer inlet is located in the geometric centre of the watershed) and the time of flow, t_f, is 5 min. The time of concentration of the entire watershed is therefore 20 min. Figure 7.14 shows the watershed with isochrones of 5, 10, 15, and 20 min. If the runoff coefficient is constant and uniform and equal to 0,85, calculate the maximum outflow following a precipitation of 50 mm/h falling for 5 min over the entire surface of the watershed.

7.7 For the parking lot in problem 7.6, plot the first 30 minutes of the runoff hydrograph obtained with a precipitation of 50 mm falling for 30 min over the entire surface of the watershed.

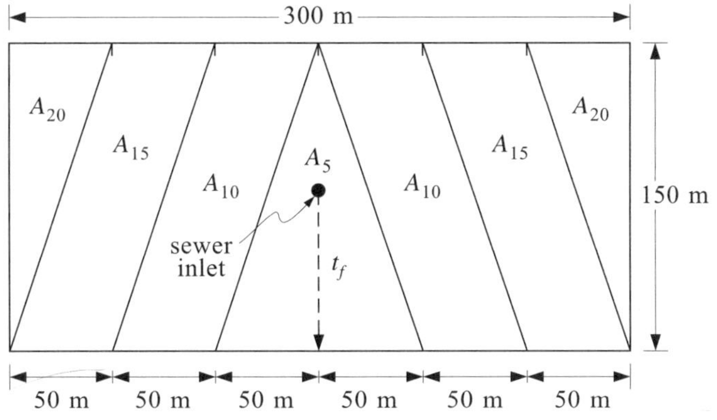

Figure 7.14 Parking lot (problem 7.6) A_{10} = area with time of concentration or inlet time, t_e, of 10 min.

7.8 Figure 7.15 shows a site located in the Montreal region that is part asphalt (runoff coefficient, *R*, of 0,8) and part grass (coefficient *R* of 0,18). Determine the inlet time, t_e, for each of the two sub-basins (t_e from *A* to *C*, and t_e from *B* to *C*):

a) using the Kerby model;
b) using the Kirpich model.

Choose the inlet time that must be used to calculate the size of the storm-sewer pipe at the discharge of the watershed; comment.

7.9 If precipitation with a return period of 10 years falls on the watershed of problem 7.8, calculate the maximum outflow of each of the two sub-basins, after evaluating the two inlet times, t_e, using the Kerby model. State which of the two inlet times is critical, and comment on your choice.

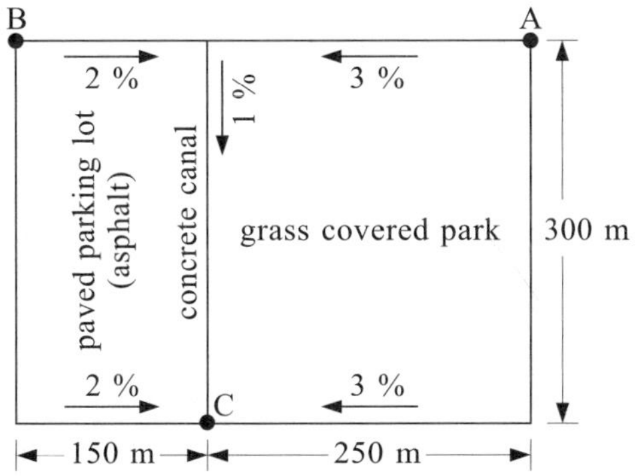

Figure 7.15 Park in the Montreal region (problem 7.8)

7.10 Solve problem 7.9 for a rainfall with a return period of 5 years.

7.11 A 50 000 m^2 (5 ha) sub-basin is located in the Montreal region in a district composed strictly of single-family dwellings (R = 0,4 and n = 0,1). If the average slope of the watershed toward the outlet is 5%, and if the water must run off over a distance of 150 m to reach the sewer inlet (discharge of the sub-basin), calculate the time of concentration (inlet time in this case) of this urban sub-basin. Using the Kerby equation, evaluate the maximum (normal) runoff at the discharge of this sub-basin following precipitation with a return period of 5 years: a) if the runoff coefficient is constant; b) if the runoff coefficient is variable over time. So-called impermeable surfaces cover 30% of the sub-basin and absorptivity of the permeable surfaces is average. State your recommendations.

7.12 A 400 000 m^2 (40 ha) building site located in the Montreal region will accomodate single-family houses on half of the area and row houses on the other half. The time of concentration of this watershed is 40 min, its runoff coefficient is 0,55, and its Manning coefficient n is 0,1. The main sewer located at the discharge of the watershed is able to carry a maximum flow of 1,27 m^3/s. Is the capacity of the main sewer sufficient to carry the runoff of rainfalls with a return period of 5 years? Comment on your results.

7.13 A 5 000 m^2 (0,5 ha) watershed with a runoff coefficient of 0,5 and a time of concentration of 22 min is located in the Montreal region. Using the rational equation and considering a return period of 5 years, calculate, in m^3/s, the following flows:

a) the maximum normal runoff at this watershed's discharge;
b) the flow 10 min after rainfall begins;
c) the flow 30 min after rainfall begins;
d) the flow 45 min after rainfall begins;
e) the maximum flow, considering a 10 min rainfall (from the onset of the rainfall, the precipitation front falls on the watershed's discharge);
f) the flow 4 min after the 10 min rainfall begins;
g) the flow 12 min after the 10 min rainfall begins;
h) the maximum flow, considering a 40 min rainfall;
i) the flow 20 min after the 40 min rainfall begins;
j) the flow 42 min after the 40 min rainfall begins.

7.14 A 4 000 m^2 (0,4 ha) watershed with a runoff coefficient of 0,6 and a time of concentration of 15 min is located beside the watershed described in problem 7.13 and flows toward the same discharge point. Calculate the maximum normal flow in the storm sewer used for both watersheds for a precipitation with a return period of 5 years.

7.15 In the Montreal region, two adjoining watersheds, A_1 and A_2, discharge into the same storm sewer. The surface area of watershed A_1 is 0,6 ha, its time of concentration is 5 min, and its runoff coefficient is 0,6. The surface area of watershed A_2 is 0,8 ha, its time of concentration is 8 min, and its runoff coefficient is 0,5. Using the rational equation, determine the maximum normal runoff (in m^3/s) in the storm sewer that receives the runoff of both watersheds. Storm sewers are designed for a 5-year return period.

7.16 In the Montreal region, two adjoining watersheds, A_1 and A_2, discharge into the same storm sewer. The surface area of watershed A_1 is 0,6 ha, its time of concentration is 5 min, and its runoff coefficient is 0,8 (paved parking lot). The surface area of watershed A_2 is 0,8 ha, its time of concentration is 8 min, and its runoff coefficient is 0,15 (flat lawn). Using the rational equation, determine the maximum runoff (in m^3/s) in the storm sewer that receives the runoff of both watersheds. (Do not perform more than two different and pertinent computations.) What is the maximum normal flow? Storm sewers are designed for a 5-year return period.

7.17 In the Montreal region, the characteristics of a watershed connected to a combined sewer system are as follows: A = 50 ha, R = 0,9, and t_c = 20 min. The combined sewage of this watershed is discharged into a reservoir, from where it is evacuated to the river by a 40 m long, reinforced-concrete outfall (n = 0,013). Determine the minimum theoretical diameter of the outfall if the water level in the reservoir must not exceed 20 cm over the highest level of the river. Minor head losses are negligible. The minimum and maximum velocities permitted in a combined outfall are 0,9 and 3 m/s. The outfall is always submerged. What is the maximum velocity of the sewage in the outfall? What are the maximum head losses in the outfall? For the design of combined sewer systems, precipitations with a return period of 10 years are used.

7.18 A rectangular parking lot in the Montreal region is made of concrete (Manning's n = 0,013) and measures 30 m by 50 m. Its sole water inlet is located at its geometric center. The corners of the 30 m side (the highest) are at 100 m elevation, and the two other corners are at 99,5 m elevation. In all directions, the surface slants toward the water inlet as gently as possible, but never falling below 2%. A return period of 5 years is considered, and the runoff coefficient of the parking lot is 0,85. Considering these data, what should be the elevation of the water inlet? What is the inlet time? And what is the maximum runoff at the inlet? Use the Kerby model. Comment.

7.19 When using the rational equation, can a rainfall of duration t be used if it is smaller that the time of concentration t_c of the watershed (for a given return period)? Comment.

7.20 When using the rational equation, can a rainfall of duration t be used if it is greater that the time of concentration t_c of the watershed (for a given return period)? Comment.

Chapter 8

DESIGN OF SEWER SYSTEMS

INTRODUCTION

A sewer system is designed to collect and transport wastewater or runoff mainly by gravity and must comply with the following requirements. It must:

- be deep enough for sewage collection (from buildings for example), preferably without pumping;
- resist dead loads due to backfill (material placed over pipes to fill trenches) and live loads, due for example to moving vehicles;
- have sufficient hydraulic capacity to transport maximum design flows;
- have sufficient slope for the flow velocity when pipes are flowing full to prevent any suspended solids from being deposited;
- have enough equipment (connections, street inlets, valves, etc.) for the collection and evacuation of wastewater and for the maintenance of the network;
- be low cost;
- be safe for maintenance crews and the public over its entire design period.

Designing a sewer system is not easy. A location must be found, the dimension and slope of each pipe length must be calculated, the pipe material (concrete, thermoplastic...) must be chosen, and the burying depth of the pipes must be determined.

When transporting wastewater by gravity, the engineer generally takes advantage of the natural slope of the terrain. However, if the slope is not sufficient to avoid the deposit of solids, pumping stations are used.

Finally, it is useful to recall that sewerage refers to the collection, treatment, and disposal of liquid waste. Sewerage or sewage works include all the physical structures required for these tasks (McGhee, 1991, p. 266).

8.1 TYPES OF COLLECTION SYSTEMS AND SEWERS

One of the three following sewer systems is chosen depending on the type of wastewater to be evacuated (fig. 8.1):

- a *sanitary sewer system* (mistakenly called a domestic system), which carries sanitary wastewater i.e. domestic and commercial wastewater, industrial raw or partially treated wastewater (depending on the capacity of the sewage treatment plant), and parasitic flow (infiltration/inflow); basement drains are connected to sanitary connections;
- a *storm-sewer system*, which carries runoff, including water from foundation drains, and, depending on local practices, water collected by roof gutters (usually, flat roofs are drained into the storm (or combined) sewer through storm (or combined) connections and inclined roofs are drained onto permeable surfaces);
- a *combined sewer system*, which carries both sanitary wastewater and runoff.

A sanitary sewer system usually works in conjunction with a storm-sewer system, which is made of storm-collection pipes or ditches. The combined sewer system, however, is not associated with any other system for the collection and evacuation of wastewater or runoff. In Quebec, combined sewers are no longer permitted, except in special cases with the permission of the Quebec Ministry of the Environment.

The expression "*pseudo-domestic* or pseudo-sanitary sewer system" is used to identify sanitary sewer systems that carry not only sanitary wastewater but also water from foundation drains, flat roofs, driveways of underground garages; basement drains are also connected to pseudo-sanitary sewer connection. Nowadays, the Quebec Ministry of the Environment forbids construction of such systems (fig. 8.1).

8.2 PLANNING CONSTRUCTION OF A SEWER SYSTEM

Planning for construction of a sewer system begins by clearly identifying watersheds and sub-basins to be drained as well as existing infrastructures. Then, a discharge site must be chosen. Thus, from the start, the engineer requires topographic maps with contour lines. In general, the direction of the system follows the slope indicated by elevation lines to avoid the need for pumping sewage or water from deep excavations.

The boundaries of the area under study are often determined by the boundaries of the natural watersheds. However, future needs and occupancy characteristics must be studied carefully. The engineer must ensure that the planned construction and the watershed's boundaries respect the regional sewer-system development plan (if there is one).

The design of a sewer system follows the stages described below: preliminary study, design, construction, and finally operation and maintenance.

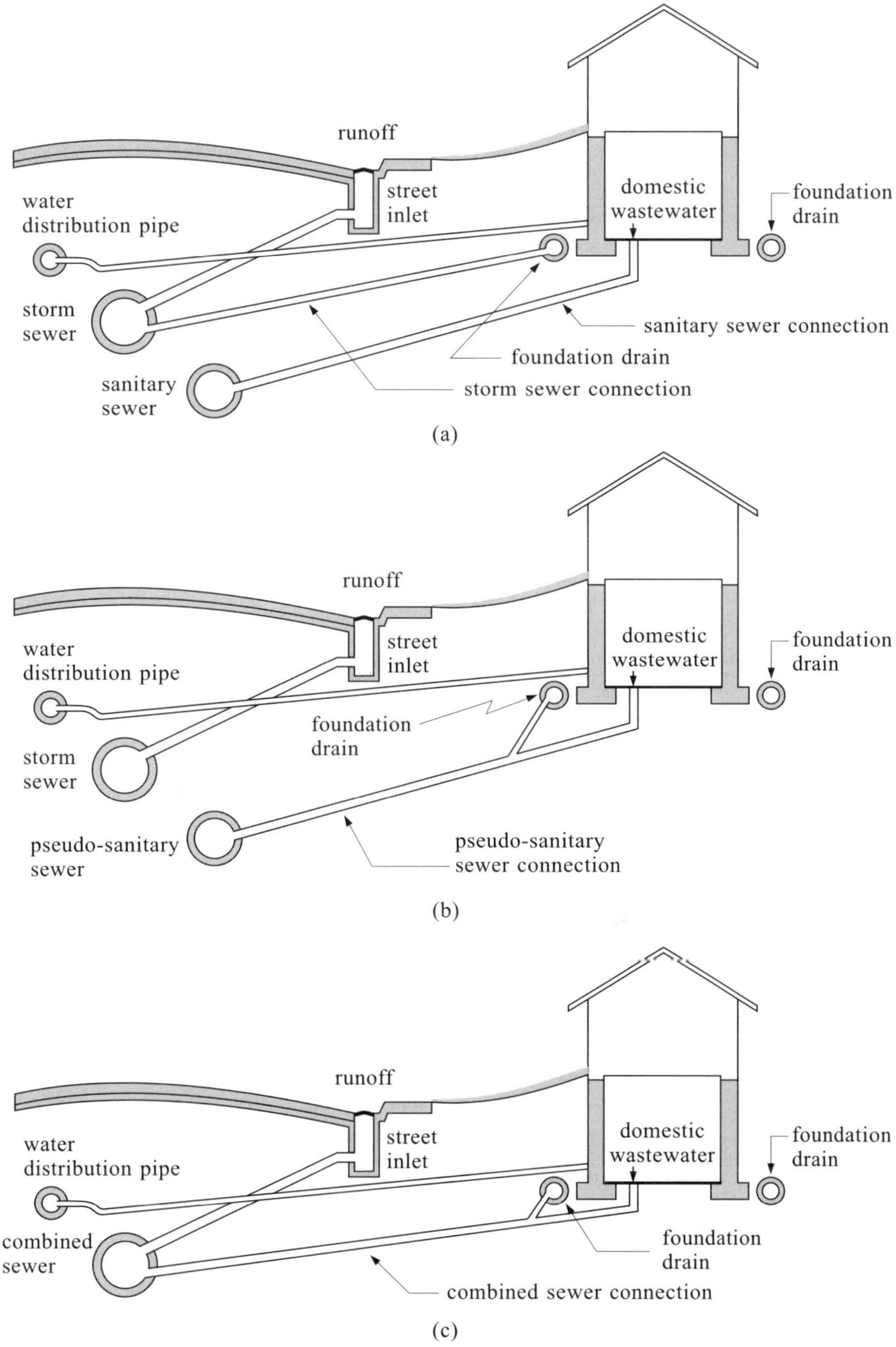

Figure 8.1 Schematic representation of service connections to different types of sewer systems: a) sanitary and storm systems; b) pseudo-sanitary and storm systems; c) combined system

Preliminary study. During this first stage, technical and economic ground rules are laid down. Subsequent decisions rely on these ground rules to determine the final design of the project. The report presented by the engineer must include the following: the objectives of the project, a detailed list of existing structures and services, a study of the physical environment (using topographic maps, aerial photographs, results from drilling and probing, etc.), a future needs study (area's occupancy and demographic predictions), a background for design criteria, identification of possible solutions, a comparative study of various possible solutions, and a recommendation of the final solution.

Design. During this stage, the specifications and drawings for construction are prepared. These documents will be used when inviting tenders for the construction of the project; thus, they must be *complete, clear,* and *concise*.

Construction. This stage includes inviting tenders, granting the contract, construction, and *surveillance* of the work, which is also an important part of the construction.

Operation and maintenance. The operation and maintenance of a sewer system is usually the responsibility of the system's owner. It is, however, the engineer's duty to advise the owner of maintenance practices.

8.3 COMPONENTS OF A SEWER SYSTEM

Sewer systems are made of a series of pipe lengths assembled in such a way that the sewage is transported from where it is produced to an evacuation point. Elements of a sewer system vary depending on the type of system, as they must be adapted to the categories of wastewater transported.

8.3.1 Types of Sewers in a Typical Collection System (fig. 6.1)

There are five types of sewer pipes: branch or lateral sewer, main sewer, trunk sewer, intercepting sewer, and outfall.

The *branch (or lateral)* sewer transports wastewater from one or several street sections; wastewater from sectors located upstream from it must not be discharged in it; it has no tributary flow except from house connections.

The *main sewer* carries wastewater from several branch sewers; the main sewer, as its name indicates, is the main evacuation axis of a small watershed.

The *trunk sewer*, a large main sewer, receives sewage from main sewers; it is the main evacuation axis of a large watershed.

The *intercepting sewer* collects all or part of the wastewater transported by the main and trunk sewers and carries it to the sewage treatment plant. Usually, it is buried deep enough to collect the wastewater by gravity. In the Montreal area, there are two intercepting sewers: one to the north, along Rivière-des-Prairies, and one to the south, along the St. Lawrence River. They transport, from west to east, wastewater from the entire Montreal area (most of the island of Montreal) to the treatment plant in the Rivière-des-Prairies district (fig. 8.2).

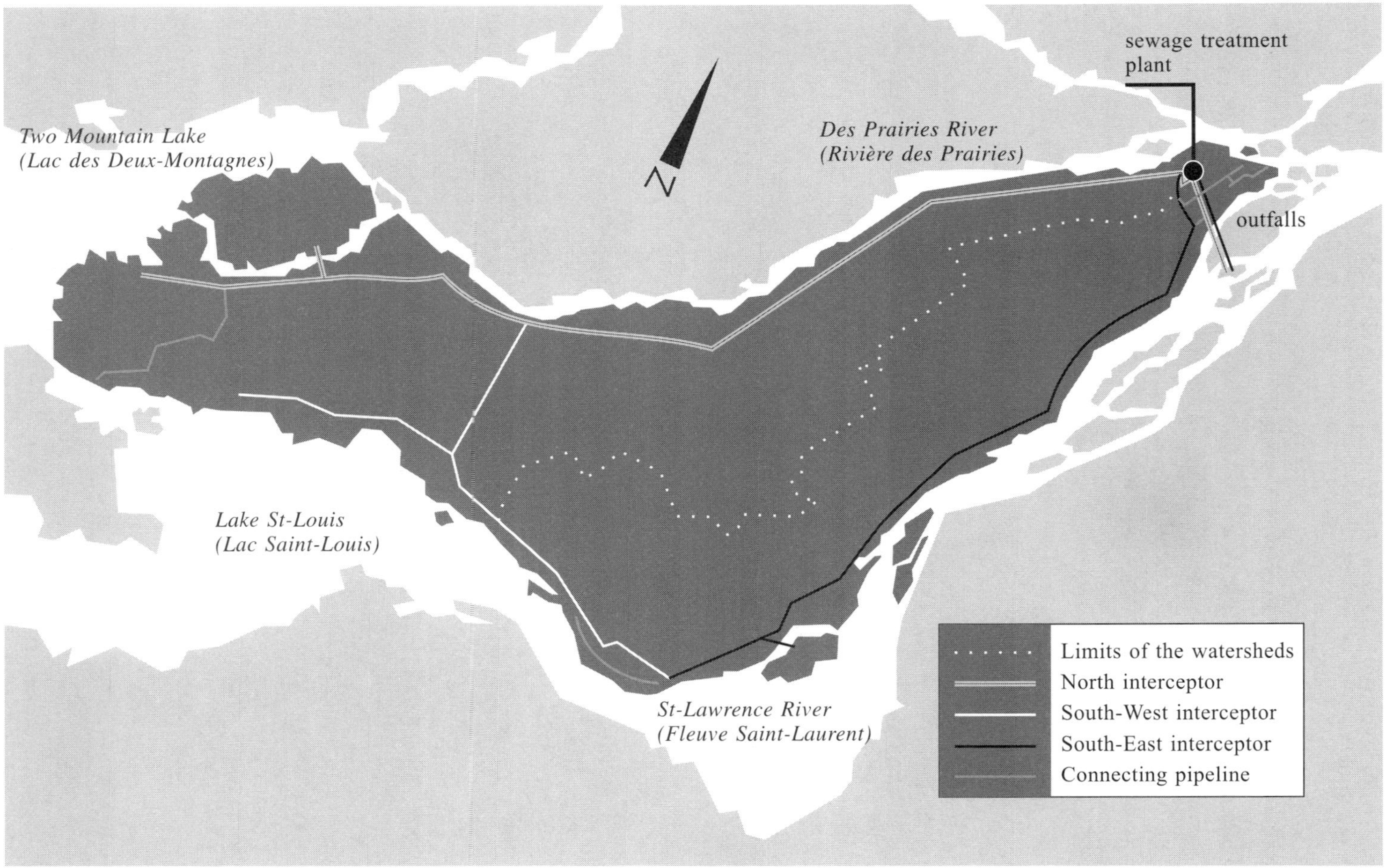

Figure 8.2 Main layout of intercepting sewers of the Montreal area.

From: Montreal Urban Community, Water Treatment Services, *Projet d'épuration des eaux, Plan directeur du système d'interception*, p. 4 (in French).

Finally, the *outfall* evacuates wastewater (raw, partially treated, or treated) into a receiving body of water.

8.3.2 Sewer Manhole and Drop Sewer Manhole

The *sewer manhole* (figs. 8.3 and 8.4) is an element of major importance in a sewer system. It provides access to sewer pipes for maintenance and ensures adequate ventilation in the system, which is essential for the evacuation of foul-smelling, toxic, and explosive gases (H_2S, NH_3, CH_4, CO_2, etc.; see chapter 10).

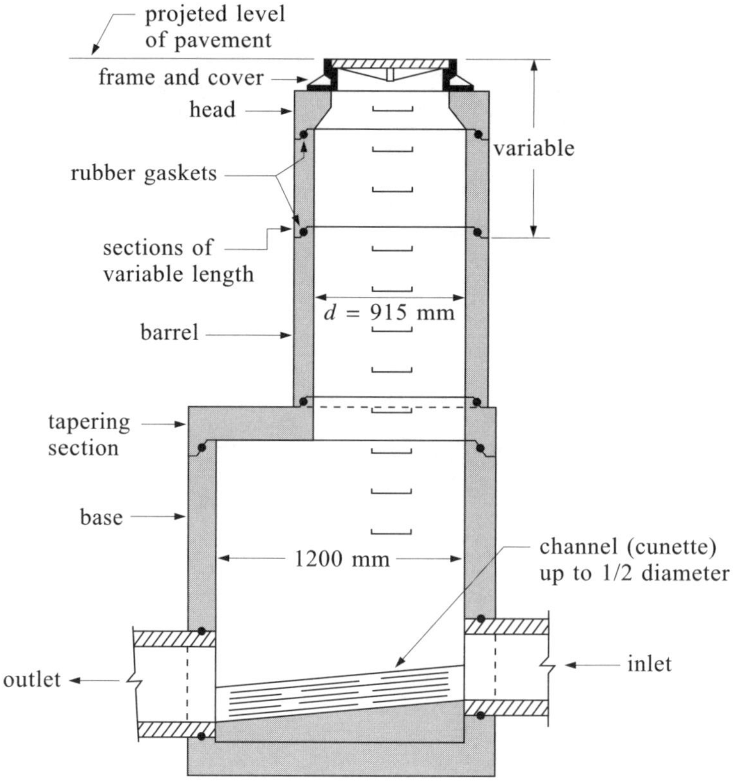

Figure 8.3 Sewer manhole

From: Bureau de normalisation du Québec, *Conduites d'eau et d'égouts* (Les Publications du Québec, HQ 1809-300, 1987) p. 110.

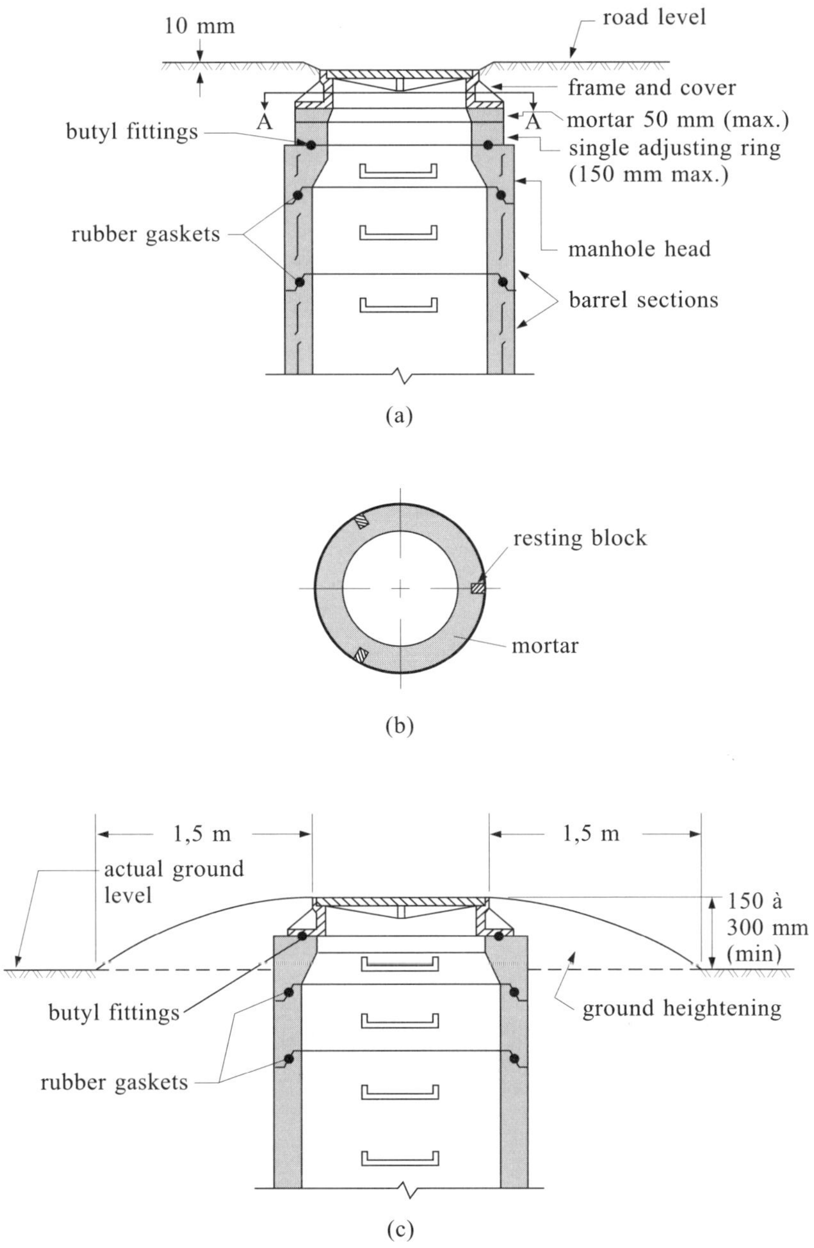

Figure 8.4 Installation of sewer manhole frame and cover: a) in a paved street; b) section A-A; c) in the field

From: Bureau de normalisation du Québec, *Conduites d'eau et d'égouts* (Les Publications du Québec, HQ 1809-300, 1987), p. 110.

A *drop sewer manhole* (fig. 8.5) is installed when the invert of the pipe carrying the wastewater to the sewer manhole is more than 600 mm above the crown (top) of the evacuating pipe. Below (section 8.4.5), we will describe the various types of sewer manholes and the rules regulating their locations. When the depth of the sewer manhole reaches 6 m, a safety platform made of two grade covers must be installed.

8.3.3 Street Inlet

A street inlet is used strictly by storm-sewer and combined sewer systems. Its function is to collect surface runoff, mostly from the street gutter, so that it can be evacuated through the storm sewer or the combined sewer (fig. 8.6).

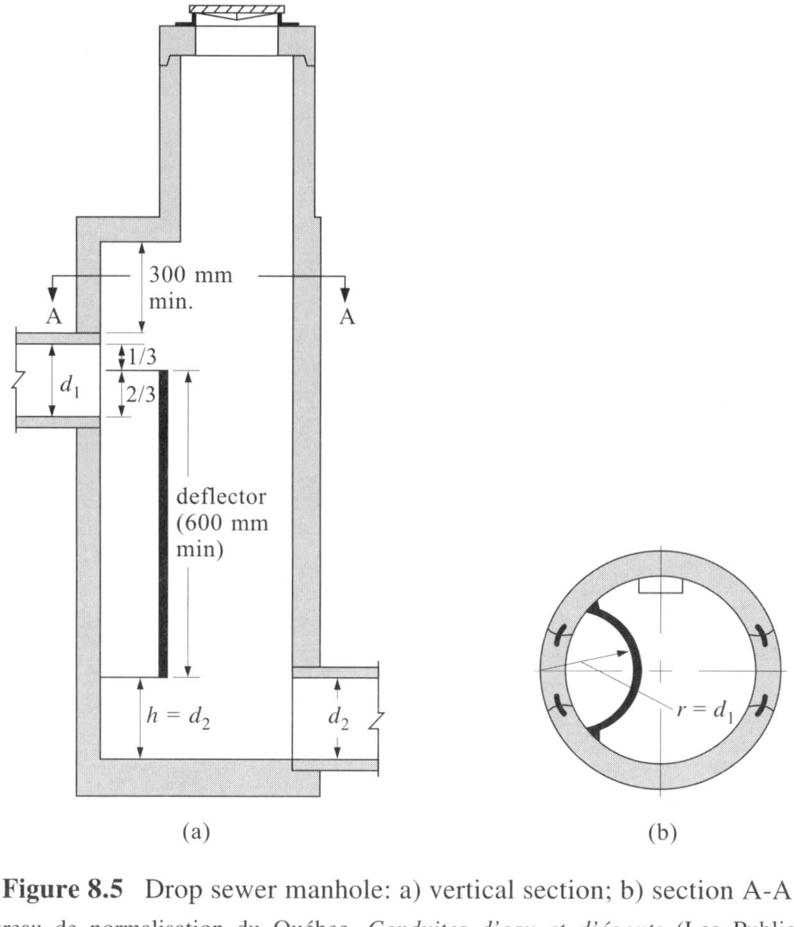

Figure 8.5 Drop sewer manhole: a) vertical section; b) section A-A

From: Bureau de normalisation du Québec, *Conduites d'eau et d'égouts* (Les Publications du Québec, HQ 1809-300, 1987), p. 110.

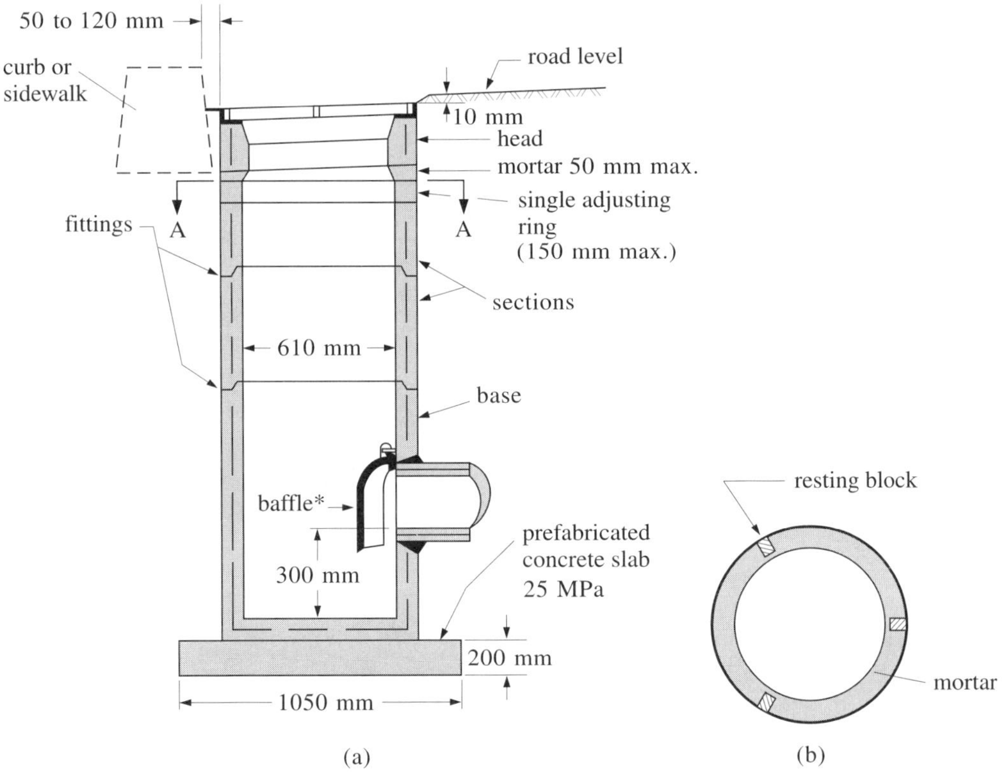

* Baffle prevents access of floating material to pipes.

Figure 8.6 Street inlet: a) vertical section; b) section A-A

From: Bureau de normalisation du Québec, *Conduites d'eau et d'égouts* (Les Publications du Québec, HQ 1809-300, 1987), p. 113.

8.3.4 Building Service Connection

Building service connections (figs. 8.7 and 8.8) evacuate wastewater from houses and other buildings. The *sanitary sewer connection* transports domestic wastewater to the sanitary sewer system (including water from basement drains), and the *storm-sewer connection* evacuates drainage water (foundation drains, garage drains, and flat-roof drains) to the storm-sewer system. The *combined sewer connection* evacuates combined sewage (domestic wastewater and runoff). The *pseudo-sanitary sewer connection* carries domestic wastewater and the water from foundation and basement drains, flat roof drains, and below-ground driveway drains to the pseudo-sanitary sewer. Figure 8.7 shows a building service connection. Figure 8.8 illustrates the position of the sanitary or storm sewer service connection in relation to house basements and street sewer pipe. The horizontal and vertical distances show how the top of the sewer pipe crown (top) is buried at least 2,0 m deep to provide a service connection slope of 2%.

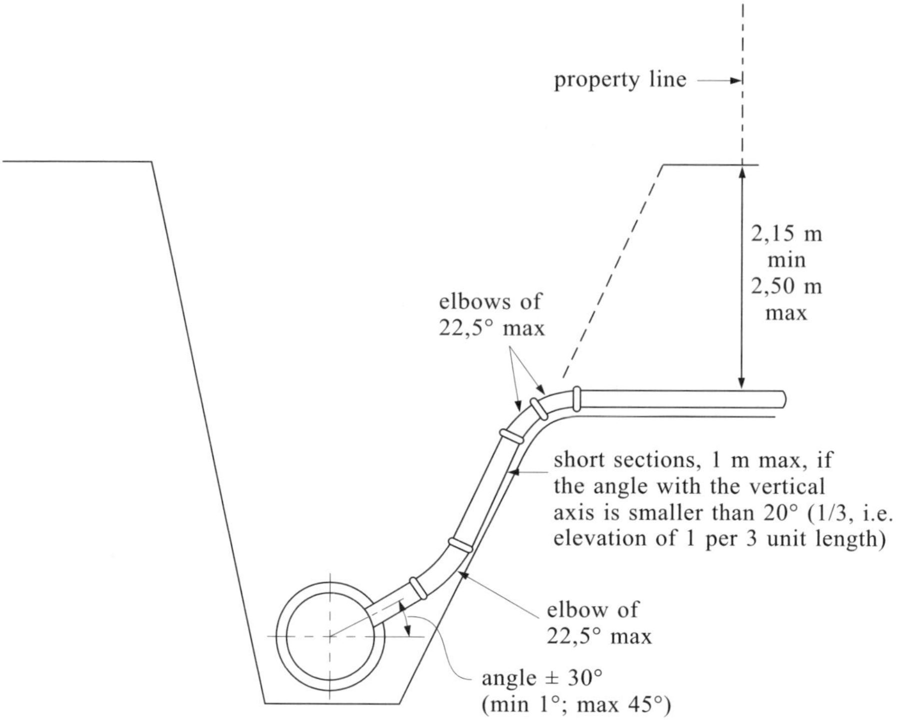

Figure 8.7 Building service connection link to the sewer

From: Bureau de normalisation du Québec, *Conduites d'eau et d'égouts* (Les Publications du Québec, HQ 1809-300, 1987), p. 114.

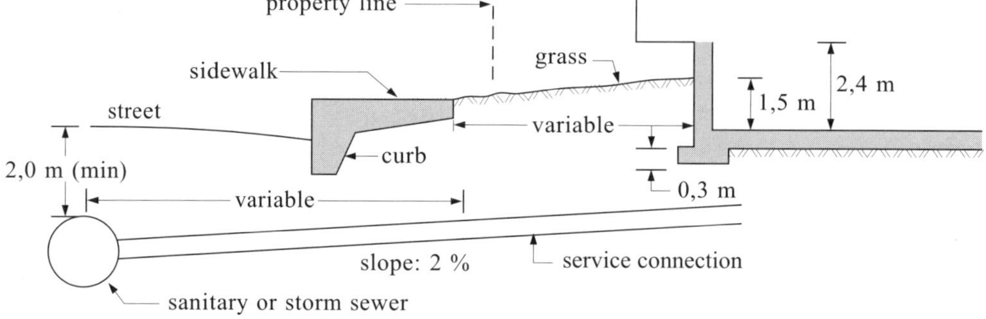

Figure 8.8 Drawing of sanitary or storm sewer service connection

8.4 DESIGN AND CONSTRUCTION OF ELEMENTS OF A SEWER SYSTEM

In the last 50 years, engineers have learnt a great deal about the design and construction of sewer systems. The sections below present basic engineering knowledge of pipe materials, the shapes of pipe sections, required slopes, flow velocity, sewer man-hole and street-inlet installation, and joint watertightness to reduce infiltration flows.

8.4.1 Sewer and Drinking-Water Pipe Materials

Additional information is given on drinking water pipes in section 4.12.4.

Sewer pipes can be made of various materials. The choice of a material is a function of:
- the roughness of the material;
- the life span of the material and the sewer system's appurtenances;
- the material's resistance to erosion, abrasion, acids, bases, gases, solvents, etc.;
- the level of complexity of handling and installation of the pipes;
- the pipe resistance to dead and live loads (also resistance to inside pressure in the case of drinking water pipe);
- the watertightness of joints;
- the availability of special connections;
- the cost of purchasing, handling, and installing the pipes.

Materials most commonly used for the manufacture of sewer pipes are asbestos-cement, unreinforced concrete, reinforced concrete (pre-cast or cast-in-place), cast iron, ductile cast iron, concrete pressure pipe, galvanized steel and thermoplastic.

Asbestos-cement. The nominal diameters of *asbestos-cement* pipes vary from 100 to 900 mm (4 to 36 in; table 8.1). These pipes can be used to evacuate wastewater by gravity or under pressure (backflow pipe of pumps), and to distribute water for consumption. They are available in five different structural classes: 1500, 2400, 3300, 4000 and 5000 (see section 8.4.10 and chapter 11 for definitions). Pipes in the 1500 and 2400 classes are rarely used because of their brittleness.

Advantages. The main advantages of asbestos-cement pipes are as follows:
- they are light weight;
- they are easy to handle and install;
- they have good watertightness at the joints between pipe lengths and between the pipe and service connections;
- the material is smooth (Manning's roughness coefficient is 0,010); however, the Manning's roughness coefficient of asbestos-cement seems to increase with the age of the pipe, to reach the roughness of concrete (0,013);
- long pipe sections (4 m) are available; thus, the number of joints is small compared to concrete pipes.

Disadvantages. The disadvantages of asbestos-cement pipes are more important:
- asbestos-cement pipes are brittle;
- they require excellent bedding (bottom of the trenches where the pipe is laid);
- the cost of large-diameter pipes is high.

Unreinforced concrete. Such pipes are rarely used because they are brittle and crack easily tending to increase infiltration rates.

Pre-cast reinforced concrete. Sewer pipes made of *pre-cast reinforced concrete* are most commonly used. Their nominal diameters vary from 150 to 3800 mm (table 8.1). They are used to evacuate wastewater by gravity and are available in five structural classes (chapter 11). The joints are waterproofed by adding a rubber gasket between pipe lengths (figs. 8.9 and 8.10). Nowadays, this practice is mandatory for sanitary as well as storm sewers.

Table 8.1 Diameters of asbestos-cement and reinforced-concrete sewer pipes available on the market

Diameter			Asbestos cement	Reinforced concrete
nominal	actual*			
(mm)	(mm)	(in)		
100	102	4	x	
125	127	5	x	
150	152	6	x	x
200	203	8	x	x
225	229	9		x
250	255	10	x	
300	305	12	x	x
350	355	14	x	
375	380	15		x
400	405	16	x	
450	455	18	x	x
500	510	20	x	
525	535	21		x
600	610	24	x	x
675	685	27		x
750	760	30	x	x
900	915	36	x	x
				(diameter may reach 3800 mm)**

* Actual diameter ≅ nominal diameter × (2,54/2,50). 2,54 and 2,50 are respectively the exact number and the approximate number of centimeters in one inch.

** The nominal diameters of reinforced concrete pipes over 900 mm increase in 150 mm (6 in) steps to 2400 mm (95 in), then in 300 mm (12 in) steps to 3800 mm (150 in). If the desired diameter is greater than 3800 mm, the pipes must be cast in place.

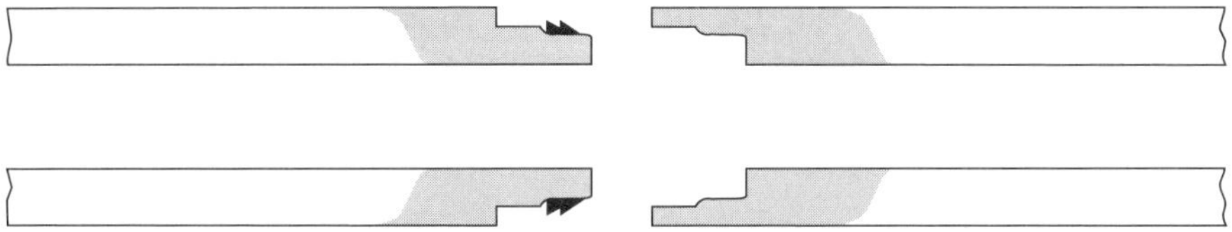

Figure 8.9 Joints between two pipe lengths

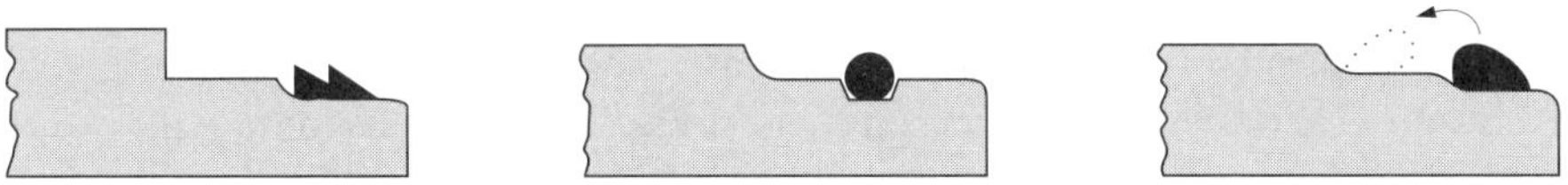

(a) (b) (c)

Figure 8.10 Types of gaskets used for joints between concrete pipe lengths: a) lock-joint (push-on); b) O-ring; c) roll-on

From: Bureau de normalisation du Québec, *Conduites d'eau et d'égouts* (Les Publications du Québec, HQ 1809-300, 1987), p. 104.

To avoid water infiltration at the joints of service connections, special connectors must be used, such as prefabricated tees and saddles. Pipes of 150 to 225 mm in diameter are available in sections of 1 m, and pipes of larger diameter are available in sections of 2,4 m. **The approximate thickness of the concrete pipe wall, in inches, equals the pipe diameter in feet, plus one. Thus, the thickness of a pipe wall with a diameter of 2 feet is 3 inches (2 + 1).**

Advantages. In general, these pipes:
- are very resistant to dead and live loads;
- are available in a wide variety of diameters and classes.

Disadvantages. However, pre-cast reinforced concrete pipes are:
- prone to acid corrosion;
- especially hard to handle in large-diameter sizes, mostly because of their great weight.

Cast-in-place reinforced concrete. For very large diameters (1,8 to 2 m and more), economically speaking, it can be advantageous to cast concrete pipes in place.

Cast iron, ductile cast iron, concrete pressure pipe, and galvanized steel. Sewer pipes made of these materials are not as common as the sewer pipes described above.

Cast iron pipes have been replaced nowadays by lighter, more resistant, and less brittle ***ductile cast iron pipes***, which are used when there is a risk of water infiltration, especially when the water table is high. They are also used for elevated pipes (e.g., under bridge decks), as sewer outfalls, and for pipelines inside sewage and water-treatment plants. These pipes are available in very long lengths that limit the number of connecting joints and the infiltration associated. They show good resistance to loads, but are prone to corrosion.

Concrete pressure pipes have a greater resistance to loads than other types of pipes, because of their steel cores. They are used mostly for water intake pipes and the large diameter pipes of water-distribution networks. They are quite costly, available in very long lengths, and permit very little water infiltration mainly because of their high resistance to loads and internal water pressure.

Galvanized steel pipes are mainly used for the drainage of highways, railroads (e.g. for culverts), and so on. These pipes are available in a wide variety of diameters and lengths. They are prone to corrosion and their hydraulic capacity is poor, as their Manning coefficient is high (up to 0,027).

Plastic (more precisely thermoplastic). Plastic pipes are used increasingly in both sewer systems and water-distribution networks. They are used mostly for small-diameter pipes of sanitary sewer systems, for service connections, in houses, and as foundation drains; they are also used for small-diameter pipes in distribution networks.

Plastic pipes are light and have watertight joints. They are also available in very long lengths. However, plastic is very flexible, and the pipes can collapse under high loads. This may occur if the bedding and backfill are not installed properly (see art. 11.2 for more information).

8.4.2 Minimum Diameters of Sewer Pipes

In Quebec, for maintenance and self-cleaning purposes, minimum authorized diameters are 200 mm for sanitary sewer pipes and 300 mm for storm or combined sewers. The 300 mm diameter is required because storm or combined sewers often receive heavy material like sand and gravel during street cleaning and heavy rainfalls. Moreover, for major runoff, pipes of diameters less than 300 mm are adequate only over short distances.

More importantly, when pipes meet at a sewer manhole, the diameter of the downstream pipe (the one evacuating wastewater) must never be smaller than the diameter of the upstream pipe(s) (which bring wastewater to the sewer manhole).

8.4.3 Sewer Pipes with Unusual Shapes

The cross section of most prefabricated sewer pipes is circular. Pipes of other cross-sectional shapes can also be built or cast in place to facilitate their installation and construction or ease water flow (fig. 8.11). Sections can be egg-shaped, semi-elliptical, horseshoe or basket handle shaped.

The horseshoe cross section is most frequently used for large cast-in-place reinforced concrete pipes because of ease of construction, especially in bedrock (see appendix A for hydraulic characteristics of some of these types of pipes).

8.4.4 Actual Inner Diameter of Sewer Pipes

In terms of hydraulics, the *inner* diameter of a pipe is the one that matters most. In fact, pipes of identical nominal diameters do not necessarily have the same inner diameters. The inner diameter is a function of the type of the pipe material, and this situation results from the following facts:

a) All materials do not have the same mechanical strength; the wall thickness therefore varies.

b) To install and join pipe lengths made of different materials, the outer diameters must often be the same, even though the pipe walls may have different thickness (especially the case for pipes in the water-distribution network).

c) Pipe manufacturers still produce pipes with nominal diameters in the imperial system (12 in, 24 in, etc.). Approximate values in the SI system are assigned to these diameters to avoid fractional numbers. Table 8.2 gives a list of actual inner diameters for 18-inch (450 mm) pipes made of various materials.

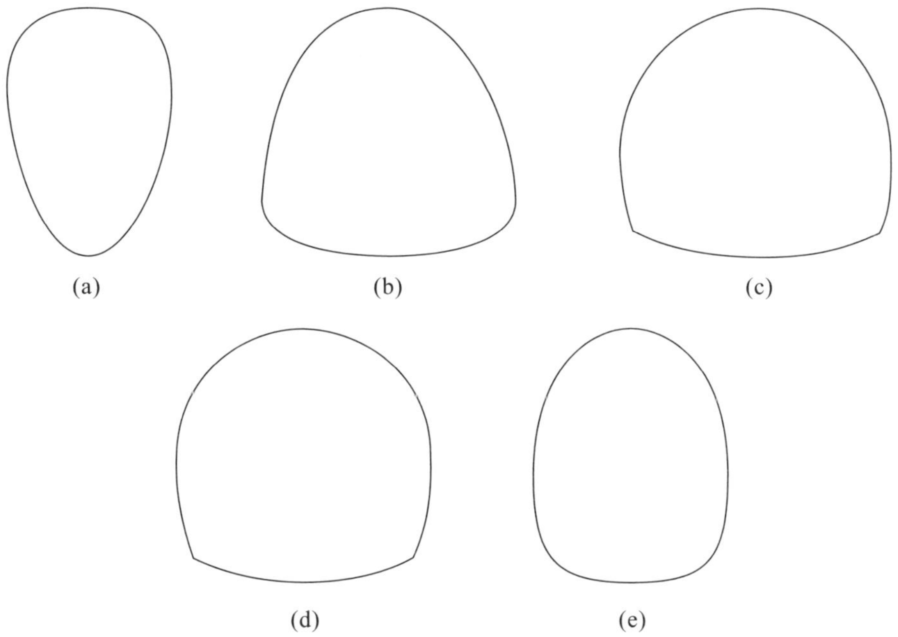

Figure 8.11 Special sections of sewer pipes: a) egg-shaped section; b) semi-elliptical section; c) horseshoe section; d) horseshoe section used in Montreal; e) basket handle section

Table 8.2 Actual inner diameters for 18-inch (450 mm) pipes, according to manufacturing materials

	Pipe material				
	Reinforced concrete	**Galvanized steel**	**Asbestos cement**	**Polyvinyl chloride**	**Polyethylene**
Nominal diameter (in)	18	18	18	18	18
Nominal diameter (mm)	450	450	450	450	450
Inside diameter (mm)	457	400	457	450	variable

8.4.5 Sewer manhole

Sewer manholes enable maintenance crews to access sewer pipes and ensure system ventilation so that toxic and explosive gases can be evacuated. They must be built so as to minimize the minor head losses they produce in the system.

The maximum distance between two sewer manholes depends upon the diameters of the pipes that connect them. For pipes with inner diameters between 200 and 1220 mm, sewer manholes should not be more than 100 m apart. For inner diameters exceeding 1220 mm, the distance should not exceed 150 m.

The installation of a sewer manhole is mandatory:

- when there are changes in diameters and/or slopes of the sewer pipes;
- when there is a junction of pipes from different directions;
- at the beginning of the system – that is, at the head the first pipe length of a local sewer draining an urban sub-basin;
- at the junction of two pipes buried at two different depths; moreover, if the difference in elevation is important, a drop manhole must be installed so that maintenance crews working at the bottom of the sewer manhole are not showered with sewage from the upstream sewer pipe (section 8.3.2). Figures 8.3 and 8.5 show an ordinary sewer manhole and a drop sewer manhole.

8.4.6 Sewer-Pipe Alignment

In general, a sewer pipe is installed in a straight line between two sewer manholes. However, a slight change in alignment may be made without the use of sewer manholes. Elbows – flexible joints that do not cause leakage – or pipe lengths with fitted ends can be used to change direction.

8.4.7 Burying Depth of Sewer Pipes

As much as possible, sewer pipes must be buried deep enough for gravity to carry wastewater from surrounding dwellings and land. However, buildings with especially deep basements must be equipped with pumping stations to evacuate wastewater and foundation drain water.

Sanitary Sewer. To drain basements, the pipe crown (top) must be at least 2 m under the ground surface (this depth is called the pipe cover). At this depth, there is no risk of frost in the Montreal region: in fact, 1,85 m is adequate for drinking water, and wastewater freezes at a lower temperature than drinking water as it contains salts in solution. When building a sewer, the inner height at the bottom of the pipe (invert) is always the measurement indicated on drawings. For example, when installing a sanitary sewer pipe with a diameter of 375 mm where the ground level is 110,5 m, the pipe invert is laid at a depth of 108,12 m (110,5 m – 2,0 m of cover – 0,38 m of actual inner diameter). Note that the thickness of the pipe wall is not taken into account as it is included in the recommended 2 m cover.

Storm Sewer. To fulfil its function efficiently, the storm sewer must meet the following conditions:

- it must be buried deep enough to evacuate foundation drain water without pumping;
- the pipe crown (top) must be at least 2,0 m under the ground surface (surface of street) (fig. 8.8); however, it is the invert that is, more often, located at this depth. The required depth may, however, be smaller in areas where foundation drains are not deep or where drain water is pumped;
- to facilitate sewer-pipe connections on both sides on the street, installation of sewer pipes and water distribution pipes on the same horizontal plane must be avoided: usually, sewer pipes are installed beneath water-distribution pipes to minimize the risk of contaminating drinking water.

The reader will find other information concerning burying depth of sewer pipes in sections 8.5.3 and 9.1.2.

Sanitary and storm sewer pipes should also not be on the same horizontal plane. When computations lead to such an arrangement, usually at the head of the system, the sanitary pipe is lowered enough so that its crown (top) is at the same level as the storm-pipe invert.

8.4.8 Slope of Sewer Pipes

Slope in itself is not a requirement in the design of sewer pipes; it must only be steep enough to generate a minimum velocity to ensure the self-cleaning of the pipes.

Slope and Minimum Flow Velocity. The evacuation of wastewater by gravity must always occur at a flow velocity that prevents suspended matters from being deposited. A good flow velocity slows, or even stops, the formation of foul-smelling, toxic, and explosive gases (H_2S, NH_3, CH_4, CO_2, etc.). The desired minimum flow velocity, taking into account the pipe flow at the beginning of the design period, varies from 0,6 to 0,75 m/s for sanitary sewer systems. For combined and storm sewers, it reaches 0,9 m/s: storm and combined wastewater carries matter heavier than the matter transported by sanitary wastewater. **According to normal engineering practices, minimum flow velocities must be respected when pipes are flowing full.**

When the design flow is much less than the capacity of the pipe – capacity being the flow when the pipe is flowing full (section 2.1) – the engineer must calculate the actual flow velocity (for the maximum daily flow of the first year of the design period for example) to check that it is acceptable and make the appropriate adjustments, if necessary. In the case of circular pipes, critical flows are less than half the capacity of the pipe since it is only then that flow velocities are smaller than when pipe is flowing full (section 2.3).

Table 8.3 shows the minimum slopes required to maintain flow velocities at 0,6 m/s in circular pipes flowing full.

Slopes greater than the ones shown in table 8.3 must be used (e.g. slope of 0,5%) for the first pipe lengths of a sewer system where flow velocities are often very low because the pipes never flow full (sect. 9.1.2): the minimum diameter (200 mm or 300 mm depending on the type of sewer) is chosen for these pipes as it facilitates maintenance.

Table 8.3 Slopes required to maintain flow velocities at 0,6 m/s (circular pipes flowing full) as a function of pipe diameters and Manning's roughness coefficient

Diameter			**Slope (m/m)***	
nominal (mm)	**actual (mm)**	**actual (in)**	**Manning's roughness coefficient n = 0,013**	**Manning's roughness coefficient n = 0,015**
150	152	6	0,004 8	0,006 3
200	203	8	0,003 2	0,004 3
250	255	10	0,002 4	0,003 2
300	305	12	0,001 4	0,002 5
375	380	15	0,001 9	0,001 9
450	455	18	0,001 4	0,001 5
525	535	21	0,000 9	0,001 2
600	610	24	0,000 75	0,001 0

* These slopes must be multiplied by 1,56 to maintain a velocity of 0,75 m/s and by 2,25 to maintain a velocity of 0,9 m/s.

Maximum Flow Velocity. The ideal velocity is usually higher than the minimum flow velocity. However, it is recommended that an upper limit be set to control noise and to avoid excessive wear and abrasive deterioration of the pipes and other elements (including sewer manholes). Flow velocities greater than 12 m/s are considered excessive – even damaging. In general, municipal and regional organizations set the maximum flow velocity at 3 m/s; this upper limit is not, however, as strict as the lower limit.

8.4.9 Various Elements

The *street inlet* collects runoff and carries it toward the combined or storm sewer. *Service connections* transport wastewater from buildings to sewer pipes.

Street Inlet. Street inlets are designed and installed to collect runoff from the streets and transport it to the combined or storm sewer. The street inlet also acts as a catch basin to hold back heavy debris (sand and gravel for example) carried by runoff. Figure 8.6 illustrates a typical street inlet cross section.

Street inlets, which have a maximum hydraulic capacity of 30 L/s, are installed on each side of the street along curbs and sidewalks. The distance between street inlets may vary from 45 to 90 m, depending on the surface drained, the occupancy of the areas and the slope of the street; street inlets are closer together in areas of high urban density and impermeability. Inlets must also be installed at street intersections, just upstream from pedestrian crossings; however, the exact layout depends on the slope of the streets and on municipal requirements (sect. 9.3).

Service Connections. There are four major categories of service connections (fig. 8.1): *sanitary sewer connections* (sanitary wastewater and basement drains), *storm sewer connections* (foundation drains, drains from driveways located below ground level, and flat roof drains), *combined sewer connections* (sanitary wastewater, foundation and basement drains, flat roof drains, and drains from driveways located below ground level), and *pseudo-sanitary sewer connections* (sanitary wastewater, foundation and basement drains, and drains of driveways located below ground level). The last is now prohibited in Quebec.

Service connections are linked to sewer pipes using special equipment that ensures a watertight fit to minimize infiltration in the system. Small-diameter sewer pipes are connected by prefabricated tees, and large-diameter pipes are connected by special saddles (section 8.4.1). Although there are differences among municipalities, the minimum diameter for service connections is generally 150 mm. **The 150 mm diameter is used more and more for the connection of sewer inlets**; in the past, pipes of 225 mm diameter were mostly used.

8.4.10 Strength Classification of Pipes

Prefabricated pipes are available in different classes, which identify their bearing capacity after installation. In practice, these classes are associated with the depth at which pipes can be buried. Asbestos-cement pipes are available in classes 1500, 2400, 3300, 4000, and 5000. These numbers identify the bearing capacity in pounds per foot length (1 lb/ft ≈ 1,50 kg/m).

There are five classes, I to V, of prefabricated reinforced-concrete pipes (section 11.8). In practice, only classes III to V are readily available; the others must be custom made.

8.5 HEAD LOSSES IN A SEWER SYSTEM

The designer of a sewer system cannot choose the slope of pipes without first determining head losses due to friction and minor head losses.

8.5.1 Head Losses in Sewer Pipes

Head losses due to friction in a sewer pipe are important parameters when determining the dimensions of sewer pipes; minor head losses and head losses due to friction enable us to determine the minimum slope of pipes necessary to ensure flow by gravity and adequate flow velocities.

8.5.2 Head Losses Due to Various Equipment

When determining the piezometric line, many engineers neglect minor head losses due to equipment, especially due to sewer manholes. This omission is generally acceptable because Manning's roughness coefficient *n* suggested for pipes used as sewers takes into account minor head losses. Other engineers systematically specify lowering the pipe invert located downstream from a sewer manhole by 1,5 to 3,0 cm. However, head losses can sometimes be greater; thus, they must be determined in the same manner as for large-diameter pipes with low slopes.

8.5.3 Elevation of Sewer Pipes at Sewer Manholes

In sewer manholes used solely as junctions between two pipes, the pipe crowns (tops) are placed at the same height; however, the downstream pipe is sometimes lowered if the minor head losses at the sewer manhole are greater than the difference between the upstream and downstream pipe diameters. In sewer manholes collecting wastewater of more than one sewer pipe, the height of the downstream pipe (evacuating pipe) is a function of the crown (top) of the most deeply buried upstream pipe. **The crown (top) of the downstream pipe cannot be higher than the lowest crown (top) of the upstream pipes; it can, however, be lower for other hydraulic or topographic considerations.** The manhole at this location may also be used as the head of another sewer network. If this is the case, the first pipe length of this new sewer network will be linked to the manhole at a greater elevation than the foregoing upstream pipes from which it does not receive any sewage. **Finally, as stated in section 8.4.2, the diameter of the downstream pipe cannot be smaller than the largest diameter of the upstream pipes.**

8.6 SEWAGE PUMPING STATIONS

When wastewater cannot be evacuated by gravity alone, a pumping station must be installed. For small flows, prefabricated units are often used, with a single well and submersible pumps. For larger flows, special pumping stations must be built on site. These stations are often equipped with two wells: a wet well, collecting wastewater, and a dry well, where the pumps are installed.

8.6.1 Pumps

To avoid development of septic conditions and production of bad odours, a wastewater should not remain at the pumping station for more than 30 minutes. The dimensions of the pumping station and the capacity of the pumps must be chosen accordingly. Even for the smallest stations, two pumps are installed, preferably with a fixed speed: each has a capacity equal to the maximum flow. When there are more than two pumps and the one with the greatest capacity breaks down, the others must be able to take over and pump the maximum flow.

To remain operational during a power outage, the pumping station should be equipped with gas motor pumps or emergency generators: the total capacity of this pumping equipment must equal to the maximum wastewater flow at the pumping station.

8.6.2 Evacuation of Excess Sewage Through a Weir

To avoid flooding a pumping station or wastewater backflow in sewer pipes, a weir is generally installed in a sewer manhole upstream from the station. This weir should have a capacity equal to the maximum flow at the station. Depending on available options, wastewater excess can be evacuated in to a stream, held in a reservoir, or disposed of in the manner least damaging to the environment.

8.6.3 Ventilation

Except in very small pumping stations, a forced-air ventilation system must be installed. It should be equipped with a manual start-up switch at the entrance to the station, and the air should be pushed from the outside in, to ensure sufficient air exchange: a) in wet wells, 6 air exchanges per hour during the winter and 12 during the summer; b) in dry wells and shelters, 3 air exchanges per hour.

8.7 INVERTED SIPHON

Figure 8.12 shows a structure used to transport wastewater under major unavoidable depressions (rivers, canals, etc.) without causing too much head loss. The weirs allow distribution of flow between pipes: when the small-diameter pipe runs full, the excess flows in to a pipe with an intermediate diameter, then in to the largest pipe. The flow velocity of the water in these pipes must be high enough for self-cleaning of the pipes. The velocity must therefore be 1 m/s for sanitary flows (minimum diameter of the pipes is 150 to 200 mm) and 1,5 m/s for combined and storm flows (minimum diameter of the pipes is 300 mm). Pipes flow under pressure, and the difference in elevation between the entrance and the exit of the siphon is calculated as a function of minor head losses and friction.

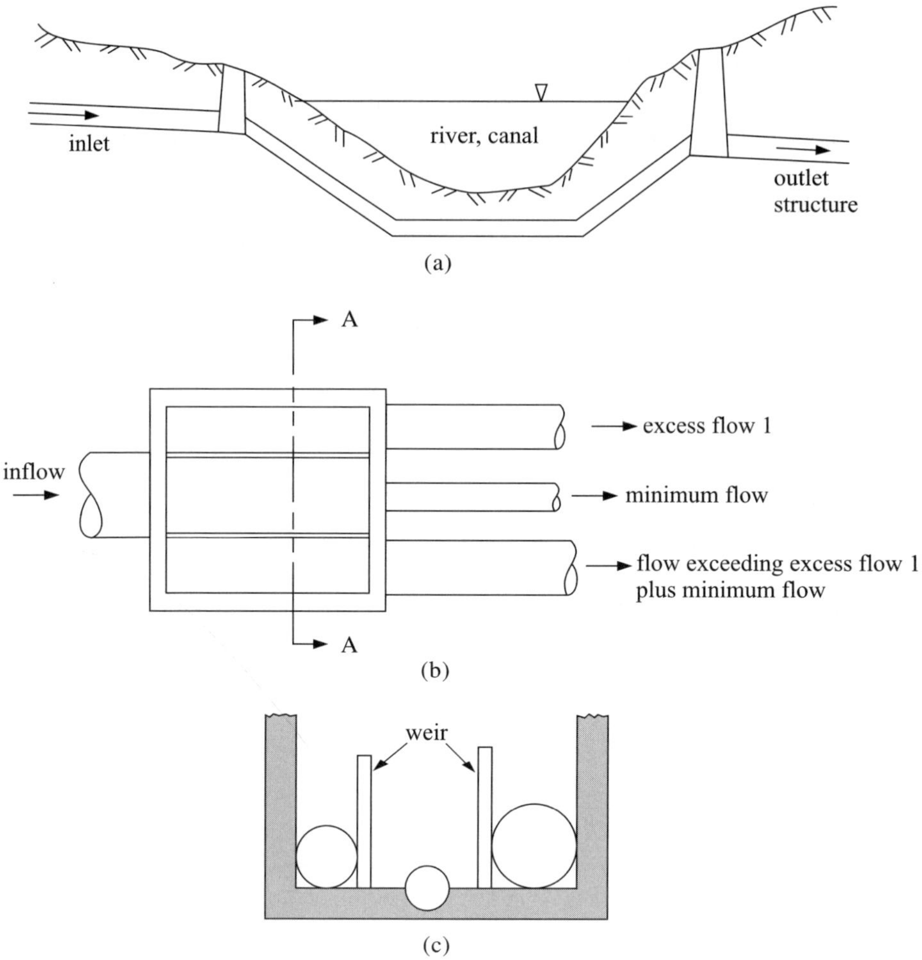

Figure 8.12 Inverted Siphon: a) plan; b) vertical section; c) section A-A: weir
From: W. Viessman Jr. and M. J. Hammer, *Water Supply and Pollution Control*, 5th ed. (Harper Collins, 1993), p. 219.

Example 8.1

An inverted siphon is made of a unique concrete pipe of inside diameter 355 mm. Its total length is 150 m and its Manning's roughness coefficient is 0,013. The design flow is 200 L/s (0,2 m^3/s). Compute the outlet elevation of this siphon if its inlet elevation is 100 m. Minor head losses are negligeable.

Solution

An inverted siphon flows under pressure and does not operate otherwise. Thus, the Manning equation may be used with hydraulic radius R_h replaced by $d/4$.

$$v = \frac{1}{n}\left(\frac{d}{4}\right)^{2/3} s^{1/2}$$

and

$$v = Q/A = 0{,}2 \div \pi\,(0{,}355)\,2/4$$
$$= 2{,}02 \text{ m/s}$$

so

$$s^{1/2} = v \times n \times (4/d)^{2/3}$$
$$= 2{,}02 \times 0{,}013 \times (4/0{,}355)^{2/3}$$
$$= 0{,}132$$

and

$$s = 0{,}017 \text{ m/m}$$

Thus, elevation of the siphon outlets, h_s, is:

$$h_s = 100 - 150 \times 0{,}017$$
$$= 97{,}45 \text{ m}$$

8.8 TUNNELS

In urban areas, a tunnel is often built when a major street crosses under an obstacle such as a railroad track. The major artery goes through the tunnel, but smaller streets do not and are constructed as dead ends. An inverted siphon (an expensive structure that requires maintenance) would not be built under the major artery; instead, the sewers would be diverted to neighbouring streets to go under the obstacle while maintaining adequate slopes to ensure flow by gravity. The water collected at the bottom of the tunnel would have to be pumped to storm or combined sewers, depending on the system used.

8.9 CULVERT

A culvert is a channel installed under a road or a railroad track and used to transport runoff from one side of the obstacle to the other. Its cross section is usually circular, rectangular, or arched. The runoff to be transported is a function of the size of the watershed and can be determined with the rational equation, as shown in chapter 7 (section 7.3.3). To avoid flooding on both sides on the road or railroad track, maximum water levels must not be exceeded (section 9.5).

8.10 PIPE INSTALLATION

Storm-sewers, sanitary-sewers, and water-distribution pipes serving houses on both sides on the street are often installed right under the pavement. The burying depth of sanitary sewer pipes follows the guidelines in section 8.4.7, and water-distribution pipes run parallel to the ground surface at sufficient depth to avoid frost (section 4.12.5). When streets are very wide, such as boulevards, one or all pipes may be doubled to facilitate connections to houses and fire hydrants.

During construction, a trench is dug deep enough to bury the deepest set of pipes, often the sanitary sewer pipes (section 8.4.7). After this first set of pipes is installed, the trench is filled up to the level of the second set of pipes, usually the storm-sewer pipes. At this level, a base is constructed and made wide enough to install the pipes. The operation is repeated to install the last set of pipes, those for water distribution (fig. 8.13). Chapter 9 describes how to determine the width of the bedding and the slopes of the trench walls.

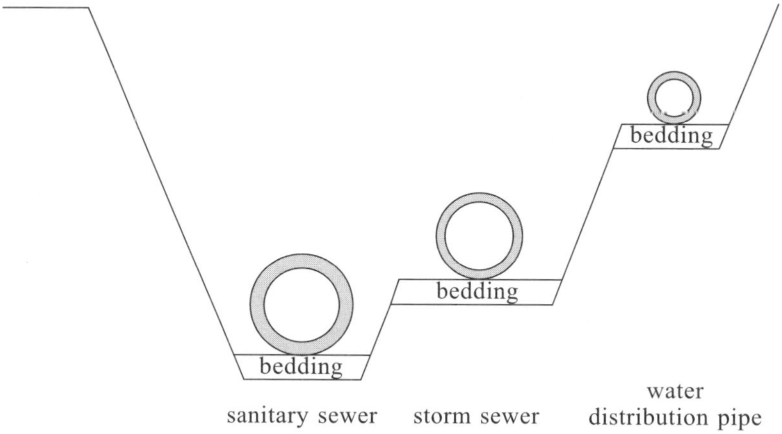

Figure 8.13 Installation of various pipes under the street

Pipe installation must also follow other guidelines. To avoid contamination of drinking water by wastewater, the water-distribution pipes must be located 300 mm distant (pipe wall to pipe wall), horizontally and vertically, from any sewer pipes (fig. 8.14). When a single trench is dug for all pipes and this requirement cannot be met (mainly in vertical distance), a separate trench is dug for the water-distribution pipes so that its horizontal distance to any sewer pipe is at least 3 m (pipe wall to pipe wall). In so doing there is an undisturbed core of natural soil between the pipes. This core need not reach the surface (see also section 9.4).

Other rules relate to the location of pipes so that backfill can be adequately compacted. Between all structures (pipes, sewer manholes, street inlets, etc.), a free distance of 600 mm (not necessarily horizontal) must be respected. Figure 8.15 illustrates this for installation of a sanitary sewer pipe parallel to a storm-sewer pipe.

Finally, at the crossing of two pipes, a minimum vertical distance of 300 mm (or 100 mm with protection of a rigid insulator between the pipes) must be allowed so the upper pipe does not crush the lower.

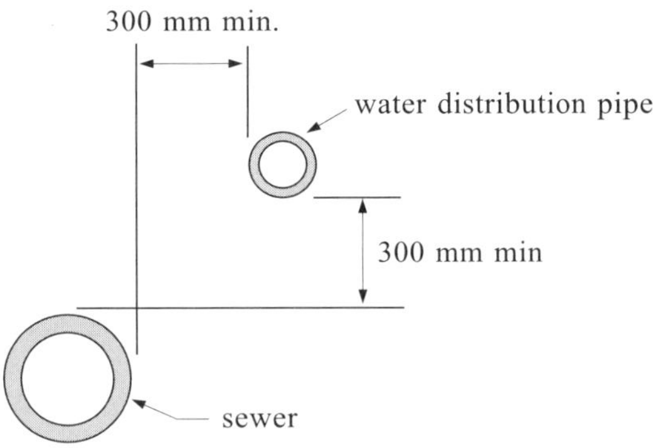

Figure 8.14 Water-distribution and sewer pipes in common trench

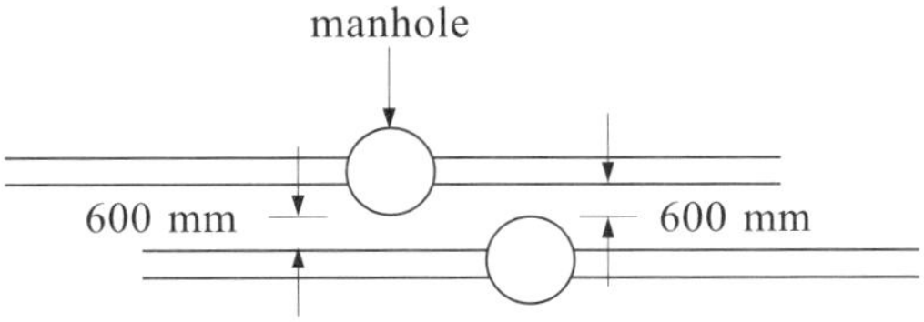

Figure 8.15 Free distance of 600 mm between two structures

8.11 SEWAGE FLOW MEASUREMENTS

Engineers must often measure sewage flows before designing and constructing sewers and interceptors, selecting pumps, designing pumping stations and estimating inflow/infiltration.

Continuous flow measurements are used at sewage treatment plants for control purposes as well as for computation of the quantity of chemical products required.

Instantaneous flow measurements. When data is needed only for exceptional events or as preliminary information, engineers rely on simple inexpensive tools:

- Measurement, with the help of a chronometer, of the time required to fill up a reservoir of known volume;
- Use of a portable rectangular or V notch weir or a combination of both;
- Measure of the flow velocity (rotating devices with helix or buckets – not appropriate for sewage as they may be easily obstructed by debris – electromagnetic sensor or doppler) and the flow section;
- Use of tracers : rhodamine VT, lithium salts, etc.

Temporary continuous flow measurements: from one week to many months. These measures are used to pick up information regarding sewage flow variations during the day, during and after rainfall or from day to day, and also to calibrate computer models like SWMM:

- Rectangular or V notch weir or a combination of both;
- Parshall or Palmer Bowlus flumes.

Permanent continuous flow measurements. Permanent measurement devices are installed at sewage treatment plants, at city limits or at industrial sewage effluents:

- Parshall flume is the most frequently used installation at sewage treatment plants;
- Weirs;
- Pumping stations : when pump capacity and useful volume are known and the duration of pump operations are monitered, it is possible to compute the flow through the station.

Registration devices. Most popular devices use bubbling, ultrasound and pressure.

Parshall flumes. The Parshall flume is one of the most successful structures developed for the measurements of flows (fig. 8.16). It has found considerable use in irrigation, water supply and sewage treatment. It is made of a converging section followed by a dropping throat section and a diverging section. The first two sections create critical flow conditions unless a hydraulic jump is formed by a high down-stream water level and the jump backs up into the throat section.

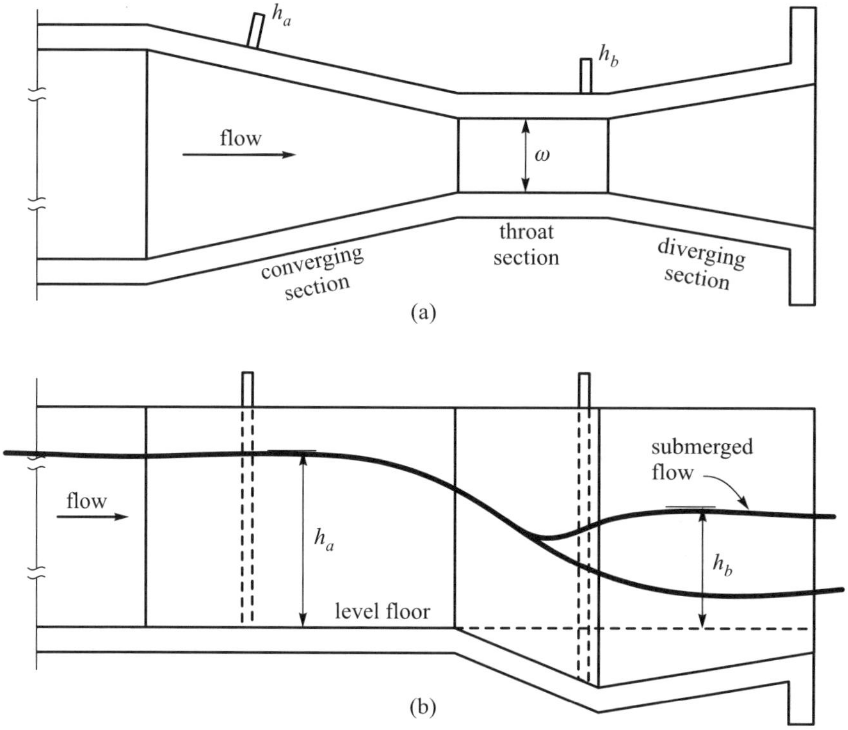

Figure 8.16 Parshall flume: a) plan; b) longitudinal section

Parshall, in 1920, developed equations to express the relationship between the flows in the flume and the depth of water, h_a (table 8.4). These equations are valid provided the depth of water, h_b, is not too high.

Table 8.4 Equations to compute discharges in various size Parshall flumes

Throat width, *w*		Equations	Conditions, h_b/h_a
ft	in		
0	3	$Q = 0{,}992(h_a)^{1{,}547}$	$\leq 0{,}6$
0	6	$Q = 2{,}06(h_a)^{1{,}58}$	$\leq 0{,}6$
0	9	$Q = 3{,}07(h_a)^{1{,}53}$	$\leq 0{,}6$
1 to 8	---	$Q = 4wh_a^{1{,}522\left(w^{0{,}026}\right)}$	$\leq 0{,}7$
10 to 50	---	$Q = (3{,}6875w + 2{,}5)h_a^{1{,}6}$	$\leq 0{,}8$

When h_b/h_a exceeds values shown in table 8.4, the flow is submerged. The effect of the submergence is to reduce the discharge compute with the value h_a. Thus the discharge computed with the above equation must be corrected by a negative quantity. The diagrams give the corrections for submergence for Parshall flumes of various sizes (fig. 8.17).

Corrections for the 1-ft flume are applied to larger flumes by multiplying the correction for the 1-ft flume by the appropriate factor in table 8.5.

Table 8.5 Correction to the 1-ft flume correction factor

Throat width, *w* (ft)	Correction factor
1	1,0
1,5	1,4
2	1,8
3	2,4
4	3,1
6	4,3
8	5,4

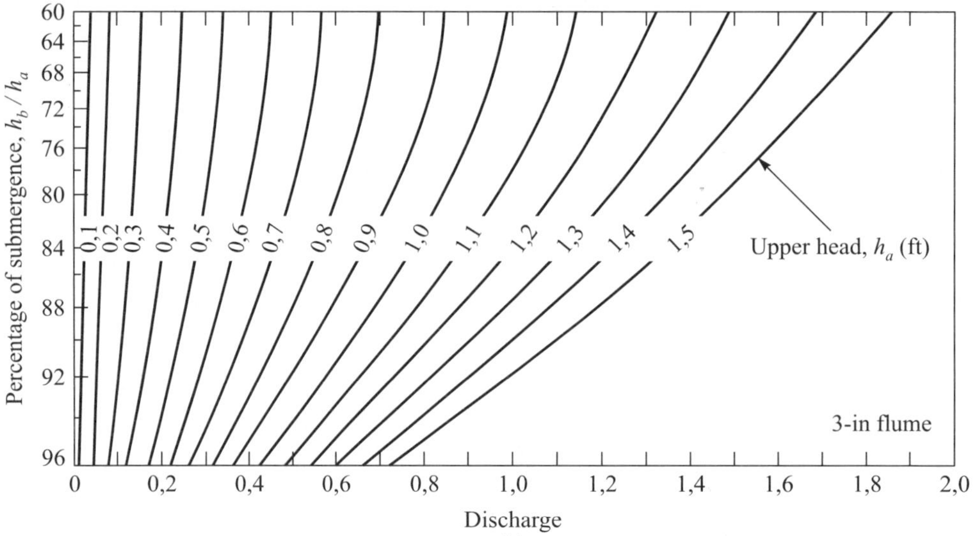

(a)

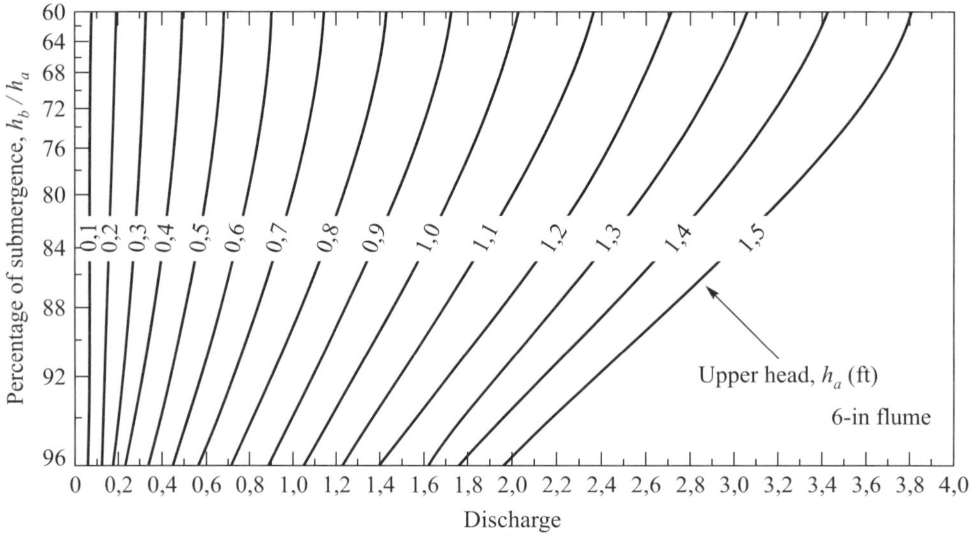

(b)

Figure 8.17 Diagrams for computing submerged flow through Parshall flumes of various sizes: a) $w = 3$ in; b) $w = 6$ in

Similarly, corrections for the 10-ft flume are applied to larger flumes by multiplying the correction for the 10-ft flume by the appropriate factor in table 8.6.

Table 8.6 Correction to the 10-ft flume correction factor

Throat width, *w* (ft)	Correction factor
10	1,0
12	1,2
15	1,5
20	2,0
25	2,5
30	3,0
40	4,0
50	5,0

Example 8.2

Hydraulic conditions in a 1 foot Parshall flume (w = 1 ft) are such that the depths of water, h_a and h_b, are 1,80 and 1,48 respectively. Compute the discharge.

Solution

$h_b/h_a = 1{,}48/1{,}8 = 0{,}82$ or 82%.

Since h_b/h_a is larger than 0,70 or 70%, the flow is submerged.

Discharge assuming no submergence:

$$Q = 4wh_a^{1{,}522\left(w^{0{,}026}\right)}$$

$$= 4 \times 1 \times (1{,}8)^{\left(1{,}522(1)^{0{,}026}\right)} = 4 \times 1 \times (2{,}446) = 9{,}79\ \text{ft}^3/\text{s}$$

Correction to take into account the submergence:

On the appropiate graph, one finds a correction of 1,22 ft^3/s.

Actual discharge through the Parshall flume:

$9{,}79 - 1{,}22 = 8{,}57$ ft^3/s

Palmer-Bowlus Flumes. The Palmer-Bowlus flume is essentially a portable restriction installed in a channel to produce a higher velocity critical flow in the throat (fig. 8.18). The flume is most often used in manholes or open round or rectangular bottom channels to measure flowrates. It is useful as a temporary installation to provide flow data. Unlike the Parshall flume, the dimensional configuration of a Palmer-Bowlus flume is not rigidly established for each flume size. Each flume must be calibrated by the manufacturer.

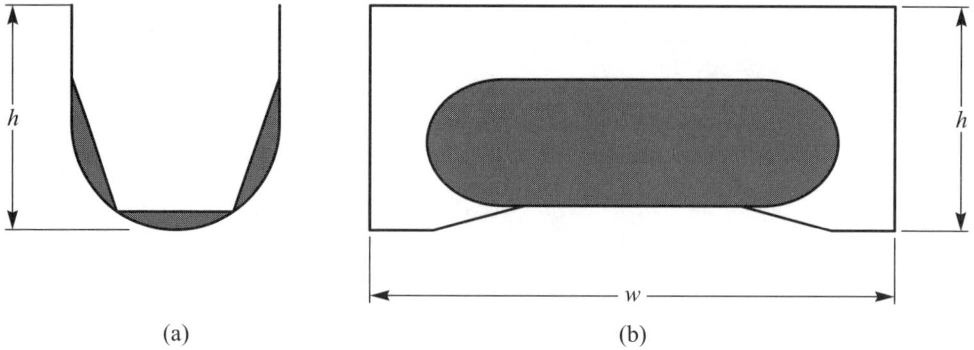

Figure 8.18 Palmer-Bowlus flumes: a) front view; b) longitudinal section

Example 8.3

In considering the construction of an interceptor, you are asked to compare maximum sanitary and runoff flows generated in a suburban residential area (suburban watershed) where all houses are multiple-family dwellings. The sewage network is entirely combined. Use the following data and information:

Surface of the watershed	165 ha
Ultimate population	25 575 (155/ha)
Maximum rainfall return period	10 years
IDF curves (Montreal)	Mitci
Runoff coefficient	0,7
Sewage flow velocity in the network pipes	2 m/s
Maximum travelling distance in the network pipes	2600 m
Inlet time	6 min
Infiltration	60 L/pers·d
Inflow	100 L/pers·d

Maximum runoff flow

Time of concentration, t_c:

The watershed is approximately equal to a 1300 m side square. The time of concentration, t_c, is thus equal to the sum of inlet time, t_e, and the flow time, t_f, over a maximum distance of the order of 2 × 1300 m = 2600 m.

$$t_c = t_e + t_f$$
$$= 6 \text{ min} + 2600 \text{ m} \div (2 \text{ m/s} \times 60 \text{ s/min}) = 27{,}6 \text{ min}$$

Maximum rainfall intensity, I, for a 10 year return period:

$$I = \frac{2743{,}2}{t_c + 14} = \frac{2743{,}2}{27{,}6 + 14} = 65{,}9 \text{ mm/h}$$

Maximum runoff flow, Q:

$$\begin{aligned} Q &= KAIR \\ &= 2{,}75 \times 10^{-3} \times 165 \text{ ha} \times 65{,}9 \text{ mm/h} \times 0{,}7 \\ &= 20{,}9 \text{ m}^3\text{/s} \end{aligned}$$

Maximum sanitary sewage flow (to be captured by the interceptor)

Average domestic sewage flow, $(Q_d)_{avg}$:

$$\begin{aligned} &= 25\,575 \text{ pers} \times 0{,}320 \text{ m}^3\text{/(pers·d)} \times (\text{d}/86\,400 \text{ s}) \\ &= 0{,}0947 \text{ m}^3\text{/s} \end{aligned}$$

Maximum peak factor, FP_{max}, for domestic sewage flow:

$$\begin{aligned} FP_{max} &= 1{,}742Q^{-0{,}1506} = 1{,}742 \times (0{,}0947)^{-0{,}1506} \\ &= 2{,}48 \end{aligned}$$

Maximum domestic sewage flow, $(Q_d)_{max}$:

$$(Q_d)_{max} = 0{,}0947 \times 2{,}48 = 0{,}235 \text{ m}^3\text{/s}$$

Infiltration:

$$\begin{aligned} &= 60 \text{ L/(pers·d)} \times 25\,575 \text{ pers} \times (\text{d}/86\,400 \text{ s}) \times 10^{-3} \text{ m}^3\text{/L} \\ &= 0{,}018 \text{ m}^3\text{/s} \end{aligned}$$

Inflow:

$$\begin{aligned} &= 100 \text{ L/(pers·d)} \times 25\,575 \text{ pers} \times (\text{d}/86\,400 \text{ s}) \times 10^{-3} \text{ m}^3\text{/L} \\ &= 0{,}03 \text{ m}^3\text{/s} \end{aligned}$$

Maximum sanitary sewage flow:

$$\begin{aligned} &= \text{maximum domestic sewage flow} + \text{infiltration} + \text{inflow} \\ &= 0{,}235 \text{ m}^3\text{/s} + 0{,}018 \text{ m}^3\text{/s} + 0{,}03 \text{ m}^3\text{/s} \\ &= 0{,}2830 \text{ m}^3\text{/s} \end{aligned}$$

This is the maximum sewage flow the interceptor will carry to the sewage treatment plant.

In summary

Maximum runoff flow: $20{,}9 \text{ m}^3\text{/s}$
Maximun sanitary flow: $0{,}2830 \text{ m}^3\text{/s}$

Example 8.4

In relation to example 8.3, you are asked to compare the annual runoff volume and the annual sanitary sewage volume. To this end, you must recall that in the Montreal area it falls an average of 100 cm of precipitation (runoff + snow). Evaluate the storm overflow volume.

Annual sanitary sewage volume:

= average sanitary sewage flow × 365 d/y
= (average domestic flow + infiltration + inflow) m^3/s × (365 × 24 × 60 × 60) s/y
= (0,0947 + 0,018 + 0,03) m^3/s × 31 560 000 s/y
= 4 500 187 m^3/y

This is the annual sewage volume the interceptor brings to the sewage treatment plant unless special measures, like the use of detention basins, are taken to increase it.

Annual runoff volume:

= height of precipitation × surface of the watershed × runoff coefficient
= 100 cm/y × 0,01 m/cm × (165 ha × 10 000 m^2 /ha) × 0,7
= 1 155 000 m^3/y

Annual storm overflow volume:

As a first approximation, the annual storm overflow volume is the difference between annual runoff volume and inflow admitted to the interceptor i.e. :

= 1 155 000 m^3/y – 100 L/(pers·d) × 25 575 pers × 365 d/y × 10^{-3} m^3/L
= 1 155 000 m^3/y – 933 487,5 m^3/y
= 221 512,5 m^3/y

Practically the annual storm overflow volume will be smaller since part of it is admitted to the interceptor when it does not carry the maximum sanitary flow.

In summary

Annual volume of sanitary sewage:	4 500 187 m^3/y
Annual volume of runoff:	1 155 000 m^3/y
Accepted inflow to the interceptor:	933 987,5 m^3/y
Excess runoff flow:	221 512,5 m^3/y

CONCLUSION

Sewer pipes must be designed and constructed according to engineering rules and practices. Table 8.7 gives a summary of the requirements related to sewer-pipe diameters and flow velocities in various types of sewers.

Except for pipes with very large diameters, pipe diameters are usually selected from what is readily available on the market, for obvious economic reasons. Pipe materials are chosen according to local conditions. Finally, at a sewer manhole, the diameter of the downstream pipe must never be smaller than the diameter of the upstream pipe and the burying depth of the downstream pipe crown (top) must never exceed the crown (top) of the upstream pipe.

Table 8.7 Engineering rules related to the diameters of sewer pipes and flow velocities of various types of sewers

Pipe	Nominal diameter		Flow velocity	
	minimum (mm)	maximum (mm)	minimum (m/s)	maximum (m/s)
Sanitary sewer	200	no limits	0,60 to 0,75	3
Storm sewer	300	no limits	0,9	3
Combined sewer	300	no limits	0,9	3

PROBLEMS

8.1 Three sewer pipes carry wastewater to the same sewer manhole. Their diameters and height of inner inverts are as follows:

$$d_1 = 305 \text{ mm}, \; h_1 = 125{,}50 \text{ m}$$

$$d_2 = 380 \text{ mm}, \; h_2 = 126{,}50 \text{ m}$$

$$d_3 = 455 \text{ mm}, \; h_3 = 125{,}50 \text{ m}$$

The wastewater brought by the three upstream pipes is evacuated from the manhole by a 610 mm pipe. Determine the maximum height at which the invert of the downstream pipe must be installed.

8.2 Three sewer pipes carry wastewater to the same sewer manhole. Their diameters and height of inner inverts are as follows:

$$d_1 = 305 \text{ mm}, \; h_1 = 30{,}34 \text{ m}$$

$$d_2 = 455 \text{ mm}, \; h_2 = 30{,}79 \text{ m}$$

$$d_3 = 380 \text{ mm}, \; h_3 = 30{,}18 \text{ m}$$

A 610 mm pipe evacuates from the manhole the wastewater brought by the three upstream pipes. Determine the maximum height at which the inner invert of the downstream pipe must be installed.

8.3 A storm-sewer pipe with an inner diameter of 915 mm, transporting a flow of 0,7 m^3/s at a velocity of 0,92 m/s, must cross under a ground depression using an inverted siphon. Compute the minimum inside diameter of the pipe necessary to carry this flow, if a single pipe is used for the siphon. The pipe is made of concrete (Manning's n = 0,013) and has a total length of 150 m. Determine the head loss. The maximum velocity of the water in the pipe is 3 m/s. (The answer is not limited to commercial diameters and minor head losses can be neglected.)

8.4 A culvert must be built under a road in the Montreal region. The watershed feeding this culvert includes 5 000 m^2 of paved roads (R = 0,8) and 0,2 km^2 of grassy sandy soil (R = 0,1), with an average slope toward the culvert of 10%. If the inlet time is 26 min, determine the flow the culvert must be able to carry if we assume a return period of 10 years.

8.5 A single-family house is located in an urban residential area and is connected to a pseudo-sanitary sewer system.

a) What is a pseudo-sanitary sewer system? When is this type of sewer system used?

b) How many service connections are there for this house? Name them.

c) What type of water (from various sources in, on, and around the house) is transported by the service connections of this house?

d) How many sewer pipes are located under the street in front of the house? Name and identify the types of water transported by each pipe.

Chapter 9

HYDRAULIC DESIGN OF SANITARY, STORM, AND COMBINED SEWER SYSTEMS

INTRODUCTION

In previous chapters we discussed various topics: hydrology, flows in pipes, evacuation of storm and domestic wastewater flows, and maximum and minimum peak factors. This chapter demonstrates, through theory and practical applications, how engineers use this knowledge to compute sewer-pipe sizes, determine their underground installation and fix the location of street inlets.

9.1 HYDRAULIC DESIGN OF STORM-SEWER SYSTEMS

To determine the dimensions of a storm sewer pipe, engineers need the following information:

- the dimensions of the tributary watershed;
- the characteristics of Intensity-Duration-Frequency curves (IDF) or equations in the watershed region;
- the characteristics of the watershed's surface, to evaluate its runoff coefficient;
- the average slope of the watershed toward its discharge;
- data to evaluate the longest path for runoff water to reach the watershed's discharge (to reach the street inlet, in the case of an urban sub-basin);
- data and equations required to calculate the longest time runoff water takes to reach the discharge;
- engineering practices to determine the choice of sewer pipe dimensions and to compute the flow velocities in these pipes.

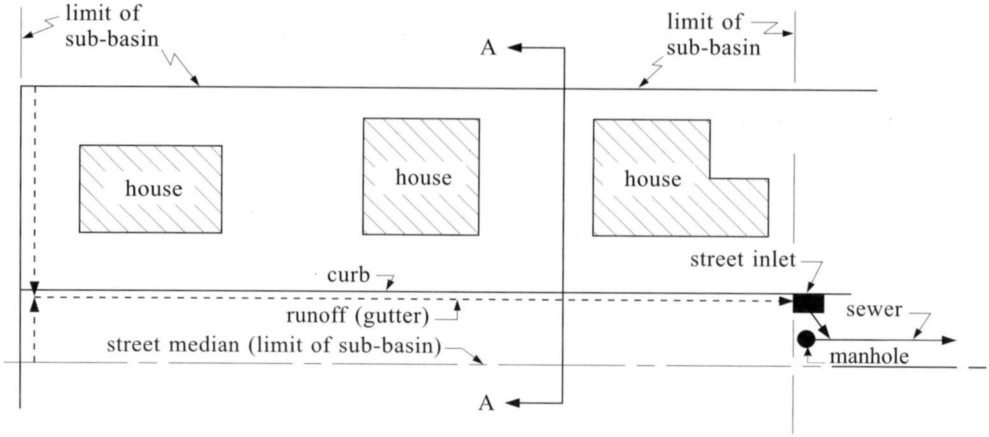

(a)

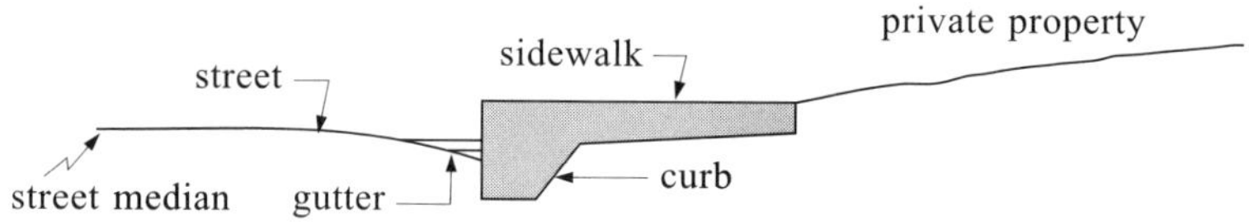

Section A-A

(b)

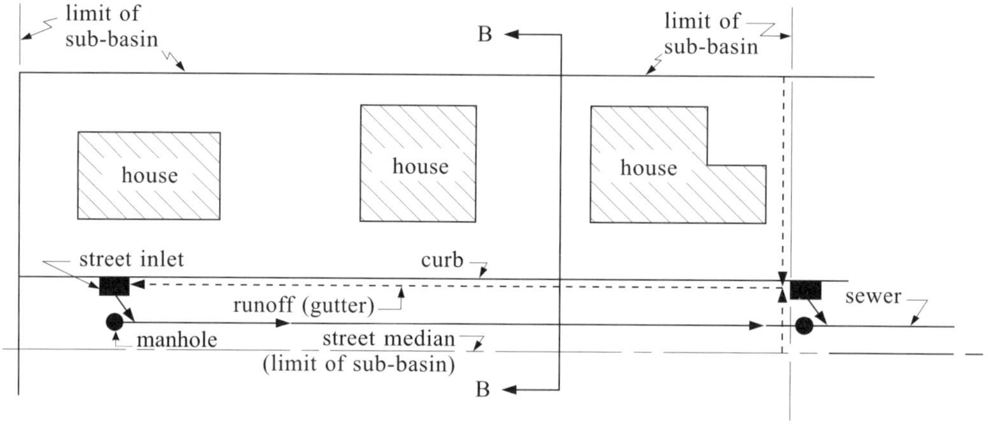

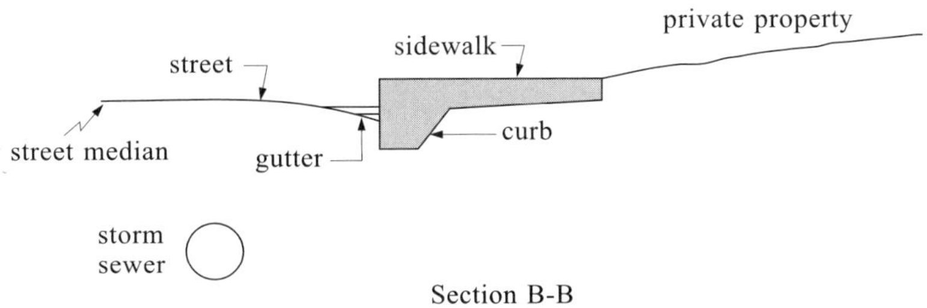

Section B-B

(d)

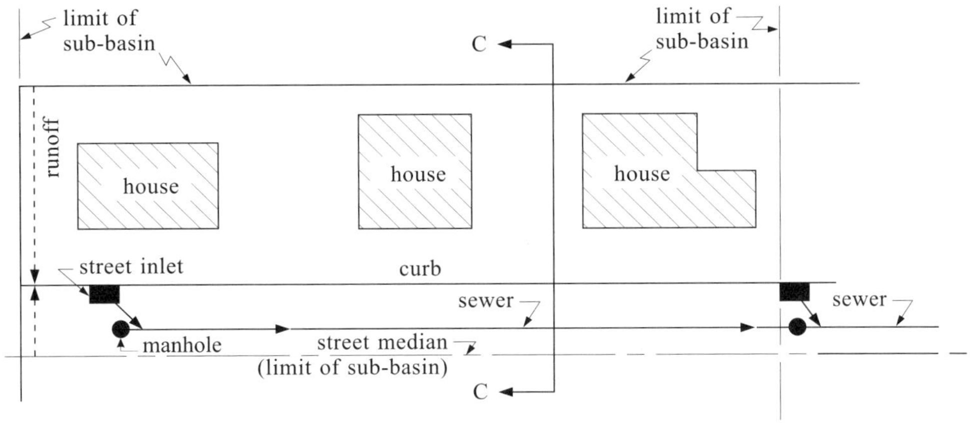

(e)

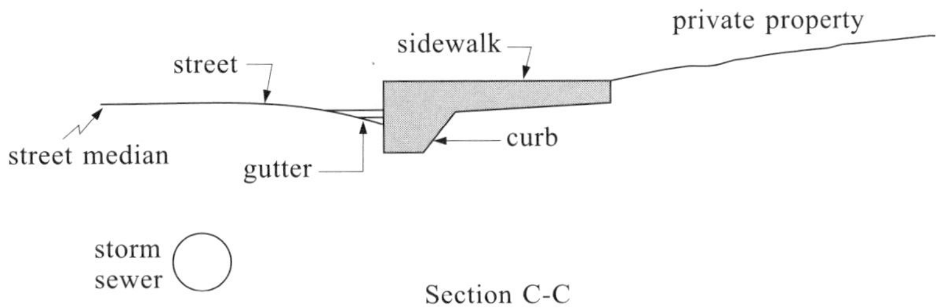

Section C-C

(f)

Figure 9.1 Computation of runoff in an urban sub-basin: a) and b) theoretical approach; c) and d) initial practical approach; e) and f) final practical approach: runoff paths for computation of inlet time

9.1.1 Computation of the Maximum Flowrate at the Discharge of an Urban Sub-basin

As seen in chapter 7, the smallest urban watershed, the urban sub-basin (for example, in a residential district), is made of a few lots and half a street section and discharges its runoff into a street inlet. Figures 9.1a and 9.1b show this type of sub-basin. It is made of three lots and includes half the width of the street in front of three lots. In this case, the water usually runs from the back of the lots toward the street (perpendicular to the street), then along the curb to the street inlet, which is connected directly to the storm-sewer pipe. When the average slope of the lots, the slope of the street, and values applicable to the equations presented in chapter 7 are known, engineers can calculate the inlet time of the sub-basin. Frequently, an identical sub-basin is located in front of it. This is also made of a few lots plus the other half of the street section. Its water is discharged into another street inlet, often located in front, along the curb on the opposite side of the street. Therefore, the sewer pipe under the street, which collects the water from both street inlets, must be able to accommodate maximum flows from two urban sub-basins.

In practice, a sewer pipe must also be installed in front of every house in order to link storm-sewer connections; thus, it is useful and hydraulically safe to consider runoff from an urban sub-basin to be captured at the head of the sub-basin (figs. 9.1c and 9.1d). The engineer uses another simplification, also hydraulically safe, by considering that the inlet time equals the inlet time of the runoff in figures 9.1e and 9.1f – in other words, the inlet time of a water droplet running on a lot located at the head of the sub-basin, perpendicularly to the street and up to the curb, the flow time along the curb not being taken into account (note that this inlet time is the same as in the previous case but without the flow time along the curb). **As we will see later (sections 9.1.2 and 9.3), the sizes of storm-sewer pipes are determined without considering the location of street inlets: for design purposes, runoff is assumed to enter the sewer from the upstream sewer manhole. The flow time on the sidewalk, if there is one, is neglected.**

N.B. It is hydraulically safe to make the previous assumptions because: a) in the first case, it is assumed that all sub-basin runoff enters the sewer at it's upstream inlet (even manhole) rather than being caught by the sewer pipe; b) in the second case, the inlet time is smaller than the actual inlet time, leading to a larger runoff when computations are made with the rational equation (smaller inlet time implies larger rain intensity and leads to larger runoff); c) in both cases, the computations lead to a larger runoff, meaning the chosen pipe will have a larger capacity than required. However, in most situations the minimum size pipe (in Quebec 300 mm diameter) still has a larger hydraulic capacity. This is why the practicing engineers are in the habit of simplifying its computations as previously indicated.

9.1.2 Design of a Reinforced Concrete Sewer System Receiving Storm Flows from Many Urban Sub-basins

Figure 9.2 is a schematic representation of a small urban watershed in the Montreal region, for which we must determine the dimensions of the various pipes. Table 9.1 lists the characteristics of the sub-basins forming this watershed and the lengths of pipes to be installed in each one.

Initial considerations and engineering practices. The diameters of various pipes in the storm-sewer system are calculated by taking into account the following rules and information (some of which were presented in chapter 8):

- Following engineering practices, the diameter of a pipe located downstream from a sewer manhole is never smaller than the diameter of the pipes located upstream (section 8.4.2).
- For sewer manholes collecting wastewater from two or three pipes, the top (interior) of the downstream pipe – the one evacuating the wastewater – is *never* located higher than the lowest top (interior) of the upstream pipes (section 8.5.3).
- To avoid freezing of the pipes and to drain the basements of houses, the pipe top must be located at least 2,0 m beneath the ground surface (in the Montreal region, a 1,85 m cover is sufficient to avoid freezing of wastewater and drinking-water pipes).
- The amount of material extracted from the ground during construction of the pipes must be minimized; thus, pipes should not be buried too deep and should be installed, when possible, parallel to the ground surface.
- It is forbidden to use a Manning's roughness coefficient smaller than 0,013, the coefficient of concrete (because of wear, the Manning's roughness coefficient of other materials – asbestos cement and thermoplastic in particular – tends towards the coefficient of concrete).
- We use the precipitation curves (IDF curves) of the region where the pipes must be installed and a minimum return period of 5 years (for a combined sewer system, a minimum return period of 10 years is suggested).
- The minimum and maximum flow velocities in the pipes must be 0,9 and 3,0 m/s, respectively, when the pipes are flowing full. In the upstream pipes of the system, flows are very low and it is suggested that a minimum slope of 0,005 m/m be used for the first 100 m of pipes. This ensures self-cleaning because the velocity is then 1,0 m/s in 300 mm diameter (min diameter) pipes with a Manning's roughness coefficient of 0,013 when they are flowing full.
- The pipes of the sanitary and storm-sewer systems are not necessarily parallel and their slopes are often different; it may even be that on the same street, storm water flows in one direction and sanitary wastewater flows in the other.

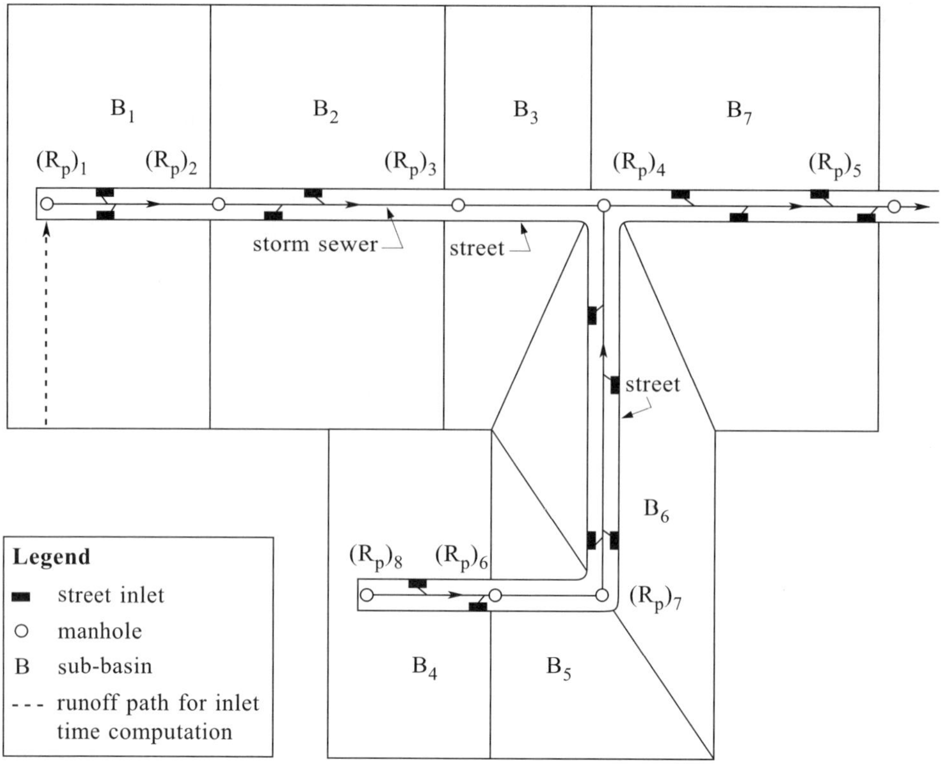

(a)

Figure 9.2 Watershed made of seven sub-basins: a) storm sewer; b) sanitary sewer

Table 9.1a Watershed shown in figure 9.2: characteristics of sub-basins

Parameter	**Sub-basin**						
	B_1	**B_2**	**B_3**	**B_4**	**B_5**	**B_6**	**B_7**
Area (ha)	1,2	2,2	2,1	1,5	1,0	2,4	4,2
Inlet time (min)	8,5	9,0	9,0	7,0	7,4	9,0	10,0
Runoff coefficient (R_T)	0,45	0,48	0,46	0,70	0,50	0,65	0,50

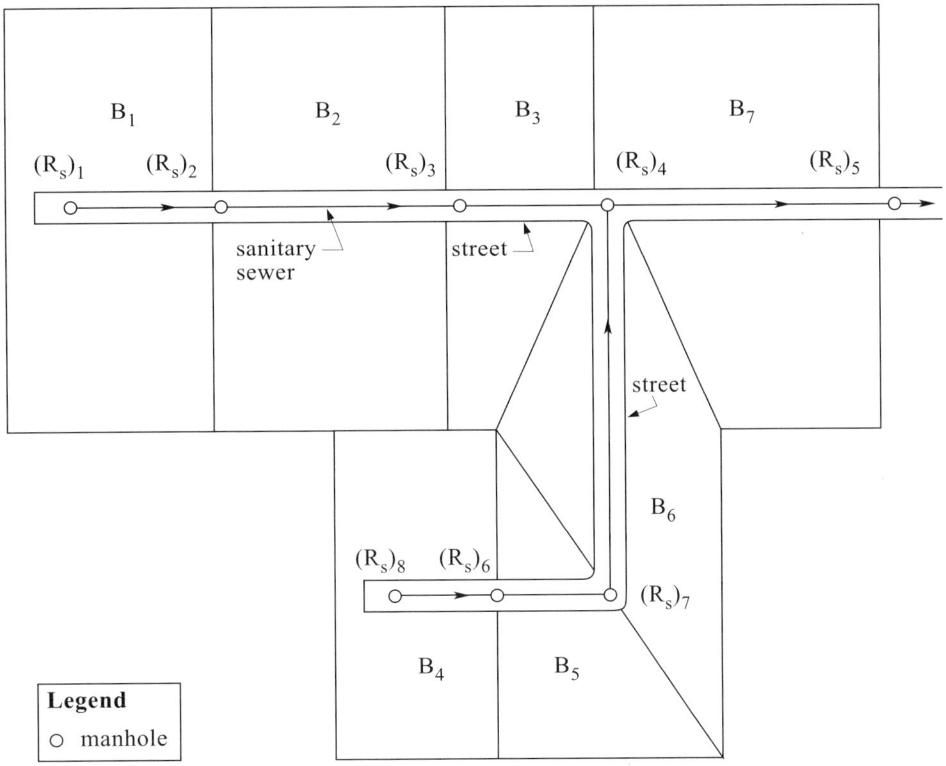

(b)

Figure 9.2 (continued)

Table 9.1b Watershed shown in figure 9.2: characteristics of the ground and pipes

Sewer line	Ground level (m)		Length (m)	Slope of natural ground in the direction of flow (m/m)
	upstream	downstream		
$(R_p)_1$–$(R_p)_2$	113,03	111,50	90	0,017
$(R_p)_2$–$(R_p)_3$	111,50	109,00	100	0,025
$(R_p)_3$–$(R_p)_4$	109,00	107,60	100	0,014
$(R_p)_8$–$(R_p)_6$	109,19	108,00	70	0,017
$(R_p)_6$–$(R_p)_7$	108,00	107,51	70	0,007
$(R_p)_7$–$(R_p)_4$	107,51	107,60	100	-0,00 8
$(R_p)_4$–$(R_p)_5$	107,60	107,20	125	0,003 2

Table 9.2 Computations for a reinforced-concrete storm-sewer system, assuming that the runoff coefficients, R_T, are constant

Computation of flow									
Sewer line (between manholes)	Time of concentration t_c (min)	Increment of tributary area (ΔA) ΔA (ha)	Runoff coefficient of (ΔA) R_T	Equivalent impermeable area of ΔA ΔAR_T (ha)	Cumulative area A (ha)	Cumulative impermeable area AR (ha)	Runoff coefficient of tributary area A R	Intensity of precipitation I (mm/h)	Maximum flow $2{,}75 \times 10^{-3}\ AIR$ Q_{max} (m^3/s)
$(R_p)_1$–$(R_p)_2$	8,50	1,2	0,45	0,54	1,2	0,54	0,45	106,56	0,158
$(R_p)_2$–$(R_p)_3$	9,22	2,2	0,48	1,06	3,4	1,60	0,47	102,94	0,451
$(R_p)_3$–$(R_p)_4$	9,80	2,1	0,46	0,97	5,5	2,57	0,47	100,20	0,710
$(R_p)_8$–$(R_p)_6$	7,00	1,5	0,70	1,05	1,5	1,05	0,70	114,97	0,332
$(R_p)_6$–$(R_p)_7$	7,49	1,0	0,50	0,50	2,5	1,55	0,62	112,08	0,478
$(R_p)_7$–$(R_p)_4$	9,00	2,4	0,65	1,56	4,9	3,11	0,64	104,02	0,897
$(R_p)_4$–$(R_p)_5$	10,44	4,2	0,50	2,10	14,6	7,78	0,53	97,34	2,083

Computation of flows and choice of pipe diameters. A number of steps are involved in computing the flows to be evacuated through different sewer lines and determining the slopes and diameters of pipes required for this evacuation. This section presents a computation method using a grid, like the one presented in table 9.2.

NOTE: In practice, the engineer installs street inlets only after designing storm-sewer pipes. Moreover, as an initial estimate, the engineer considers increasing the diameter of the pipes every 100 to 200 m for diameters smaller than 750 mm (30 in), and every 300 m for diameters between 750 and 1 200 mm (48 in).

Pipe $(R_p)_1$–$(R_p)_2$: Computation of the maximum storm flow to be evacuated. Pipe $(R_p)_1$–$(R_p)_2$ must be able to evacuate the maximum flow originating from sub-basin B_1, for which the time of concentration (also the inlet time in this case) is 8,5 min. Let us now explain how to find values related to this sewer line and presented in table 9.2.

t_c: time of concentration of sub-basin B_1, time also equal to inlet time, t_e. In table 9.1, we find that t_e = 8,5 min.

Table 9.2 (continued)

Computation of sewer characteristics						**Ground elevations**		**Inner invert elevations (minimum cover of 2 m)**	
Slope of sewer s (m/m)	Nominal diameter of sewer d_{nom}* (mm)	Flow velocity in sewer flowing full v_f (m/s)	Length of sewer line L (m)	Flowing time in sewer line - pipe flowing full t_f (min)	Hydraulic capacity of sewer flowing full Q_f (m^3/s)	upstream (m)	downstream (m)	upstream (m)	downstream (m)
0,017	375	2,09	90	0,72	0,237	113,03	111,50	110,65	109,12
0,025	450	2,86	100	0,58	0,464	111,50	109,00	109,05	106,55
0,014	600	2,60	100	0,64	0,759	109,00	107,60	106,39	104,99
0,017	450	2,36	70	0,49	0,383	109,19	109,19	106,74	105,55
0,007	600	1,84	70	0,63	0,537	108,00	108,00	105,39	104,90
0,025	900	1,44	110	1,27	0,946	107,51	107,51	104,60	104,33
0,007	1050	2,67	125	0,78	2,403	107,60	107,60	104,18	103,31

* All computations made with actual diameters (table 8.1).

ΔA: area of the watershed drained by the sewer line under study [area where pipe $(R_p)_1$–$(R_p)_2$ is located]. In this case, ΔA is equal to the area of sub-basin B_1, thus 1,2 ha.

R_T: runoff coefficient of area ΔA, thus 0,45 (table 9.1a).

ΔAR_T: equivalent impermeable area (of sub-basin B_1) supplying runoff in the sewer line in question plus the equivalent impermeable area(s) supplying runoff in the upstream sewer line(s). The equivalent impermeable area is the impermeable area supplying the same runoff as the area it represents. Here, $\Delta AR_T = 0{,}54$ (= 1,2 × 0,45).

A: area of the watershed supplying runoff to the sewer line in question – that is, the sum of the areas supplying runoff to the upstream sewer line(s) and of ΔA. In this case, $A = (\Delta A)_{B_1} = 1{,}2$ ha.

AR: total equivalent impermeable area supplying runoff to the sewer line in question. Here, $AR = \Sigma \Delta AR_T = 0{,}54$ (= 1,2 × 0,45).

R: runoff coefficient of the entire area of the watershed supplying runoff to the sewer line in question. Here, $R = AR \div A$, thus 0,45 (= 0,54 ÷ 1,2).

I: intensity of the precipitation, computed with the equation of the appropriate IDF curve. In the Montreal region, for a rainfall with a return period of 5 years, the equation proposed by Mitci (table 7.3) is used, hence:

$$I = \frac{2184{,}4}{t_c + 12} = \frac{2184{,}4}{8{,}5 + 12}$$
$$= 106{,}56 \text{ mm/h}$$

According to the assumptions presented in chapter 7 on computation of the maximum flow using the rational equation, we looked for the precipitation intensity of a rainfall of duration equal to the time of concentration of the watershed tributary to the sewer line in question, thus 8,5 min.

Q_{max}: flow calculated with the rational equation (equation 7.3):

$$Q_{max} = 2{,}75 \times 10^{-3} AIR$$
$$= 2{,}75 \times 10^{-3} \times 1{,}2 \times 106{,}56 \times 0{,}45$$
$$= 0{,}158 \text{ m}^3/\text{s}$$

Pipe*$(R_p)_1$–$(R_p)_2$: *Computation of the diameter and slope of the pipe. The following procedure describes the steps involved in choosing the diameter, slope, and elevation of a pipe carrying the flow calculated above.

s: slope of the sewer pipe. As an initial estimate, the engineer chooses a slope equal to the slope of the ground (0,017 for the present case); if this value is inappropriate, it must be changed (often, the value is changed to account for the minimum and maximum velocities discussed in section 8.4.8).

d: diameter of the considered sewer pipe. For the first sewer line, the smallest diameter accepted in engineering practices is usually chosen as an initial estimate, hence 300 mm for storm pipe. For the present case, a 300 mm pipe is inadequate: checking of this statement is left to the reader; therefore, a diameter of 375 mm is chosen.

v_f: flow velocity of runoff in a sewer of diameter d and of slope s flowing full (section 8.4.8); when the Manning roughness coefficient is known, this value can be obtained using the Manning equation or directly, using the Manning nomograph; for the present case, v_f = 2,09 m/s (computation with actual diameter).

L: length of the sewer line in question; here, L = 90 m.

t_f: time taken by wastewater to flow through the considered sewer line when the pipe is flowing full ($t_f = L/v_f$); here, t_f = 0,72 min.

Note: In most cases, the maximum design flow lies somewhere between 50% and 100% of pipe capacity. Thus actual flow velocity is always greater than 100% flow velocity when pipe is flowing full (v_f). This means that actual t_f is smaller than calculated. However, it is assumed that the error on the computed runoff flow is negligible. If in doubt, the engineer should make a more precise computation with the actual t_f.

Q_f: flow carried by the sewer pipe $(R_p)_1-(R_p)_2$ of diameter d and slope s when it is flowing full (the velocity v_f times the inside cross section of the pipe; Q_f can also be obtained using the appropriate Manning nomograph). If flow Q_f is smaller than flow Q_{max}, a modification must be made; the diameter and/or the slope must be increased, a pipe with a smaller Manning coefficient may be chosen, if it is permitted, and so on. The chosen pipe has a hydraulic capacity Q_f of 0,237 m^3/s, larger than the computed maximum flow Q_{max} (0,158 m^3/s).

Pipe $(R_p)_1-(R_p)_2$: Elevations of the inverts at the pipe's ends. A contractor installing a sewer pipe with the previously computed characteristics needs to know the elevations of the inner inverts at each end of the pipeline. These elevations are important for hydraulic reasons. The values are obtained by subtracting from the ground level over the pipe the burying depth (here 2 m) and the inner required diameter of the pipe (section 8.4.7). This computation is accurate only if the slope of the pipe is the same as the slope of the ground level. When they differ, the computation uses only the slope of the pipe [see pipe $(R_p)7-(R_p)4$].

Pipe $(R_p)_2-(R_p)_3$: Computation of the maximum storm flow. Pipe $(R_p)_2-(R_p)_3$ must be able to carry the maximum flow originating from a watershed made of sub-basins B_1 and B_2. To compute this, the time of concentration of the watershed (sum of sub-basins B_1 and B_2) is evaluated at sewer manhole $(R_p)_2$ (the discharge of the watershed under study), as required by the rational equation. Two values must be compared, the inlet time of sub-basin B_2 (9,0 min) and the time of concentration of sub-basin B_1, which, at the sewer manhole $(R_p)_2$, is equal to the sum of its inlet time and the flow time in pipe $(R_p)_1-(R_p)_2$ (8,5 min + 0,72 min = 9,22 min). The greater of the two values is the required time of concentration. Then, the maximum flow is calculated using the rational equation.

The following procedure shows the steps necessary to compute the maximum storm flow.

t_c: time of concentration of the watershed (sub-basin B_1 + sub-basin B_2) tributary to the sewer line $(R_p)_2-(R_p)_3$, time equal to the sum of the inlet time, t_e, of sub-basin B_1 and flow time of the wastewater, t_f, in pipe $(R_p)_1-(R_p)_2$ (see previous step); it therefore yields:

$$\begin{aligned} t_c &= t_e + t_f \\ &= 8{,}5 + 0{,}72 \\ &= 9{,}22 \text{ min} \end{aligned}$$

ΔA: increment of tributary area to be considered for sewer line $(R_p)_2-(R_p)_3$ [area where $(R_p)_2-(R_p)_3$ is located]. Area ΔA is therefore equal to the area of sub-basin B_2, thus 2,2 ha.

R_T: runoff coefficient of the area ΔA of sub-basin B_2, hence 0,48.

ΔAR_T: equivalent impermeable area (of B_2) tributary to the considered sewer line in addition to the equivalent impermeable area(s) tributary to the sewer line(s) located upstream, hence 1,056 (= 2,2 × 0,48).

A: total area of the watershed tributary to sewer line $(R_p)_2$–$(R_p)_3$, thus A = 3,4 ha (= 1,2 + 2,2).

AR: total equivalent impermeable area tributary to sewer line $(R_p)_2$–$(R_p)_3$, that is, the sum of ΔAR_T of the above sewer line and ΔAR_T of the considered sewer line; AR = 1,596 ha (= 0,54 + 1,056), thus ≅ 1,60 ha.

R: runoff coefficient of the total area tributary to the considered sewer line; $R = AR \div A$ = 0,469 (= 1,596 ÷ 3,4),

I: intensity of precipitation with a duration equal to the time of concentration of the watershed, considering a return period of 5 years, hence:

$$I = \frac{2184{,}4}{t_c + 12} = \frac{2184{,}4}{9{,}22 + 12}$$
$$= 102{,}94 \text{ mm/h}$$

Q_{max}: flow computed using the rational equation; it yields:

$$Q_{max} = 2{,}75 \times 10^{-3} AIR$$
$$= 2{,}75 \times 10^{-3} \times 3{,}4 \times 102{,}94 \times 0{,}469$$
$$= 0{,}451 \text{ m}^3/\text{s}$$

Pipe $(R_p)_2$–$(R_p)_3$: computation of the diameter and slope of the pipe. The following procedure is used to choose the diameter, slope, and elevation of the pipe $(R_p)_2$–$(R_p)_3$ which carries the flow calculated above.

s: slope of the sewer pipe. As an initial estimate, the engineer chooses a slope equal to the slope of the ground surface; in the present case, s = 0,025.

d: diameter of the considered sewer line $(R_p)_2$–$(R_p)_3$. According to engineering practices, this diameter cannot be smaller than the diameter of the above pipe line $(R_p)_1$–$(R_p)_2$; in this case, a pipe of d = 450 mm is required to meet identified needs.

v_f: flow velocity of wastewater in pipe $(R_p)_2$–$(R_p)_3$ when flowing full; in this case, v_f = 2,86 m/s.

L: length of the considered sewer pipe $(R_p)_2$–$(R_p)_3$; here, L = 100 m.

t_f: time required for wastewater to flow through the sewer line $(R_p)_2$–$(R_p)_3$ when it is flowing full ($t_f = L/v_f$); here, t_f = 0,58 min.

Q_f: flow carried by the sewer pipe $(R_p)_2$–$(R_p)_3$ of diameter d and slope s when flowing full; for the chosen pipe, Q_f = 0,464 m³/s, which is greater than the computed maximum flow Q_{max}.

Then, the elevations of the pipe are chosen using the same procedure as for pipe $(R_p)_1$–$(R_p)_2$.

Pipe $(R_p)_3$–$(R_p)_4$. The characteristics of this pipe are computed the same way as for pipe $(R_p)_2$–$(R_p)_3$.

Pipe $(R_p)_8$–$(R_p)_6$. The characteristics of this pipe are computed in the same manner as for pipe $(R_p)_1$–$(R_p)_2$; pipe $(R_p)_8$–$(R_p)_6$ collects the water from an urban sub-basin.

Pipe $(R_p)_6$–$(R_p)_7$. The characteristics of this pipe are calculated the same way as for pipe $(R_p)_2$–$(R_p)_3$.

Pipe $(R_p)_7$–$(R_p)_4$. In this case, the inlet time in sub-basin B_6 (9,0 min) is greater than the sum of the inlet time of sub-basin B_4 and the flow time of the wastewater in pipes $(R_p)_8$–$(R_p)_6$ and $(R_p)_6$–$(R_p)_7$, that is, 8,12 min (= 7,0 + 0,49 + 0,63). Therefore, the entire watershed contributes to runoff only after 9,0 min, and it is only at this time that pipe $(R_p)_7$–$(R_p)_4$ collects the maximum flow, that is, 0,890 m^3/s. After 8,12 min of rainfall, the proportion of the watershed's area contributing to runoff is only:

$$\begin{aligned} A &= \frac{8{,}12}{9{,}0} \times 4{,}9 \\ &= 4{,}42 \text{ ha} \end{aligned}$$

where 4,9 is the sum of the surface areas of sub-basins B_4, B_5, and B_6.

The intensity of the precipitation would be:

$$\begin{aligned} I &= \frac{2184{,}4}{8{,}12 + 12} \\ &= 108{,}57 \text{ mm/h} \end{aligned}$$

The runoff would then be only:

$$\begin{aligned} Q &= 2{,}75 \times 10^{-3} \times 4{,}42 \times 108{,}57 \times 0{,}635 \\ &= 0{,}84 \text{ m}^3/\text{s} \left(< 0{,}890 \text{ m}^3/\text{s}\right) \end{aligned}$$

NOTE*:* It was assumed here that the runoff coefficient of the watershed's surface contributing to runoff was equal to the runoff coefficient of the watershed formed by B_4, B_5, and B_6; if this were not the case, additional computations would be required.

In this case, the pipe slope must differ from the slope of the ground surface, which rises in the direction of flow. The pipe slope, *s*, is positive, but the slope of the ground surface is negative. The elevation, *h*, of the inner invert at sewer manhole $(R_p)_4$ is then calculated as follows:

$$\begin{aligned} h \text{ of inner invert at sewer manhole } (R_p)_4 &= h \text{ of inner invert at sewer manhole } (R_p)_7 - s \times L \\ &= 104{,}60 - 0{,}0025 \times 110 \\ &= 104{,}33 \text{ m} \end{aligned}$$

Pipe $(R_p)_4$–$(R_p)_5$. As seen in chapter 7, the maximum flow at the discharge of a watershed is generally the flow caused by a rainfall as long as the time of concentration of the watershed, since then the entire watershed's surface contributes to runoff. However, there are exceptions to this rule when the hypotheses of the rational equation are not met (irregularly shaped watershed, surfaces with runoff coefficients distributed unevenly, etc.). These exceptions must be identified to avoid installing pipes that are too small, therefore causing flooding and inviting lawsuits. The engineer now has the information necessary to compute the maximum flow carried by pipe $(R_p)_4$–$(R_p)_5$, thus:

t_c: time of concentration of the watershed formed by sub-basins B_1, B_2, B_3, B_4, B_5, B_6, and B_7, that is, 10,44 min, which is equal to the longest of the following times:

- t_e of sub-basin B_7, thus 10,0 min;
- t_e of sub-basin B_1 + t_f of sub-basins B_1, B_2, and B_3, thus 10,44 min (= 8,5 + 0,72 + 0,58 + 0,64);
- t_e of sub-basin B_4 + t_f of sub-basins B_4, B_5, and B_6, thus 9,39 min (= 7,0 + 0,49 + 0,63 + 1,27);
- $(t_e + t_f)$ of sub-basin B_6, thus 10,27 min (= 9,0 + 1,27).

ΔA: 4,2 ha

R_T: 0,5

ΔAR_T: 2,1 ha

A: total area of the watershed, thus 14,6 ha

AR: AR of sub-basins B_1, B_2, and B_3 + AR of sub-basins B_4, B_5, and B_6 + ΔAR_T, thus:

$$\begin{aligned} AR &= 2{,}57 + 3{,}11 + 2{,}1 \\ &= 7{,}78 \text{ ha} \end{aligned}$$

R: $$AR \div \left(\sum_j (A)_{B_j} + \Delta A \right)$$

thus:

$$\begin{aligned} R &= 7{,}78 \div (1{,}2 + 2{,}2 + 2{,}1 + 1{,}5 + 1{,}0 + 2{,}4 + 4{,}2) \\ &= 0{,}53 \end{aligned}$$

I: $$\begin{aligned} I &= \frac{2184{,}4}{t_c + 12} = \frac{2184{,}4}{10{,}44 + 12} \\ &= 97{,}34 \text{ mm/h} \end{aligned}$$

Q_{max}: $$\begin{aligned} Q_{max} &= 2{,}75 \times 10^{-3} \times 14{,}6 \times 97{,}34 \times 0{,}53 \\ &= 2{,}07 \text{ m}^3/\text{h} \end{aligned}$$

s: 0,007 m/m (since the slope of the ground surface is too small)

d: 1 050 mm (actual diameter = 1 070 mm)

v_f: 2,67 m/s

L: 125 m

t_f: 0,78 min

Q_f: 2,403 m^3/s ($Q_f > Q_{max}$ = 2,07 m^3/s)

Pipe $(R_p)_4$–$(R_p)_5$ collects water from the two pipes meeting at sewer manhole $(R_p)_4$, that is pipes $(R_p)_3$–$(R_p)_4$ and $(R_p)_7$–$(R_p)_4$. In this case (section 8.5.3), the inner crown (top) of pipe $(R_p)_4$–$(R_p)_5$ at sewer manhole $(R_p)_4$ cannot be higher than the lowest inner top of the two pipes located upstream from sewer manhole $(R_p)_4$:

h of the inner top of pipe $(R_p)_3$–$(R_p)_4$ at sewer manhole $(R_p)_4$ = h (of the inner invert) + d (inside of pipe)

= 104,99 + 0,610

= 105,60 m

h of the inner top of pipe $(R_p)_7$–$(R_p)_4$ at sewer manhole $(R_p)_4$ = h (of the inner invert) + d (inside of pipe)

= 104,33 + 0,915

= 105,25 m

Therefore, the inner top of pipe $(R_p)_4$–$(R_p)_5$ at sewer manhole $(R_p)_4$ is placed at a height of 105,25 m. As the diameter of this pipe is 1070 mm, its inner invert, at sewer manhole $(R_p)_4$, is located at a height of 104,18 m (= 105,25 – 1,07; fig. 9.3). At this location, the (inside) crown of the sewer pipe is at 107,60 – 105,25 = 2,35 m underground – more than the 2,0 m minimum required. Otherwise, the pipe could have been lowered.

The elevation of inner invert of pipe $(R_p)_4$–$(R_p)_5$ at $(R_p)_5$ is calculated the same way as the elevation of the inner invert of pipe $(R_p)_7$–$(R_p)_4$ at $(R_p)_4$, because the pipe slope, s, differs from the slope of the ground surface; hence:

h of the inner invert of pipe $(R_p)_4$–$(R_p)_5$ at sewer manhole $(R_p)_5$ = h of the inner invert at sewer manhole $(R_p)_4 - s \times L$

= 104,18 – 0,007 × 125

= 103,31 m

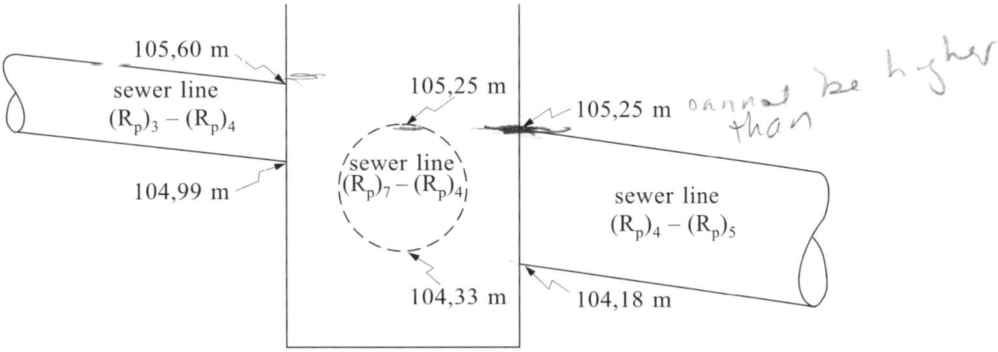

Figure 9.3 Sewer manhole $(R_p)_4$ of the storm sewer system; the numbers indicate inner elevations of inverts and crowns (tops) of pipes

9.2 DESIGN OF SANITARY SEWER SYSTEMS

To design a sanitary sewer system, we need information about:

- the area tributary to this sewer system (in comparison with the storm sewers, we refer to this area as the watershed);
- urban planning for the area;
- current and future population of the area;
- characteristics and quality of the existing sanitary sewer system, when appropriate;
- actual and future average domestic wastewater flow per inhabitant;
- actual and future infiltration;
- actual and future inflow.

9.2.1 Average, Maximum, and Minimum Sanitary Flowrates in a Sanitary Sewer

To illustrate the computation of sanitary flows, let us study the district shown in figures 9.2 and 9.4, assuming that the population density is very high – 600 persons/ha. As for storm sewers, we assume that sanitary wastewater of a sub-basin is collected at a sewer manhole located in the upper part of this sub-basin – $(R_s)_1$ for sub-basin B_1. For infiltration and inflow, we assume the values of 60 and 50 L/(person·d), respectively (see table 6.6); the domestic flow is 320 L/(person·d). Reinforced-concrete pipes (n = 0,013) are used; their nominal diameter cannot be smaller than 200 mm, according to Quebec standards. The pipes must be buried deep enough for their crowns (tops) to be at least 2 m under the ground surface ("cover" refers to the depth of the ground covering the pipe) to ensure that the wastewater will not freeze and, more importantly, that wastewater from neighbouring houses can be collected without pumping. To ensure self-cleaning of the pipes, we plan a minimum flow velocity of 0,6 m/s or more when the pipes are flowing full.

Computation of Average and Maximum Sanitary Flows. To compute maximum and minimum sanitary flows, the average domestic flow transported by the sanitary sewer must be known. This is also useful when designing sewage treatment plants.

The **average domestic flow** is obtained by multiplying the number of users by the average domestic sewage per inhabitant. Thus, for pipe $(R_s)_1$–$(R_s)_2$, which collects sewage only from watershed B_1:

$$\begin{aligned}\left(Q_{dom}\right)_{avg} &= 1{,}2 \text{ ha} \times 600 \text{ persons/ha} \times 320 \text{ L/(person·d)} \times \text{d/86 400 s} \\ &= 2{,}67 \text{ L/s}\end{aligned}$$

Average sanitary flow can be determined by adding infiltration [60 L/(person·d)] and inflow [50 L/(person·d)] to the average domestic flow (2,67 L/s). The average sanitary flow in pipe $(R_s)_1$–$(R_s)_2$ is therefore 3,59 L/s (= 2,67 + 0,5 + 0,42).

In a pipe line, the maximum sanitary flow is equal to the sum of the maximum domestic flow, infiltration, and inflow. The maximum domestic flow is equal to the average domestic flow multiplied by the appropriate peak factor (see section 6.2.1).

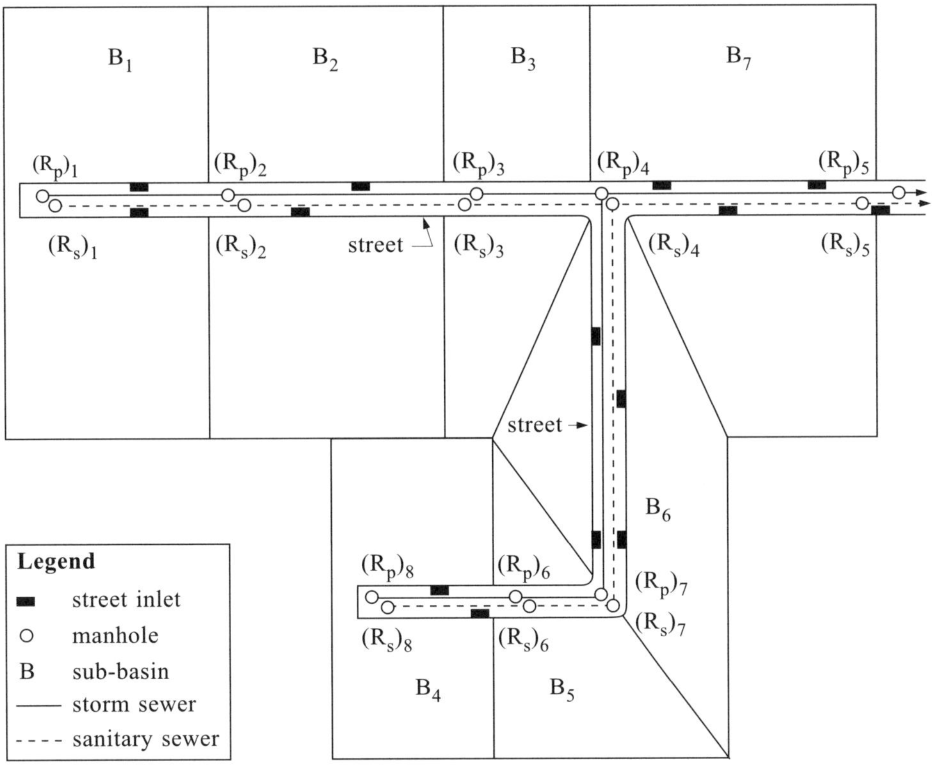

Figure 9.4 Storm sewer and sanitary sewer installed side by side in the watershed shown in figure 9.2

Pipe $(R_s)_1$–$(R_s)_2$: Computation of the maximum sanitary flow to be evacuated. Pipe $(R_s)_1$–$(R_s)_2$ must be able to evacuate the maximum domestic flow, plus infiltration and inflow originating from sub-basin B_1. The following steps determine the maximum flow transported by a sanitary pipe, more specifically for sewer $(R_s)_1$–$(R_s)_2$ (table 9.3).

ΔA: area of the watershed tributary to the considered pipe line – area of the sub-basin where pipe $(R_s)_1$–$(R_s)_2$ is located in addition to the area of upstream sub-basins. In this case, ΔA is equal to the area of sub-basin B_1, thus 1,2 ha.

ΔP: population of watershed B_1 tributary to the considered pipe in addition to the populations tributary to pipes upstream from the considered pipe. Here, ΔP is the population of sub-basin B_1, thus 720 persons (1,2 ha × 600 persons/ha).

P: total population tributary to the considered pipe, that is, the sum of populations related to pipes upstream and ΔP, thus 0 + 720 = 720 persons.

$(Q_{dom})_{avg}$: average *domestic* flow, obtained by multiplying the number of inhabitants by the average domestic flow per person; this yields:

$$\left(Q_{dom}\right)_{avg} = 720 \text{ persons} \times 320 \text{ L/(person·d)} \times \text{d/86 400 s}$$

$$= 2{,}67 \text{ L/s} \left(0{,}0027 \text{ m}^3\text{/s}\right)$$

Table 9.3 Computations related to the sanitary sewer system shown in figure 9.4

Average of tributary population									
Sewer line (between manholes)	Increment of tributary area (ΔA)	Increment of tributary population	Total population tributary	Average domestic flow	Maximum peak factor	Maximum domestic flow	Increment of parasitic flow	Total parasitic flow	Maximum sanitary flow
	ΔA (ha)	ΔP	P	$(Q_{dom})_{avg}$ (L/s)	FP_{max}	$(Q_{dom})_{max}FP$ (L/s)	ΔQ_{par} (L/s)	Q_{par} (L/s)	$(Q_{san})_{max}$ (L/s)
$(R_s)_1$–$(R_s)_2$	1,2	720	720	2,67	4,00	10,67	0,92	0,92	11,59
$(R_s)_2$–$(R_s)_3$	2,2	1320	2040	7,56	3,64	27,47	1,68	2,60	30,07
$(R_s)_3$–$(R_s)_4$	2,1	1260	3300	12,22	3,38	41,33	1,60	4,20	45,53
$(R_s)_8$–$(R_s)_6$	1,5	900	900	3,33	4,00	13,33	1,15	1,15	14,48
$(R_s)_6$–$(R_s)_7$	1,0	600	1500	5,56	3,81	21,16	0,76	1,91	23,07
$(R_s)_7$–$(R_s)_4$	2,4	1440	2940	10,89	3,44	37,47	1,83	3,74	41,21
$(R_s)_4$–$(R_s)_5$	4,2	2520	8760	32,44	2,92	94,71	3,21	11,15	105,86

FP: maximum peak factor (see section 6.2.1 and table 6.2 for computations); in this example, it yields:

$$\begin{aligned} FP &= 1{,}742\left[\left(Q_{dom}\right)_{avg}\right]^{-0{,}1506} \\ &= 1{,}742\,(0{,}0027)^{-0{,}1506} \\ &= 4{,}25 \end{aligned}$$

Therefore, 4,0 is chosen because FP must not exceed 4,0.

$(Q_{dom})_{avg} \cdot FP$: maximum *domestic* flow, thus:

$$\begin{aligned} \left(Q_{dom}\right)_{max} &= 2{,}67 \times 4{,}0 \\ &= 10{,}67 \text{ L/s} \end{aligned}$$

ΔQ_{par}: parasitic flow originating from sub-basin B_1 ($\Delta Q_{par} = Q_{inf} + Q_i$); it yields:

$$\begin{aligned} \Delta Q_{par} &= \left[60 \text{ L/(person·d)} + 50 \text{ L/(person·d)}\right] \\ &\quad \times 720 \text{ persons} \times \text{d}/86\,400 \text{ s} \\ &= 0{,}92 \text{ L/s} \end{aligned}$$

Table 9.3 (continued)

Computation of sewer characteristics					Ground elevations		Inner invert elevations (minimum cover of 2 m)	
Slope of sewer s (m/m)	Nominal diameter of sewer d_{nom}* (mm)	Flow velocity in sewer flowing full v_f (m/s)	Length of sewer line L (m)	Hydraulic capacity of sewer flowing full Q_f (m^3/s)	upstream (m)	downstream (m)	upstream (m)	downstream (m)
0,017	200	1,38	90	0,044 5	113,03	111,50	110,83	109,30
0,025	200	1,67	100	0,054 0	111,50	109,00	109,30	106,80
0,014	225	1,35	100	0,055 7	109,00	107,60	106,77	105,37
0,017	200	1,38	70	0,044 5	109,19	108,00	106,99	105,80
0,007	200	0,88	70	0,028 6	108,00	107,51	105,80	105,31
0,0077	225	1,00	110	0,041 3	107,51	107,60	105,28	104,43
0,0110	300	1,45	125	0,106	107,60	107,20	104,36	102,99

* All computations made with actual diameters (see table 8.1).

Q_{par}: total parasitic flow (including parasitic flow originating from upstream sub-basins) evacuated by the pipe in question. In this case, as there are no pipes located upstream:

$$Q_{par} = \Delta Q_{par} = 0{,}92 \text{ L/s}$$

$(Q_{san})_{max}$: maximum sanitary flow evacuated by the considered pipe, hence:

$$\left[\left(Q_{dom}\right)_{avg} \times \text{FP}\right] + Q_{par} = 10{,}67 + 0{,}92$$
$$= 11{,}59 \text{ L/s}$$

To compute the characteristics of the pipe, the same procedure is used as for storm sewers.

Maximum flows are computed in the same way for other sanitary pipes.

Pipe $(R_s)_7$–$(R_s)_4$. As for the storm-sewer system, when the ground surface slope rises in the direction of the flow, the slope of the pipe, *s*, must be different. The chosen slope (0,0077) is the minimum slope required for a 225 mm pipe, flowing full, to have a hydraulic capacity barely equal to the maximum flow it must carry. The height of the inner invert of pipe $(R_s)_7$–$(R_s)_4$ at sewer hole $(R_s)_4$ is computed as follows:

h of the inner invert of pipe $(R_s)_7$–$(R_s)_4$ at sewer manhole $(R_s)_4$ = h of the inner invert at sewer manhole $(R_s)_7 - s \times L$

$$= 105{,}28 - 0{,}0077 \times 110$$

$$= 104{,}43 \text{ m}$$

Pipe $(R_s)_4$*–*$(R_s)_5$. This pipe collects wastewater transported to sewer manhole $(R_s)_4$ by the pipes $(R_s)_3$–$(R_s)_4$ and $(R_s)_7$–$(R_s)_4$. In this case (section 8.5.3), the crown (top) of pipe $(R_s)_4$–$(R_s)_5$ at sewer manhole $(R_s)_4$ cannot be higher than the lowest crown (top) of the two upstream pipes at $(R_s)_4$:

h of the crown (top) of pipe $(R_s)_3$–$(R_s)_4$ at sewer manhole $(R_s)_4$ = h (of inner invert) + d (inside of pipe)

$$= 105{,}37 + 0{,}229$$

$$= 105{,}60 \text{ m}$$

h of the crown (top) of pipe $(R_s)_7$–$(R_s)_4$ at sewer manhole $(R_s)_4$ = h (of inner invert) + d (inside of pipe)

$$= 104{,}43 + 0{,}229$$

$$= 104{,}66 \text{ m}$$

Thus, the inner crown (top) of pipe $(R_s)_4$–$(R_s)_5$ at sewer manhole $(R_s)_4$ is 104,66 m high. As the inner diameter of this pipe is 305 mm, its inner invert, at sewer manhole $(R_s)_4$, is 104,36 m (= 104,66 – 0,305; fig. 9.5).

The chosen slope (0,0110) for pipe $(R_s)_4$–$(R_s)_5$ is the minimum slope required for a 305 mm pipe, flowing full, to ensure a hydraulic capacity barely equal to the maximum flow it must carry.

The elevation of the inner invert of downstream pipe $(R_s)_4$–$(R_s)_5$ at sewer manhole $(R_s)_5$ is computed in the same way as for the elevation of the inner invert of pipe $(R_s)_7$–$(R_s)_4$ at $(R_s)_4$, because the pipe slopes differs from the ground surface slope; thus:

h of the inner invert at sewer manhole $(R_s)_5$ = h of the inner invert at sewer manhole $(R_s)_4 - s \times L$

$$= 104{,}36 - 0{,}0110 \times 125$$

$$= 102{,}99 \text{ m}$$

Note: It is important to avoid locating the sanitary and storm-sewer pipes in the same horizontal plan to ensure connections to houses on each side of the street (section 8.4.7). For this reason, the sanitary pipe may be lowered until its crown (top) is at the same elevation as the invert of the storm pipe.

Table 9.3 lists the results of computations related to the sanitary system shown in figure 9.4.

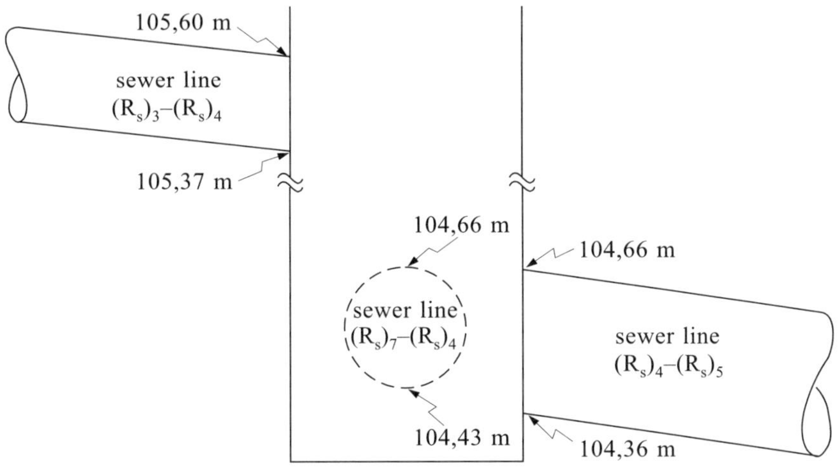

Figure 9.5 Sewer manhole $(R_s)_4$ of the sanitary sewer system; numbers correspond to the inner elevations of inverts and crowns (tops) of the pipes

Computation of Minimum Sanitary Flows. Computation of the minimum flows is useful when designing pumping stations and choosing pumps. It is also advantageous to know sewage velocities in pipes at minimum flows so as to evaluate the risk of deposited solids and resulting bad odors, especially in summer. The minimum sanitary flow is equal to the sum of the minimum domestic flow and infiltration (the minimum flow occurs when it does not rain, thus in absence of inflow). The following steps are used to compute the sanitary flow in pipe $(R_s)_1$–$(R_s)_2$.

Peak factor (table 6.2):

$$\begin{aligned} \text{FP} &= \left[\left(Q_{dom}\right)_{avg}\right]^{0{,}1506} \Big/ 1{,}742 \\ &= (0{,}0027)^{0{,}1506} / 1{,}742 \\ &= 0{,}24 \end{aligned}$$

Infiltration:

$$\begin{aligned} Q_{inf} &= 60 \ \text{L}/(\text{person·d}) \times 720 \ \text{persons} \times \text{d}/86\,400 \ \text{s} \\ &= 0{,}5 \ \text{L/s} \ \left(0{,}0005 \ \text{m}^3/\text{s}\right) \end{aligned}$$

Minimum sanitary flow:

$$\begin{aligned} \left[\left(Q_{dom}\right)_{avg} \times \text{FP}\right] + Q_{inf} &= (2{,}67 \times 0{,}24) + 0{,}5 \\ &= 1{,}14 \ \text{L/s} \ \left(0{,}00114 \ \text{m}^3/\text{s}\right) \end{aligned}$$

9.2.2 Design of a (Sanitary) Interceptor for Two Sanitary Sewer Systems

Figure 9.6 shows an intercepting sewer collecting sanitary wastewater from the two sanitary sewer systems of watersheds C and D. In general, the design of major sanitary pipes, including the intercepting sewer, depends on both actual and future flows (in 15 and 30 years). The following section shows how to compute future wastewater flows (in 30 years) in the two intercepting pipe lines shown in figure 9.6.

Characteristics of Watersheds. Wastewater from the 100 ha watershed C is already being collected by a sanitary sewer system: in the future, the remaining 150 ha will also be collected by sanitary sewers. The current population of 5000 will ultimately reach 20 000. Current infiltration is 1000 m^3/d, and it is estimated to reach 2500 m^3/d. Inflow is, and will remain, 50 m^3/d. The 200 ha of watershed D are not yet developed, but we estimate the population will reach 15 000, inflow will remain zero (though it is too optimistic – table 6.6), and infiltration will be 1120 m^3/d. Table 9.4 shows a summary of the two watersheds' characteristics.

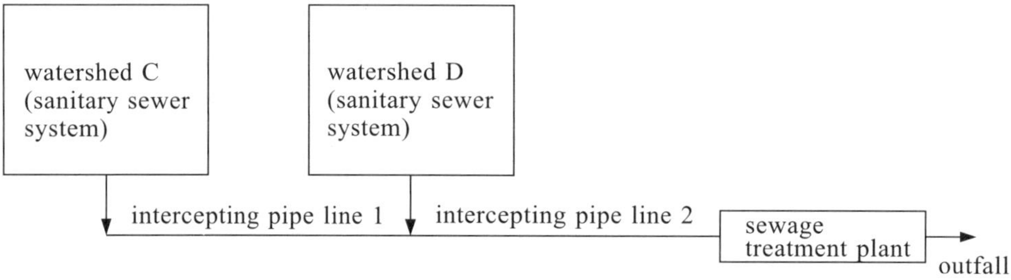

Figure 9.6 Intercepting sewer collecting sanitary wastewater from two sanitary sewer systems

Table 9.4 Characteristics of two watersheds where wastewater is collected by sanitary sewer systems (fig. 9.6)

	Watershed C (250 ha) (Sanitary sewer system)		**Watershed D (200 ha) (Sanitary sewer system)**	
	present	**in 30 years**	**present**	**in 30 years**
Population	5 000	20 000	0	15 000
Domestic flow [L/(person·d)]	320*	320*	320*	320*
Service area (ha)	100	250	0	200
Infiltration flow (m^3/d)	1000	2 500**	0	1 120**
Inflow (m^3/d)	50***	50***	0	0***

* See section 6.2.1.

** See table 6.6; because of the sandy nature of the soil, we assumed 74,7 L/person·d.

*** More reallistically, we should use 50 L/(person·d).

Average Flows in 30 Years. Average sanitary flows are equal to the sum of average domestic flows, infiltration, and inflow. For example, we will demonstrate how to compute future average flows (in 30 years) in the first intercepting pipe line collecting wastewater from watershed C.

The average **domestic flow** is equal to the number of inhabitants multiplied by the average domestic flow per inhabitant, thus:

$$\begin{aligned}(Q_{dom})_{avg} &= 20\,000 \text{ persons} \times 320 \text{ L/(person·d)} \times \text{d/86 400 s} \\ &= 74{,}07 \text{ L/s } (0{,}074 \text{ m}^3/\text{s})\end{aligned}$$

Infiltration and inflow add the following flow to the average domestic flow:

$$\begin{aligned}Q_{inf} + Q_i &= (2500 + 50) \text{ m}^3/\text{d} \times 10^3 \text{ L/m}^3 \times \text{d/86 400 s} \\ &= 29{,}51 \text{ L/s } (0{,}030 \text{ m}^3/\text{s})\end{aligned}$$

The average sanitary flow is therefore:

$$\begin{aligned}(Q_{san})_{avg} &= 74{,}07 \text{ L/s} + 29{,}51 \text{ L/s} \\ &= 103{,}6 \text{ L/s } (0{,}104 \text{ m}^3/\text{s})\end{aligned}$$

Maximum and Minimum Flows in 30 Years. The minimum and maximum flows transported by the intercepting sewers must be known for the design of interceptors and sewage-treatment plants, and to manage the operation of weirs and retention-detention basins located along the interceptors.

a) The **first intercepting pipe line** must be able to handle the flows originating from watershed C. The following steps compute average maximum and minimum domestic and sanitary flows.

- *Domestic flows* (i.e. without infiltration and inflow)
  - future average domestic flow originating from watershed C:

$$\begin{aligned}(Q_{dom})_{avg} &= 20\,000 \text{ persons} \times 320 \text{ L/(person·d)} \times \text{d/86 400 s} \\ &= 74{,}07 \text{ L/s } (0{,}074 \text{ m}^3/\text{s})\end{aligned}$$

  - maximum peak factor for Quebec:

$$\text{FP}_{max} = 1{,}742\, Q^{-0{,}1506} = 1{,}742\ (0{,}074)^{-0{,}1506} = 2{,}58$$

  - maximum domestic flow:

$$(Q_{dom})_{max} = 0{,}074 \times 2{,}58 = 0{,}191 \text{ m}^3/\text{s } (191{,}0 \text{ L/s})$$

  - minimum peak factor for Quebec:

$$\text{FP}_{min} = Q^{0{,}1506}/1{,}742 = (0{,}074)^{0{,}1506}/1{,}742 = 0{,}388$$

  - minimum domestic flow:

$$(Q_{dom})_{min} = 0{,}074 \times 0{,}388 = 0{,}029 \text{ m}^3/\text{s } (28{,}7 \text{ L/s})$$

- *Sanitary flows* (domestic flows, infiltration, and inflow)

  To compute sanitary flows, we must add infiltration plus inflow in the case of average and maximum sanitary flows. For this *first intercepting pipe line*, infiltration (2500 m^3/d) and inflow (50 m^3/d) add 0,0295 m^3/s (29,5 L/s) to the domestic flows. In 30 years it yields:

  – average sanitary flow, $(Q_{san})_{avg}$: 0,104 m^3/s (103,6 L/s)
  – maximum sanitary flow, $(Q_{san})_{max}$: 0,221 m^3/s (220,6 L/s)
  – minimum sanitary flow, $(Q_{san})_{min}$: 0,058 m^3/s (58,2 L/s) (inflow is not and must not be included)

b) The **second intercepting pipe line** must be able to transport the flows originating from watersheds C and D. The following steps compute average, maximum, and minimum domestic and sanitary flows.

- *Domestic flows*

  – future average domestic flow originating from watersheds C and D:

  $$\left(Q_{dom}\right)_{avg} = \left(20\,000 + 15\,000\right) \text{ persons} \times 320 \text{ L/(person·d)} \times \text{d/86 400 s}$$
  $$= 129{,}6 \text{ L/s } \left(0{,}130 \text{ m}^3/\text{s}\right)$$

  – maximum peak factor for Quebec:

  $$\text{FP}_{max} = 1{,}742\,Q^{-0{,}1506} = 1{,}742\ (0{,}130)^{-0{,}1506} = 2{,}37$$

  – maximum domestic flow:

  $$\left(Q_{dom}\right)_{max} = 0{,}130 \times 2{,}37 = 0{,}307 \text{ m}^3/\text{s } \left(307{,}2 \text{ L/s}\right)$$

  – minimum peak factor for Quebec:

  $$\text{FP}_{min} = Q^{0{,}1506}/1{,}742 = (0{,}130)^{0{,}1506}/1{,}742 = 0{,}422$$

  – minimum domestic flow:

  $$\left(Q_{dom}\right)_{min} = 0{,}130 \times 0{,}422 = 0{,}055 \text{ m}^3/\text{s } \left(54{,}71 \text{ L/s}\right)$$

- *Sanitary flows (domestic flows + infiltration and inflow)*

  For this *second pipe line*, infiltration (2500 m^3/d + 1120 m^3/d) and inflow (50 m^3/d) contribute 0,043 m^3/s (42,5 L/s) to the sanitary flows. In 30 years it yields:

  – average sanitary flow, $(Q_{san})_{avg}$: 0,173 m^3/s (172,5 L/s)
  – maximum sanitary flow, $(Q_{san})_{max}$: 0,350 m^3/s (349,5 L/s)
  – minimum sanitary flow, $(Q_{san})_{min}$: 0,097 m^3/s (96,6 L/s) (inflow is not and must not be included)

Table 9.5 displays the values of the various flows (in m^3/s) in the two intercepting pipe lines of the intercepting sewer shown in figure 9.6.

Note that when watershed D contributes to the flow in the second intercepting pipe line of the interceptor, the maximum flow in that reach is not the sum of the two maximum flows from both watersheds. The decreasing maximum peak factor (FP_{max})

together with the increase in average flow (fig. 6.4) due to added average flows from both watersheds explains this result. One can also understand it by the fact that the maximum flows from each of the two watersheds do not reach the second intercepting sewer line at the same time: wastewater originating from watershed C is carried a greater distance than the wastewater of watershed D. Watershed C's maximum flow reaches the intercepting sewer after watershed D has already reached maximum flow. These same facts also justify similar computation for the minimum flow related to the second intercepting pipe line.

Table 9.5 Flows in the two intercepting pipe lines of the interceptor shown in figure 9.6

	First pipe line (Watershed C: sanitary)		**Second pipe line (Watershed C: sanitary) (Watershed D: sanitary)**	
	at present	**in 30 years**	**at present**	**in 30 years**
Population	5 000	20 000	Watershed D is not contributing 5 000	20 000 + 15 000
Domestic flow				
maximum (m^3/s)	0,059	0,191	0,059	0,307
average (m^3/s)	0,019	0,074	0,019	0,130
minimum (m^3/s)	0,006	0,029	0,006	0,055
Sanitary flow				
maximum (m^3/s)	0,071	0,221	0,071	0,350
average (m^3/s)	0,031	0,104	0,031	0,173
minimum (m^3/s)	0,017	0,058	0,017	0,097

9.2.3 Design of a Sanitary Interceptor for a Sanitary Sewer System and a Combined Sewer System

Figure 9.7 illustrates an interceptor collecting wastewater from a sanitary and a combined sewer system. The following section describes the computation method used to determine future wastewater flows (in 30 years) in the two intercepting pipelines.

Characteristics of the watersheds. The characteristics of watershed E are the same as watershed C, which was described in section 9.2.2. Watershed F has a population of 7000 and will ultimately reach 9500. The domestic flow of watershed F is 320 L/(person·d), and its surface area is 200 ha. Infiltration is and will remain 3000 m^3/d, because the system is already complete. It is inappropriate to consider inflow here, as the combined sewer system must collect all of runoff. Table 9.6 summarizes the characteristics of the two watersheds.

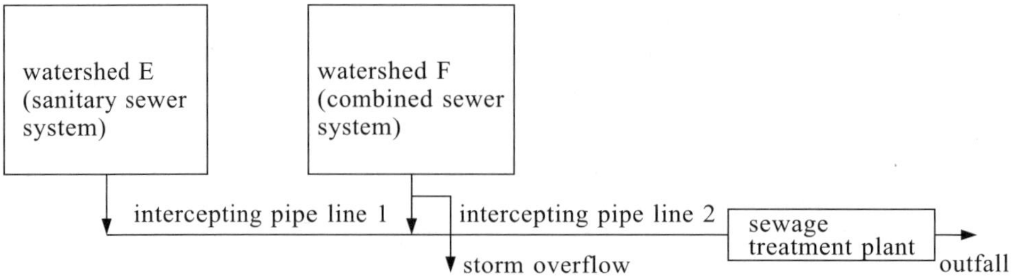

Figure 9.7 Intercepting sewer collecting sanitary wastewater from watershed E and combined wastewater from watershed F

Table 9.6 Characteristics of two watersheds, one with a combined sewer system, the other with a sanitary sewer system (fig. 9.7)

	Watershed E (250 ha) (Sanitary sewer system)		**Watershed F (200 ha) (Combined sewer system)**	
	at present	**in 30 years**	**at present**	**in 30 years**
Population	5 000	20 000	7 000	9 500
Domestic flow [L/(person·d)]	320*	320*	320*	320*
Surface area (ha)	100	250	150	150
Infiltration (m^3/d)	1 000	2 500	3 000	3 000
Inflow (m^3/d)	50**	50**	__***	__***

* For a combined sewer system, the domestic flow is obtained from *in situ* dry-weather flow (DWF) measurements (see note below).

** More reallistically, we should use 50 L/(pers·d).

*** There is no inflow in combined sewers since they are designed to transport all runoff water. However, to avoid occurrence of many stormwater overflows, it is common to accept from 0,1 to 0,5 m^3/(pers·d) of rainwater in the interceptor.

Maximum and Minimum Flows in 30 Years. This section lists the steps for evaluating the various flows in the two intercepting pipe lines collecting the wastewater from watersheds E and F.

a) The **first intercepting pipe line** must handle the flows from watershed E. The following steps compute average maximum and minimum domestic and sanitary flows in 30 years.

- *Domestic flows (i.e. without infiltration and inflow).* For the *first pipe line*, the computation is the same as in section 9.2.2; it yielded:
  - average domestic flow from watershed E, $(Q_{dom})_{avg}$: 0,074 m^3/s (74,07 L/s)
  - maximum peak factor for Quebec, FP_{max}: 2,58

- minimum peak factor for Quebec, FP_{min}: 0,388
- maximum domestic flow, $(Q_{dom})_{max}$: 0,191 m^3/s (191,0 L/s)
- minimum domestic flow, $(Q_{dom})_{min}$: 0,029 m^3/s (28,7 L/s)

- *Sanitary flows (domestic flows + infiltration and inflow).* To compute sanitary flows, infiltration must be added, as must inflow in the case of average and maximum sanitary flow. For this *first pipe line,* infiltration (2500 m^3/d) and inflow (50 m^3/d) add 0,0295 m^3/s (29,5 L/s) to the domestic flows. The computation is the same as in section 9.2.2; it yielded:
  - average sanitary flow, $(Q_{san})_{avg}$: 0,104 m^3/s (103,6 L/s)
  - maximum sanitary flow, $(Q_{san})_{max}$: 0,221 m^3/s (220,6 L/s)
  - minimum sanitary flow, $(Q_{san})_{min}$: 0,058 m^3/s (58,6 L/s) (inflow is not and must not be included)

b) The **second pipe line** of the interceptor must be able to handle the maximum flow originating from watersheds E and F. First, let us compute the *maximum domestic flow originating from the combined watershed F itself.*

- average domestic flow originating from watershed F:

$$\left(Q_{dom}\right)_{avg} = 9500 \text{ persons} \times 320 \text{ L/(person·d)} \times \text{d}/86\,400 \text{ s}$$
$$= 35{,}19 \text{ L/s } \left(0{,}035 \text{ m}^3/\text{s}\right)$$

- maximum peak factor for Quebec:

$$FP_{max} = 1{,}742\,Q^{-0{,}1506} = 1{,}742\ (0{,}0352)^{-0{,}1506} = 2{,}88$$

- maximum domestic flow originating from watershed F:

$$\left(Q_{dom}\right)_{max} = \left(Q_{dom}\right)_{avg} \times FP_{max} = 0{,}0352 \times 2{,}88$$
$$= 0{,}101 \text{ m}^3/\text{s } (101{,}4 \text{ L/s})$$

- minimum peak factor for Quebec:

$$FP_{min} = Q^{0{,}1506}/1{,}742 = (0{,}0352)^{0{,}1506}/1{,}742 = 0{,}347$$

- minimum domestic flow originating from watershed F:

$$\left(Q_{dom}\right)_{min} = \left(Q_{dom}\right)_{avg} \times FP_{min} = 0{,}0352 \times 0{,}347$$
$$= 0{,}012 \text{ m}^3/\text{s } (12{,}21 \text{ L/s})$$

NOTE: The fundamental goal of a sewage-treatment plant is to treat all sanitary sewage. In a combined sewer system, sanitary sewage is the wastewater flowing in the system when there is no precipitation: engineers usually refer to this flow as **dry-weather flow (DWF)**. In wet weather, runoff is much greater than the sanitary flow.

Under these conditions, we usually transport to the treatment plant a maximum flow equal to the maximum sanitary wastewater flow (maximum DWF) and evacuate the excess (often called the storm overflow) into the receiving stream. If this overflow causes an unacceptable level of pollution, the interceptor capacity can be increased or reservoirs may be designed specifically to store excess wastewater until the intercepting sewer is able to take it to the sewage treatment plant.

However, as it may rain at any time of day or night, combined watershed F may transport its maximum flow toward the interceptor at any time, which means that the interceptor collects the rainwater flow until it reaches its capacity; it is therefore possible that the maximum flow from watershed E reach the second intercepting pipe line of the interceptor at the same time as the maximum flow from watershed F. In preparation for such an event, the second intercepting pipe line of the interceptor is designed to handle the sum of the maximum sanitary flows from both watersheds E and F.

Moreover, the flow is minimum in the second pipe reach of the interceptor when there is no precipitation. Thus watersheds E and F behave as a single watershed with a single sanitary sewer system. The minimum flow is thus computed as in the above case (section 9.2.2).

These steps compute the average, maximum, and minimum domestic and sanitary flows in the **second pipe line** of the interceptor in 30 years.

- Domestic flows (i.e. without infiltration and inflow)
  - average domestic flow from watersheds E and F together:

$$\begin{aligned}\left(Q_{dom}\right)_{avg} &= \left(20\,000 + 9500 \text{ persons}\right) \times 320 \text{ L/(person·d)} \times \text{d}/86\,400 \text{ s} \\ &= 109{,}3 \text{ L/s } \left(0{,}109 \text{ m}^3/\text{s}\right)\end{aligned}$$

  - future maximum domestic flow; taking into account the above note, the flow is:

$(Q_{dom})_{max} = (Q_{dom})_{max}$ from watershed E + $(Q_{dom})_{max}$ from watershed F, thus:

$$\left(Q_{dom}\right)_{max} = 0{,}191 + 0{,}101 = 0{,}292 \text{ m}^3/\text{s } \left(292{,}5 \text{ L/s}\right)$$

  - minimum peak factor for Quebec:

$$\text{FP}_{min} = Q^{0{,}1506}/1{,}742 = (0{,}109)^{0{,}1506}/1{,}742 = 0{,}411$$

  - minimum domestic flow; taking into account the previous note, the flow is:

$$\left(Q_{dom}\right)_{min} = 0{,}109 \times 0{,}411 = 0{,}045 \text{ m}^3/\text{s } \left(44{,}9 \text{ L/s}\right)$$

For the *second pipe line,* infiltration (3000 m^3/d) from watershed F adds 0,035 m^3/s (34,7 L/s) to the domestic flows (inflow is not considered since watershed F is a combined sewer system). As a result, in 30 years, sanitary flows will be:

– maximum sanitary flow:

$$(Q_{san})_{max} = (Q_{san})_{max} \text{ from sanitary watershed E}$$
$$+ (Q_{dom})_{max} \text{ from combined watershed F}$$
$$+ Q_{inf} \text{ from combined watershed F}$$
$$= 0{,}221 + 0{,}101 + 0{,}035 = 0{,}357 \text{ m}^3/\text{s}$$

– average sanitary flow:

$$(Q_{san})_{avg} = (Q_{san})_{avg} \text{ from sanitary watershed E}$$
$$+ (Q_{dom})_{avg} \text{ from combined watershed F}$$
$$+ Q_{inf} \text{ from combined watershed F}$$
$$= 0{,}104 + 0{,}035 + 0{,}035 = 0{,}174 \text{ m}^3/\text{s}$$

– computation principle of the minimum sanitary flow: in this case, the average domestic flow must be computed for both watersheds together, the corresponding minimum peak factor must be found, the minimum domestic flow must be calculated, and infiltration but not inflow must be added. Hence:

– average domestic flow for the two watersheds:

$$\left(Q_{dom}\right)_{avg} = \left(20\,000 + 9500 \text{ persons}\right) \times 320 \text{ L/(person·d)} \times \text{d}/86\,400 \text{ s}$$
$$= 109{,}3 \text{ L/s} \left(0{,}109 \text{ m}^3/\text{s}\right)$$

– minimum peak factor for Quebec:

$$\text{FP}_{min} = Q^{0,1506}/1{,}742 = (0{,}109)^{0,1506}/1{,}742 = 0{,}411$$

– minimum domestic flow for the two watersheds:

$$\left(Q_{dom}\right)_{min} = 0{,}109 \times 0{,}411 = 0{,}045 \text{ m}^3/\text{s} \left(44{,}9 \text{ L/s}\right)$$

– minimum sanitary flow:

$$(Q_{san})_{min} = (Q_{dom})_{min} + Q_{inf} \text{ for the 1st pipe from watershed E}$$
$$+ Q_{inf} \text{ from watershed F}$$
$$= 0{,}045 + 0{,}029 + 0{,}035 = 0{,}109 \text{ m}^3/\text{s (108,5 L/s)}$$

Table 9.7 presents the various flows (in m^3/s) flowing in the two pipe lines of the interceptor.

Table 9.7 Flows in the two pipe lines of the interceptor shown in figure 9.7

	First pipe line (Watershed E: sanitary)		**Second pipe line (Watershed E: sanitary + watershed F: combined)**	
	at present	**in 30 years**	**at present**	**in 30 years**
Population	5 000	20 000	5 000 + 7 000	20 000 + 9 500
Domestic flow				
maximum (m^3/s)	0,059	0,191	0,137	0,292
average (m^3/s)	0,019	0,074	0,044	0,109
minimum (m^3/s)	0,006	0,029	0,016	0,045
Sanitary flow				
maximum (m^3/s)	0,071	0,221	0,184	0,357
average (m^3/s)	0,031	0,104	0,091	0,174
minimum (m^3/s)	0,017	0,058	0,062	0,109

N.B. Because it is a long, large structure, the interceptor adds an important flow of infiltration. Engineers suggest increasing infiltration by 225 L/(cm-km-d) (section 6.2.6; table 6.6).

9.3 LOCATION OF STREET INLETS

In section 9.1, we assumed the location of street inlets. The next section will list the criteria on which engineers generally rely when choosing the location of street inlets along street curbs. According to engineering practices, the pipes are usually designed first, then the locations of the street inlets are chosen. **Usually we limit to 0,03 m^3/s (30 L/s)** the maximum flow in a street inlet. To make sure that runoff flows easily toward inlets, the **slope of the gutters is at least 0,5%**.

9.3.1 Maximum Distance Between Two Street Inlets when Longitudinal Street Slope is Equal to or Greater than 0,5%

Street inlets collect water from an urban sub-basin (fig. 9.8). For the sake of demonstration, we assume the district shown in figure 9.8 is located in a residential area of the Montreal region and that all houses are single-family dwellings. When the longitudinal slope of the street is at least 5%, the curb height is constant for any value between 10 and 20 cm.

The time of concentration of this watershed (which is the inlet time, t_e) is equal to the sum of the time required by a water droplet to run off route, L_1, perpendicular to the street, and the time this same droplet takes to run off the street, along the curb, to the street inlet. The side of the street along the sidewalk is called the gutter (or curb). Assume the following:

L_1 = 40 m
L_2 = variable to be evaluated

s_1 = 1,2%
s_2 = 0,6% (> 0,5%: critical slope we have discussed previously)
n_1 = 0,1 (Manning's roughness coefficient for grass)
n_2 = 0,02 (Manning's roughness coefficient for asphalt)
R = 0,6 (runoff coefficient)
l = 22 m (width of the street)

The actual values of L_1, s_1, *and* s_2 are known and the value of L_2 is determined by trial and error. As the street inlets are part of the storm-sewer system, we have to compute a rainfall intensity with appropriate Intensity-Duration-Frequency equations. In Quebec, a return period of 5 years is suggested. The value of L_2 must be calculated, and we first assume it to be 60 m.

– Computation of the flow time on grass, $(t_L)_1$, using the Kerby equation (equation 7.6):

$$(t_L)_1 = \left(\frac{2{,}187 Ln}{s^{0{,}5}}\right)^{0{,}467} = \left(\frac{2{,}187 \times 40 \times 0{,}1}{0{,}012^{0{,}5}}\right)^{0{,}467}$$
$$= 7{,}73 \text{ min}$$

where $L = L_1$
n = Manning's roughness coefficient

– Computation of the flow time along the curb, $(t_L)_2$, using the Kerby equation:

$$(t_L)_2 = \left(\frac{2{,}187 \times 60 \times 0{,}02}{0{,}006^{0{,}5}}\right)^{0{,}467}$$
$$= 5{,}18 \text{ min}$$

– Computation of the time of concentration (which, here, is equal to the inlet time):

$$t_c = t_e = 7{,}73 + 5{,}18$$
$$= 12{,}9 \text{ min}$$

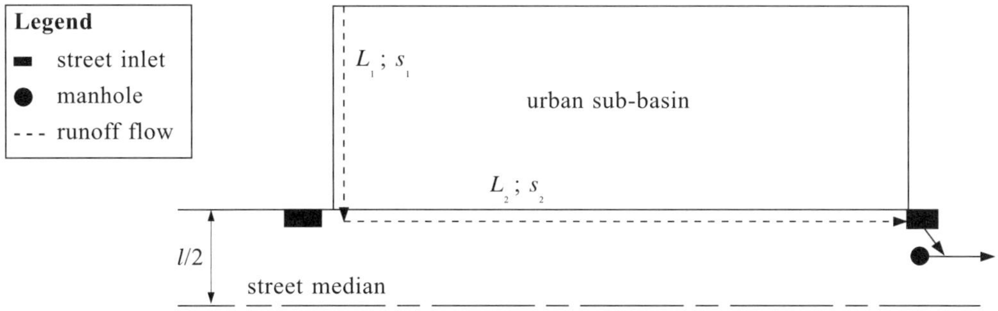

Figure 9.8 Urban sub-basin for which the maximum distance, L_2,between two street inlets is to be evaluated. The slope, s_2 (of the sidewalk, the street, and the gutter), is greater than 0,5%.

– Computation of the rainfall intensity, I, corresponding to a time of concentration of 12,9 min:

  according to the hypothesis on which the rational equation is based, we must find the intensity of a 5-year rainfall (storm-sewer network) in the Montreal region, with a duration equal to the time of concentration of 12,9 min; it yields:

$$I = \frac{2184{,}4}{t_e + 12} = \frac{2184{,}4}{t_c + 12} = \frac{2184{,}4}{12{,}9 + 12}$$
$$= 87{,}7 \text{ mm/h}$$

– Computation of the runoff flow, Q, using the rational equation:

$$Q = 2{,}75 \times 10^{-3} AIR$$
$$= 2{,}75 \times 10^{-3} \frac{(40 + 11) \times 60}{10\,000} \times 87{,}7 \times 0{,}6$$
$$= 0{,}044 \text{ m}^3/\text{s}$$

Since the calculated flow is greater than the maximum flow a street inlet can handle, this computation is repeated using a smaller value for L_2.

9.3.2 Maximum Distance between Two Street Inlets when Longitudinal Street Slope is Less than 0,5%

In this case, the slope of the gutter (the slope of the street along the curb) cannot equal the slope of the street median, which is also equal to the slope of the sidewalk: the runoff velocity of the water would be too low and would leave too much debris, the water would not be collected fast enough, and puddles would start to form. In figure 9.9, we can clearly see that the slope of the street along the curb is different than the slope of the sidewalk.

In this case, to determine the distance between two street inlets, the computations must take into account the following constraints:

– the minimum height of the curb is 10 cm to stop cars from mounting the sidewalk;
– the maximum height of the curb is 20 cm, enabling the doors of parked cars to be opened easily;
– the slope of the gutter is 0,5%.

Case 1: The slope of the street and the slope of the sidewalk are zero (fig. 9.9a). For these conditions, we can clearly see that the maximum distance between two street inlets must be calculated in the following manner:

$$s = 0{,}005 = \frac{0{,}2 - 0{,}1}{L_3/2}$$

where

$$L_3 = 40 \text{ m}$$

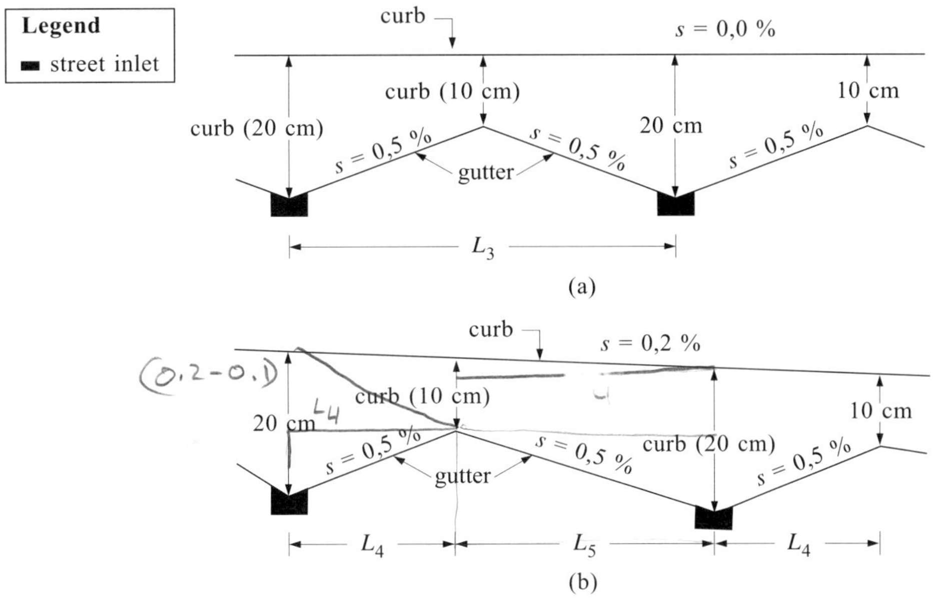

Figure 9.9 Longitudinal cross sections representing the slopes of the curb and the gutter in two streets where the maximum distance is to be determined between two street inlets: a) street slope zero; b) street slope between zero and 0,5%.

Afterward, the Kerby equation is used to verify that the flow collected at each street inlet does not exceed the 0,03 m^3/s limit. If it does, the distance between inlets must be shortened but the slope of the gutter is maintained at 0,5%.

Case 2: The street slope (which is equal to the sidewalk slope) is between 0 and 0,5% (fig. 9.9b). Let us assume, for the sake of this demonstration, that the slope of the sidewalk (of the street, and of the curb) is 0,2%. Geometrically speaking, for these conditions, we can say:

$$\frac{0,2 - 0,1}{L_4} = \frac{0,5 + 0,2}{100}$$

so

$$L_4 = 14,3 \text{ m}$$

Moreover:

$$\frac{0,2 - 0,1}{L_5} = \frac{0,5 - 0,2}{100}$$

so

$$L_5 = 33,3 \text{ m}$$

Thus, the maximum distance between two street inlets is 47,6 m (33,3 + 14,3). Then, we must make sure that the flow collected at each street inlet falls below the 0,03 m^3/s limit. If it does not, the distance between inlets must be shortened but the slope of the gutter is maintained at 0,5%.

9.3.3 Other Criteria to Be Considered for Street Inlet Location

In addition to the hydraulics practices discussed in sections 9.3.1 and 9.3.2, the street inlets must be positioned to allow free access to pedestrian crossings at street intersections. Figure 9.10 clearly shows the best location for street inlets at street intersections.

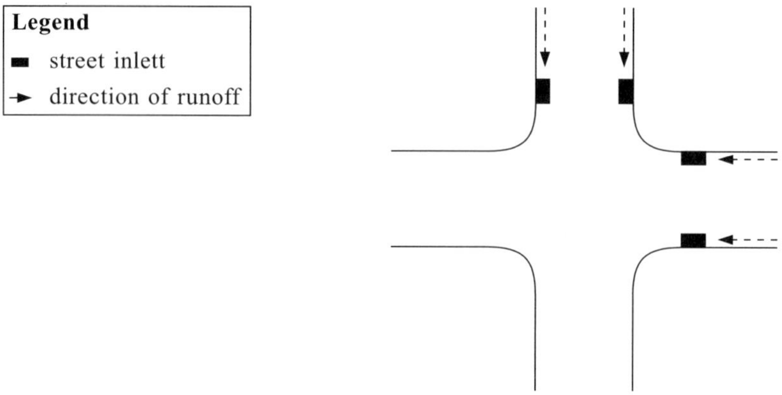

Figure 9.10 Street inlet location at a street intersection

9.4 LOCATION OF PIPES IN STREETS

Various types of pipe must be installed under the pavement; it is the engineer's role to show the contractor where the pipes should be installed relative to one another.

9.4.1 Location of Sanitary and Storm Sewer Pipes in Streets

In general, sanitary and storm sewers are installed at the same time, in the same trench, and by the same contractor. The pipes are installed as close as possible to one another to minimize excavation. Information used and engineering rules of practice are as follows:

- the thickness of the concrete pipe walls, expressed in inches, is equal to the nominal diameter of the pipe, expressed in feet, plus one. Thus, the thickness of the pipe wall for a pipe with nominal diameter of 2 ft (24 in) is 3 in (2 + 1) (section 8.4.1); other pipe wall thicknesses are given by manufacturers. This information makes it possible to compute the outside diameter;
- the minimum distance between any two exterior walls must be at least 600 mm, so that backfill can be properly compacted: this is the minimum distance between, for example, two sewer pipes or a sewer manhole and a nearby sewer pipe (fig. 8.15). This distance is often considered the horizontal distance. Moreover, the minimum vertical distance between two pipes, located one above the other – that is, crossing over – is 100 mm. This distance may be even smaller as long as rigid insulation is

installed between the pipes to avoid any load concentration; this rule, based on resistance of materials, is also valid for drinking water pipes.

– information concerning bedding thickness and width as well as the slopes of trench walls are given in section 11.9 and table 11.8.

Note: On this subject, see also section 8.10.

9.4.2 Location of Drinking-Water Pipes in Streets

Water-distribution pipes are often installed at the same time as sewer pipes. They may even be installed in the same trench, provided (section 8.10) the water-distribution pipe is at least 300 mm above (outside distance) the highest sewer pipe and at a minimum horizontal distance of 300 mm (outside distance) from any sewer pipe (figs. 8.14, 9.11, 9.12). The vertical distance is the most difficult to respect when trying to minimize excavation. These requirements, however, prevent contamination of drinking water by wastewater. If these rules cannot be followed, the water-distribution pipe must be installed in a separate trench and the minimum horizontal outside distance between the sewer and the water-distribution pipe walls must be at least 3 m (fig. 9.12). This enables us to maintain an unaltered soil core between the two pipes. This soil core, however, does not have to extend all the way to the natural ground surface.

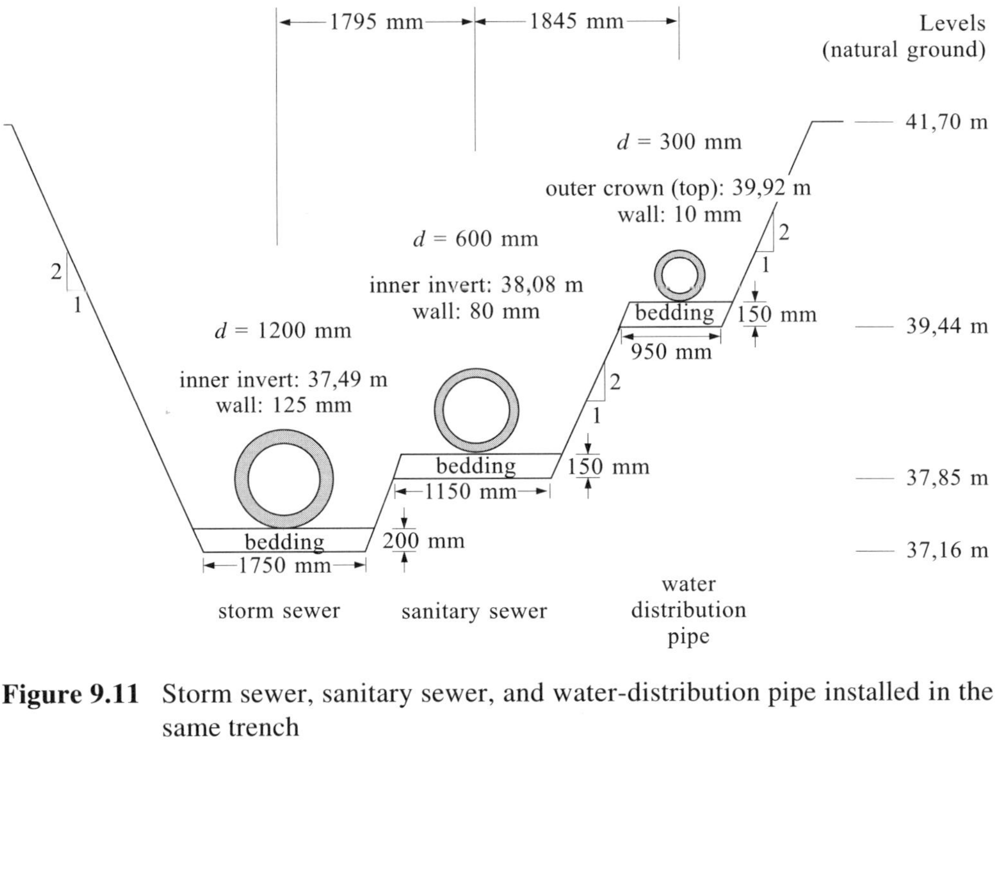

Figure 9.11 Storm sewer, sanitary sewer, and water-distribution pipe installed in the same trench

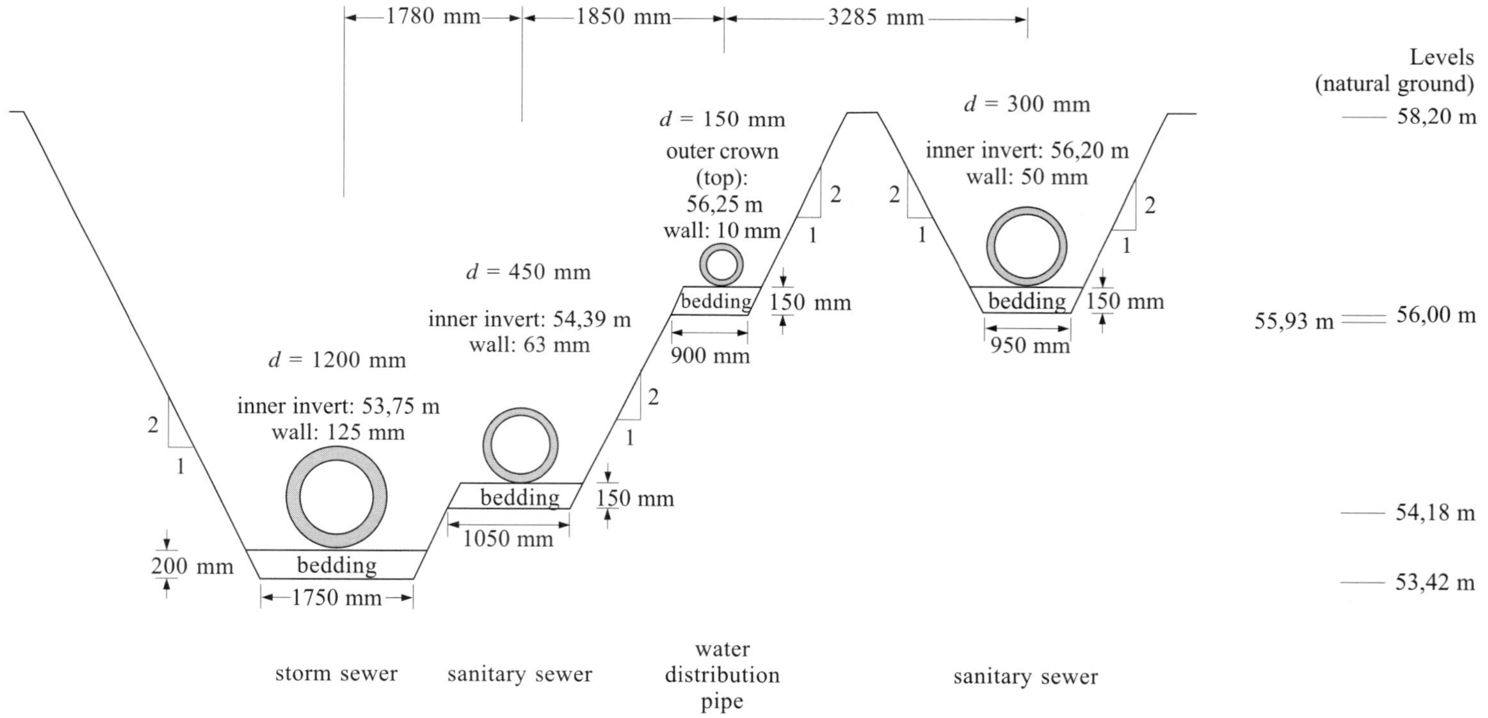

Figure 9.12 Sanitary sewer and storm sewer installed in the same trench; water distribution pipe installed in a separate trench

9.5 CULVERTS

A culvert is a storm-sewer pipe enabling runoff to cross under a road or railway track, or a channel to join two ditches, for example. The maximum flow in a culvert is calculated as for a storm sewer: by considering the size of the watershed's surface area flowing to it, its runoff coefficient, and adequate rainfall, usually a precipitation with a 5-year return period. Depending on the level of protection required for construction, the precipitation chosen may reach a return period of 25 years (ministère du Transport du Québec). In general, this computation must take into account the maximum allowable water level at the inlet (*HW* for Head Water) and outlet (*TW* for Tail Water) of the culvert to avoid flooding. The vertical cross section of a culvert is usually rectangular, circular, or elliptical. Typical culvert material is concrete or galvanized steel with or without an inside coating. Figure 9.13 shows the longitudinal cross-section of a culvert.

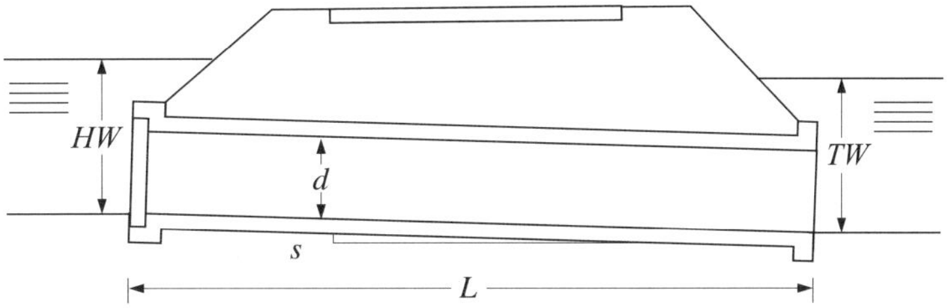

Figure 9.13 Longitudinal vertical cross section of a culvert under a road; *HW* and *TW*: maximum allowable water levels

9.5.1 General Hydraulic Characteristics

Culverts may have the following characteristics: the inlet or the outlet, or both ends, may be submerged; the flow may be laminar, critical, or turbulent; and the culvert may be too short for the water to reach uniform and stable conditions. Figure 9.14 shows various flow conditions. Usually, the culvert is assumed to be long enough for the flow to reach equilibrium, since this is generally the case and the computation results are then accurate. (If not, a complete hydraulic computation must be made.) A culvert will flow full when the outlet is submerged or when the outlet is not submerged but the headwater is high and the culvert is long. According to laboratory investigations, the entrance of an ordinary culvert will not be submerged if the headwater is less than a certain critical value, 1,5 × d for a preliminary analysis (Chow, 1959).

The computations are made in two separate steps. First, we find the entrance water level required to overcome the head losses at the inlet (in this case, the culvert acts as a weir at its inlet and its hydraulic capacity is independent of its length, slope, or Manning's n coefficient). The water level, h_e, must not be higher than the level specified by the problem (*HW*); otherwise, a culvert with a larger cross section must be chosen (or a second culvert be installed in parallel). In the second step, the computations determine whether head losses within the culvert and at its outlet govern its hydraulic capacity. If so, the water level at the culvert entrance will be higher than in the first step for the same flow.

Depending on whether the water level at the culvert entrance is higher as a consequence of inlet head losses or of total head losses related to friction within the culvert plus head losses at its outlet, we speak of inlet control or outlet control. It is interesting to recall at this stage that **in most cases the hydraulic capacity of culverts is inlet controlled**.

Culverts can be designed with the help of nomographs prepared by hydraulics engineers (TUBECON 1995; ministère du Transport du Québec, 1994) that consider previous simplifications (mainly, the culvert being long enough to reach uniform flow conditions). In the following example, we use the nomograph prepared by TUBECON, in which outlet-control results give inlet water level (h_e) plus culvert uneven level between its ends (SL). (Note: Other nomographs, such those of the ministère du Transport du Québec, give a value *H,* which is the difference of the water level at the inlet and outlet of the culvert.) It is therefore possible to compute water level at the outlet of the culvert (h_s). If h_s is higher than the maximum allowable water level at the outlet (*TW*), a larger culvert must be chosen (or a second culvert be installed in parallel).

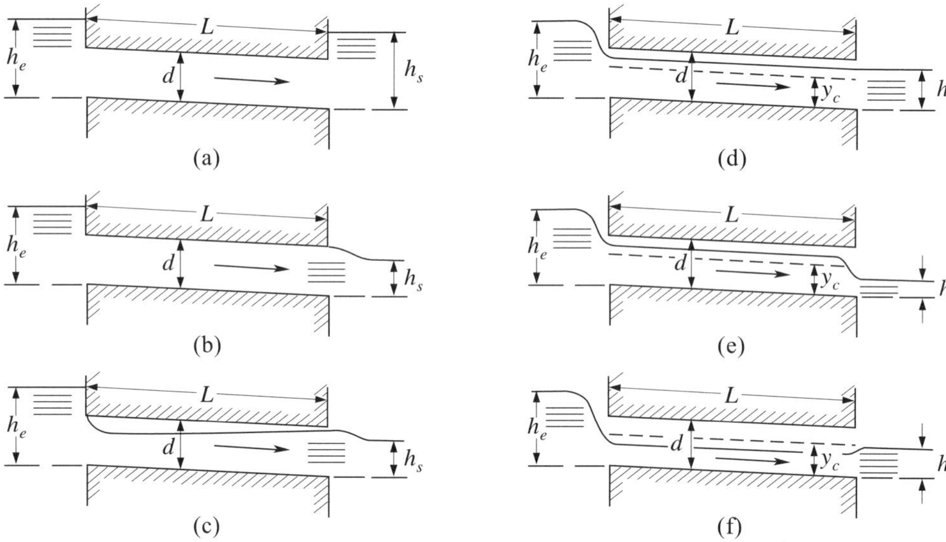

Figure 9.14 Types of culvert: a) outlet submerged. Outlet unsubmerged – headwater greater than the critical value: b) culvert hydraulically long and c) culvert hydraulically short. Outlet unsubmerged – headwater less than the critical value: d) tailwater higher than the critical depth; tailwater lower than the critical depth: e) slope subcritical and f) slope supercritical.

Adapted from: V.T. Chow, *Open-Channel Hydraulics* (New York: McGraw-Hill, 1959), p. 497.

9.5.2 Water Velocities Within Culverts for Maximum Flow (Design Flow)

When a culvert is inlet controlled it acts as a weir: in this case the culvert generally flows partly full and the water depth and velocity are computed as explained in chapter 2, using the graphs in figure 2.11 for sewers flowing partly full. A minimum self-cleaning velocity is 0,75 m/s.

When the culvert is outlet controlled, it usually flows full and water velocity is equal to the ratio of flow to the inside cross section of the culvert.

Curves such as the TUBECON curves (figs. 9.15 and 9.16) can be used for these computations. Let us illustrate this procedure with an example.

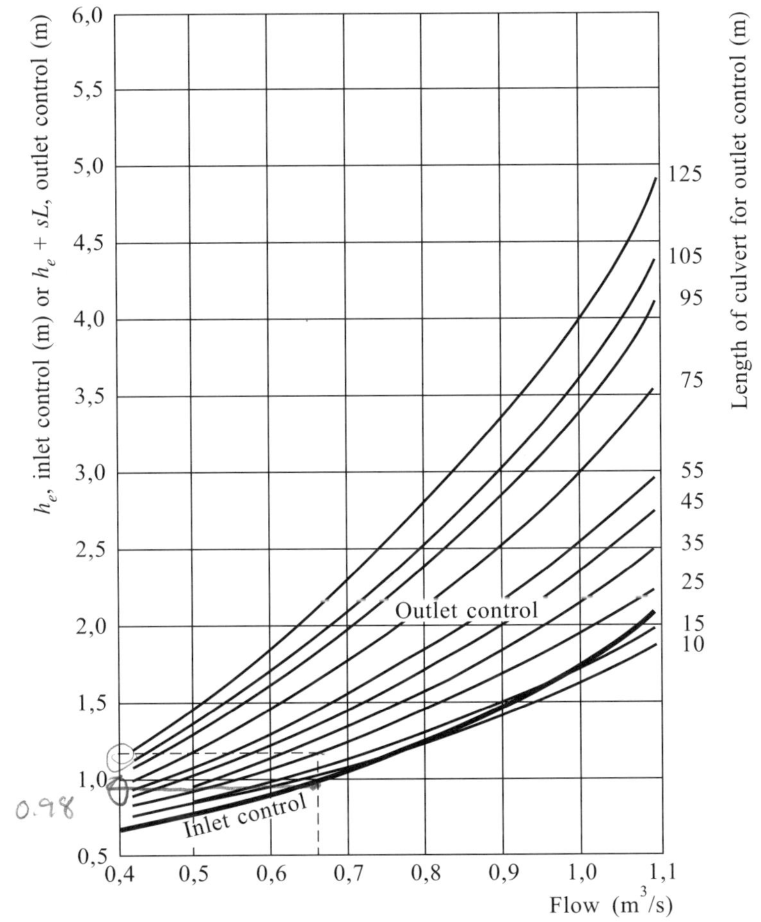

Figure 9.15 Hydraulic capacity of a concrete culvert with an inside diameter of 600 mm and of various lengths; Manning's roughness coefficient, n, is 0,012 and it is a projecting culvert inlet

From: TUBECON, *Canalisations et éléments préfabriqués en béton* (Association québécoise des fabricants de tuyaux de béton, 1995), chapter 4.

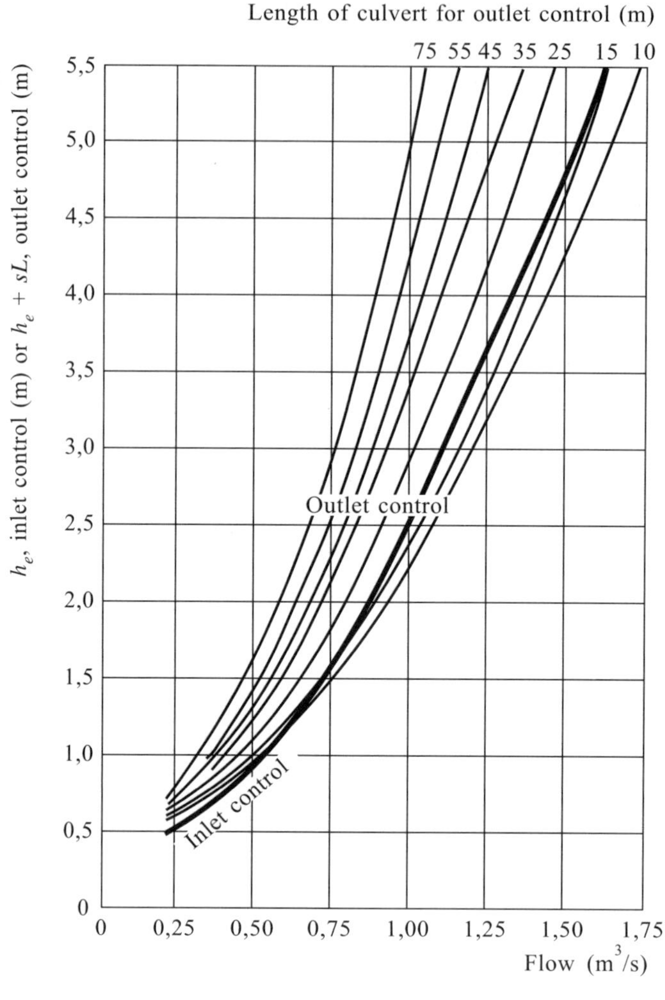

Figure 9.16 Hydraulic capacity of concrete culverts with an inside diameter of 525 mm and of various lengths

From: TUBECON, *Canalisations et éléments préfabriqués en béton* (Association québécoise des fabricants de tuyaux de béton, 1995), chapter 4.

EXAMPLE 9.1

A circular culvert crosses under a road and links two ditches located on each side of the road. Hydraulic computations have shown that the flow, Q, in the culvert is 0,66 m^3/s. The heights of the water levels allowed at the inlet, HW, and at the outlet, TW, of the culvert are 1,0 m and 0,75 m, respectively. The slope of the culvert, s, is 2%, its length, L, is 25 m and its Manning's coefficient, n, is 0,012.

a) Initial choice of the culvert's diameter: a diameter equal to or slightly greater than half the height allowed at the inlet is suggested; thus, in this case, 0,5 m (1,0 m ÷ 2); a 600 mm diameter culvert is chosen because of its market availability.
b) Checking of the water level at the inlet assuming that the inlet conditions of the culvert controls the flow: on the vertical axis of figure 9.15 (600 mm culvert), we find with the "inlet control" curve, for a flow of 0,66 m^3/s, that the height of the water level at the inlet, h_e, on the vertical axis is 0,98 m. Assuming inlet controls the flow, this 600 mm culvert is a good choice since the inlet water level is smaller than 1,0 m the maximum acceptable headwater (*HW*) level.
c) Checking of the water level at the outlet assuming that the outlet of the culvert controls the flow: in figure 9.15 (600 mm culvert), we find with the "outlet control" curve drawn for a 25 m culvert, and for a flow of 0,66 m^3/s, on the vertical axis, the value $h_e + s_L$ i.e. 1,16 m. As the culvert is considered long enough to have a stable and uniform flow, the water level at the inlet with respect to the invert, h_e, is assumed equal to the water level at the outlet with respect to the invert, h_s, which yields:

$$\begin{aligned} h_e + sL &= 1{,}16 \\ h_e &= 1{,}16 - sL \\ &= 1{,}16 - 0{,}2 \times 25 \\ &= 0{,}66 \text{ m} \end{aligned}$$

given

As the value of ($h_s = h_e$) is less than 0,75 m, the requirement at the outlet is respected. The 600 mm diameter culvert meets the requirements of the problem. Moreover, since the water level at the inlet is higher for inlet controlled than for outlet controlled computations, we conclude that this culvert is inlet controlled. As we have said, this is usually the case. Would a 525 mm diameter culvert meet these requirements (fig. 9.16)?

CONCLUSION

When designing sewer pipes, the engineer may be faced with difficult problems. For storm sewers, the rational equation is used by carefully computing the inlet time, the time of concentration, and the runoff coefficient of the watersheds, and then choosing the appropriate IDF (intensity-duration-frequency) equations and curves. For sanitary sewers, several facts must be known, such as population, domestic flow per person, maximum and minimum peak factors, infiltration, and inflow. For the intercepting sewer transporting wastewater to the sewage-treatment plant, various situations arise depending on the type of sewer systems in the tributary watersheds. In the case of combined sewer systems, the interceptors must collect a maximum flow equal to the maximum sanitary flow (which can occur at any time as it can be caused by the sanitary flow or by runoff) plus infiltration. Any rainwater excess is evacuated into the receiving stream (storm overflow) or is stored in retention/detention basins. The maximum and minimum flows in these situations (and many others) are presented in table 9.8.

Table 9.8 Maximum and minimum flows in sewers, main sewers and interceptors

Situation	**Maximum flow (Q_{max})**	**Minimum flow (Q_{min})**
1 sanitary sewer (or main sewer)	Q_{max} = population × Q_{avg}/person × FP* + Q_{inf} + Q_i	Q_{min} = population × Q_{avg}/person × FP + Q_{inf}
1 storm sewer (or main sewer)	Q_{max} = $KAIR$; 5 year return period rainfall	Q_{min} = 0
1 combined sewer (or main sewer)	Q_{max} = $KAIR$; 10 year return period rainfall	Q_{min} = population × Q_{avg}/person × FP + Q_{inf}
1 combined interceptor (1 watershed) 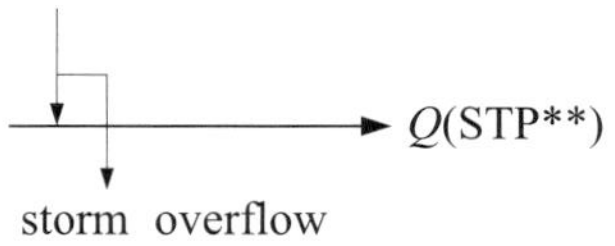	Q_{max} = population × Q_{avg}/person × FP + Q_{inf} + extra runoff flow (see N.B. at the bottom of this table)	Q_{min} = population × Q_{avg}/person × FP + Q_{inf}
Interception of two sanitary sewer systems in series (A and B) 	population (A + B) × Q_{avg}/person × FP_{AB} + $(Q_{inf})_A$ + $(Q_i)_A$ + $(Q_{inf})_B$ + $(Q_i)_B$	Q_{min} = population (A + B) × Q_{avg}/person × FP_{AB} + $(Q_{inf})_A$ + $(Q_{inf})_B$
Interception of two storm sewer systems in series (A and B) 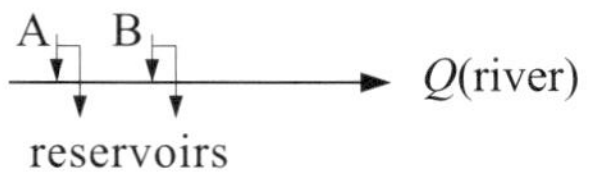	$Q_{max} = K(A_A + A_B)\, I_{AB}\, R_{AB}$; 5 year return period rainfall	Q_{min} = 0
Interception of two combined sewer systems in series (A and B) 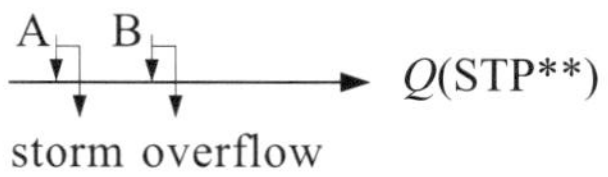	1) Interceptor (ΣQ_{max}) Q_{max} = population (A) × Q_{avg}/person × FP_A + population (B) × Q_{avg}/person × FP_B + $(Q_{inf})_A$ + $(Q_{inf})_B$ (see N.B. at the bottom of this table) 2) Main sewer $Q_{max} = K(A_A + A_B)\, I_{AB}\, R_{AB}$; 10 year return period rainfall	Interceptor or main sewer Q_{min} = population (A + B) × Q_{avg}/person × FP_{AB} + $(Q_{inf})_A$ + $(Q_{inf})_B$

* FP = maximum or minimum peak factor, accordingly. ** STP = sewage treatment plant.

N.B. In the case of combined sewer interception, it is usual to accept a certain (extra) runoff flow to avoid frequent storm overflows to the river: e.g. 0,05 m^3/person·d.

Table 9.8 (continued)

Situation	Maximum flow (Q_{max})	Minimum flow (Q_{min})
Interception of a sanitary sewer system (A) and a combined sewer system (B) 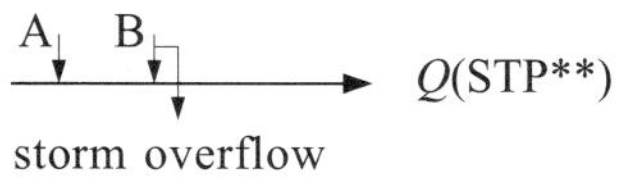 	1) Interceptor (ΣQ_{max}) Q_{max} = population (A) × Q_{avg}/person × FP_A + $(Q_{inf})_A$ + $(Q_i)_A$ + population (B) × Q_{avg}/person × FP_B + $(Q_{inf})_B$ (see N.B. at the bottom of this table) 2) Main sewer Q_{max} = population (A) × Q_{avg}/person × FP_A + $(Q_{inf})_A$ + $(Q_i)_A$ + $KA_B\, I_B\, R_B$ (10 year return period rainfall)	Q_{min} = population (A + B) × Q_{avg}/person × FP_{AB} + $(Q_{inf})_A$ + $(Q_{inf})_B$
Interception of a sanitary sewer system (A) and a pseudo-sanitary sewer system (B)  	Q_{max} = population (A + B) × Q_{avg}/person × FP_{AB} + $(Q_{inf})_A$ + $(Q_i)_A$ + $(Q_{inf})_B$ + (authorized runoff)$_B$	Q_{min} = population (A + B) × Q_{avg}/person × FP_{AB} + $(Q_{inf})_A$ + $(Q_{inf})_B$
Interception of two sanitary sewer systems (A and B) and a combined sewer system (C) 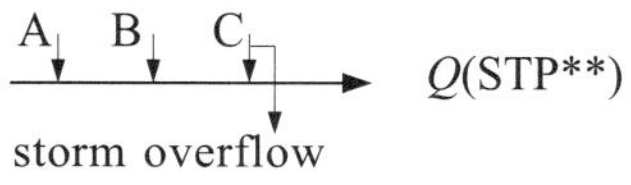 	1) Interceptor Q_{max} = population (A + B) × Q_{avg}/person × FP_{AB} + $(Q_{inf})_A$ + $(Q_{inf})_B$ + $(Q_i)_A$ + $(Q_i)_B$ + population (C) × Q_{avg}/person × FP_C + $(Q_{inf})_C$ (see N.B. at the bottom of this table) 2) Main sewer Q_{max} = population (A + B) × Q_{avg}/person × FP_{AB} + $(Q_{inf})_A$ + $(Q_{inf})_B$ + $(Q_i)_A$ + $(Q_i)_B$ + $KA_C\, I_C\, R_C$ (10 year return period rainfall)	Interceptor or main sewer Q_{min} = population (A + B + C) × Q_{avg}/person × FP_{ABC} + $(Q_{inf})_A$ + $(Q_{inf})_B$ + $(Q_{inf})_C$

* FP = maximum or minimum peak factor, accordingly.

** STP = sewage treatment plant.

N.B. In the case of combined sewer interception, it is usual to accept a certain (extra) runoff flow to avoid frequent storm overflows to the river: e.g. 0,05 m^3/person·d.

Finally, street inlets are installed along sidewalk curbs in such a way that they can collect a maximum flow of 0,03 m^3/s while keeping pedestrian crosswalks free of runoff. When the street slope is low (less than 0,05%), the slope of the gutter must nevertheless be set at 0,05%.

PROBLEMS

9.1 A storm-sewer system was designed with an inlet time, t_e, of 10 min and a rainfall return period of 5 years (Montreal region). The plans were completed and the storm sewer system was installed before streets and houses were constructed. Now, major changes have been proposed to the urban plans. These modifications affect only the inlet time, which will be 20 min. The surface areas of every sub-basin and runoff coefficients remain the same. Explain and comment on the influence of these modifications on the overall performance of the storm-sewer system already built.

a) Will the flows running off the system be greater, smaller, or unchanged? Explain.

b) Compare the computed flow with an inlet time, t_e, of 10 min and the flow obtained with a t_e of 20 min. Can you calculate the difference between the two flows in the first pipe line?

c) Would the difference computed in b) be the same for the upstream and downstream parts of the system? If not, how does it change?

9.2 In a combined-sewer system, what is the dry-weather flow (DWF)? Does this flow include infiltration and inflow? If so, can infiltration and inflow be quantified? How?

9.3 An intercepting sewer and a trunk sewer collect the wastewater of the combined sewer system of the same watershed located in the Montreal region. Explain how the design of these two pipes varies. Would your answer be the same if the interceptor and the trunk sewer were collecting the wastewater of a sanitary sewer system? Explain.

9.4 Design the storm-sewer system shown in figure 9.17. The pipes are made of concrete (Manning's n = 0,013) and the characteristics of the system are presented in table 9.9. Determine the elevation of the inner invert for every pipe and compute their diameter and slope.

At all times, make sure that the cover is at least 2,2 m over the pipe crown (top) (to compute this, neglect the thickness of the pipe walls; see section 8.4.7). In the proposed solution, take into account the following constraints:

a) Sewer manholes cannot be moved.

b) The flow velocity in the pipes, when they flow full, must remain between 0,9 and 3,0 m/s.

c) The minimum nominal pipe diameter is 300 mm.

d) Installation of a pipe with a diameter smaller than the diameter of the pipe located upstream is not permitted.

e) Excavation must be minimized by choosing the smallest possible slopes.

f) There are no constraints for slope.

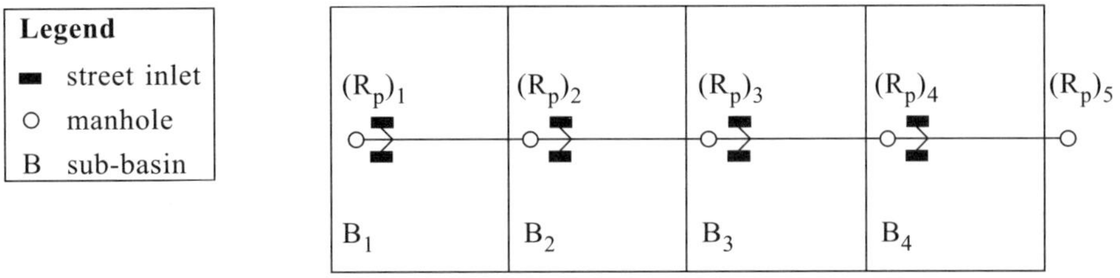

Figure 9.17 (problem 9.4) Storm-sewer system

Table 9.9a Storm sewer system (problem 9.4): characteristics of the sub-basins

Parameters	Sub-basin			
	B_1	**B2**	**B3**	**B4**
Area (ha)	1,0	1,2	1,4	1,6
Inlet time (min)	15	15	15	15
Runoff coefficient (R_T)	0,45	0,50	0,55	0,60

Table 9.9b Storm-sewer system (problem 9.4): characteristics of the pipes

Pipes	Ground elevation (m)		Length (m)
	upstream	**downstream**	
$(R_p)_1$–$(R_p)_2$	200,00	200,00	150
$(R_p)_2$–$(R_p)_3$	200,00	198,50	150
$(R_p)_3$–$(R_p)_4$	198,50	198,50	150
$(R_p)_4$–$(R_p)_5$	198,50	197,50	150

Chapter 10

DISEASE AND OTHER RISKS RELATED TO WATER

INTRODUCTION

Water can be contaminated at any time and may be unsuitable for some uses, if not all. Water is potable only when its quality is adequately controlled and, more importantly, after it has been disinfected. Municipalities responsible for supplying drinking water must meet certain standards. Sewer workers must take special precautions to protect their health and lives when maintaining and repairing sewer systems. They come into almost daily contact with soiled water and work at times in toxic or explosive environments. This chapter discusses the most common diseases that can be transmitted by bacteria, viruses, and other organisms living in water or in the sewer systems. It also presents ways to protect against these diseases.

10.1 RISKS RELATED TO CONSUMPTION OF CONTAMINATED DRINKING WATER AND TO SKIN CONTACT WITH POLLUTED WATER

Many bacterial and viral diseases are transmitted to humans when they drink water containing pathogenic organisms. Since the end of the nineteenth century, many micro-organisms have been identified and methods have been developed to either destroy them (e.g., by chlorine disinfection) or to protect humans against them (by vaccines, antibiotics, etc.).

10.1.1 Risks Related to Consumption of Contaminated Drinking Water

First, we will list the most common microbial diseases that threaten humans who drink contaminated water. For each disease, we identify the agent (bacterium, virus, protozoa, or amoeba), describe the symptoms, and evaluate the risk of death.

First of all, note that the main symptoms of waterborne diseases are the following: diarrhoea or (rarely) constipation, fever, often violent abdominal cramps, and vomiting. The similarity of these symptoms makes it virtually impossible to make a definite diagnosis. For this reason, the treating physician must be informed of every possible clue to perform an accurate diagnosis: countries visited recently, persons met, type of food ingested, professional exposure, and type of precaution taken.

Cholera. Contagious bacterial disease causing acute intestinal infection. Its symptoms are frequent diarrhoea, uncontrollable vomiting, intense thirst, and rapid dehydration. This disease may cause death in 50% of severe cases that remain untreated.

Typhoid fever. Bacterial infection causing fever, headaches, weight loss, slowed heartbeat, enlargement of the spleen, formation of pink spots on the body, dry cough, and constipation. This disease is sometimes benign and without any noticeable symptoms; however, it may cause death in 1% of cases.

Paratyphoid fever. Viral disease clinically similar to typhoid fever. The death rate, however, is much lower.

Dysentery. Generic term used to characterize diseases causing severe diarrhoea with passage of blood, accompanied by colic, nausea, and vomiting: bacillary dysentery or shigellosis (caused by various bacteria), balantidiasis (caused by ciliated protozoa), and amoebic dysentery (a serious chronic infection that diminishes quality of life and cuts life short). Shigellosis is the only type that may cause death, and the mortality rate may reach 20%.

Infectious diarrhoea. Diseases caused by various bacteria and producing the following symptoms: liquid bowel movements, vomiting, and fever. In general, these diseases do not cause death.

Table 10.1 presents all of these diseases and other related data.

10.1.2 Risks Related to Skin Contact with Polluted Water

A number of diseases are caused by pathogenic organisms capable of penetrating the skin. Individuals coming into contact with contaminated water may contract these diseases. Sewer workers and laboratory workers, students, and technicians specializing in the study and treatment of wastewater are at higher risk, as they are frequently in contact with contaminated water. This section lists the most common diseases, identifies their agents (bacterium, virus, or worm), describes their symptoms, and evaluates the potential risk of death.

Type A hepatitis. Viral infection causing fever, weight loss, nausea, abdominal pain, and finally jaundice. This disease can be severe and may become disabling over a period of several months; recovery is often very long. This disease causes death in 0,1% of cases.

Tetanus. This disease is caused by the toxin of the tetanic bacillus, which develops in wounds under anaerobic (without oxygen) conditions. Tetanus causes painful tonic spasms in the cheeks, neck, and torso. The death rate can reach 90%, depending on the age of the patient.

Bilharziasis (schistosomasis). Disease caused by worms living in the bloodstream. The symptoms are diarrhoea, abdominal cramping, liver fibrosis and cirrhosis, hypertension of the portal vein (vein transporting blood from the small intestine, pancreas, spleen, and stomach to the liver), obstruction of the urinary tract, bladder infection, and cancer of the bladder. The larvae of this parasite develop mostly in fresh water in warm-climate countries. They live in freshwater molluscs (mainly *Bulinus, Biomphalaria, Oncomelania* and *Tricula*), which act as temporary hosts. Larvae penetrate the skin and enter the bloodstream. This disease rarely causes death.

10.1.3 Risks Related to Breathing Polluted Water Droplets

It has been only a few years since legionnaire's disease (legionellosis) was linked to breathing water droplets contaminated by a certain bacterium. This bacterium usually reproduces in warm water (39° to 40,5° C) and can survive for months in tap water and distilled water.

10.1.4 Drowning

Finally, sewer workers are at risk for drowning, especially when they work in storm or combined sewer systems. The flow may rise very quickly at the beginning of or during a rainfall or when floodgates are opened to discharge a retention/detention reservoir, for example.

10.2 GASES AND VAPOURS ENDANGERING THE HEALTH AND LIFE OF SEWER WORKERS*

Water workers, particularly sewer workers, may be exposed to toxic or explosive gases or to volatile toxic substances hazardous to health and life. They also run the risk of asphyxiation (lack of oxygen). This section lists the main gases and substances to guard against.

Methane (CH_4). Colourless and odourless, methane gas is produced by the anaerobic decomposition (absence of molecular O_2) of organic matter accumulated where sewer wastewater flows too slowly. Methane is explosive (responsible for firedamp explosions in coal mines) in volume concentrations between 5% and 15%. As it is lighter than air, it is usually found in the upper parts of equipment.

Hydrogen sulphide (H_2S). This gas is colourless, toxic, and flammable. Hydrogen sulphide acid is lethal for humans in contact with it for more than a half-hour, at a volume concentration of as low as 0,05%. It is produced by the anaerobic decomposition of organic matter. This gas is explosive in volume concentrations between 4% and 46%. At high concentration, it diminishes the sense of smell which increases the risk.

---

* Adapted from: Water Pollution Control Federation, *Safety and Health in Wastewater Systems, Manual of Practice I* (Washington, 1983), pp. 8.5–8.6; Syndicat professionnel des distributeurs d'eau et Exploitants de réseaux d'assainissement, *Hygiène et Sécurité, Assainissement* (Paris, 1979).

Table 10.1 Most common diseases transmitted to humans by contaminated drinking water and wastewater

Disease	Type of pathogenic organisms	Distribution of disease	Main mode of transmission to humans	Most common means of protection	Carrier of infectious pathogenic organism
Cholera	bacterium	worldwide	water and food contaminated by faeces; wastewater; faecal-oral tract	vaccine has little efficiency; boiling and disinfection of drinking water; avoiding raw vegetables	human
Type "A" hepatitis	virus	worldwide	contact with wastewater, faecal-oral tract	injection of gamma-globulin	human
Typhoid fever	bacterium	worldwide	water and food contaminated by wastewater; shellfish	vaccination; boiling and disinfecting drinking water; pasteurising dairy products	human
Paratyphoid fever	bacterium	worldwide	water and food contaminated by wastewater; shellfish	boiling or disinfecting drinking water; pasteurising dairy products	human
Tetanus	bacterium	worldwide	spores of the bacterium introduced in a wound by dirt or rust, etc.; contact with wastewater; injection with contaminated needles	vaccination	human animals soils
Giardiasis	protozoa	worldwide	water and food contaminated by wastewater and faeces; faecal-oral tract	filtration of drinking water, hygiene	human animals

Amoebic dysentery	amoeba	worldwide	drinking water contaminated by wastewater; food contaminated by faeces; faecal-oral tract	treatment of drinking water by filtration and disinfection; hygiene	human
Balantidiasis	protozoa	worldwide	contaminated water, food and objects; faecal-oral tract	hygiene; cooking food; boiling water; water treatment including filtration	pig
Schistosomiasis (Bilharziasis)	worm	Africa; South America; West Indies	physical contact with the worm larvae found in contaminated fresh water	avoid swimming in fresh water; workers should wear boots and gloves	human
Infectuous diarrheas	various bacteria	worlwide; mostly in warm-climates with poor hygiene	water and food contaminated by wastewater; shellfish	hygiene; water disinfection; cooking of food; pasteurisation of dairy products	human and animals (pigs, cattle, sheep, chickens, cats, dogs, etc.)
Legionnaire's disease (legionellosis)	bacterium	worldwide	breathing contaminated water droplets	disinfection	water (mostly warm water)
Cryptosporidiosis	protozoa	worldwide	faecal-oral tract; drinking water and food contaminated by wastewater	filtration of drinking water, hygiene	human, pets and wild animals

Carbon dioxide (CO_2). Colourless and odourless, this gas is lethal for humans after more than half an hour of exposure when the volume concentration exceeds 6%. It is produced by the aerobic and anaerobic microbial transformation of organic matter and by several chemical transformations.

Oxygen (O_2). This gas is colourless, odourless and tasteless. It can fuel combustion but is not explosive. The atmosphere normally contains 20,9% oxygen (volume concentration). In general, a concentration of less than 19% is insufficient for humans, a concentration of under 13% is hazardous to health, and a concentration below 7% causes death. A low concentration of oxygen is often the result of a high concentration of other gases.

Several other gases can be found in sewer pipes; they originate from infiltration or discharge of industrial wastewater. These include gasoline fumes (C_5H_{12} to C_9H_{20}), carbon monoxide (CO), industrial solvents (acetone, formaldehyde, phenol, nitric acid, etc.), chlorine (Cl_2), ammonia (NH_3), and ethane (C_2H_6 or natural gas). Table 10.2 lists gases and vapours that may endanger the health of water workers.

10.3 OTHER RISKS ENCOUNTERED BY WATER-MANAGEMENT WORKERS

Workers constructing or operating sewer systems, water-distribution networks, wastewater-treatment plants, and water-treatment plants face the same risks as workers on any other construction site: falls, electrocution, collision with moving objects, and so on. For this reason, workers must wear a hard hat, steel-toed boots, safety goggles, and a safety vest. When working in a confined atmosphere, such as the sewer system or pumping stations, additional protection is required: watertight anti-slip boots, watertight suit for the entire body (to avoid skin contact with pathogenic micro-organisms), low-tension anti-explosive electrical equipment, and so on. Because toxic and explosive gases may be present, sewer workers must never go into the sewer without a gas detector (mainly for CO, CH_4 and H_2S). Sewer workers must always be attached to the surface by safety harness, with a co-worker at the ready. However, this co-worker must not attempt a rescue without being properly harnessed to the surface where a third co-worker is also ready to intervene. In pumping stations, a ventilation system is usually installed, and it must be activated before entry (art. 8.6.3).

10.4 PERSONAL HYGIENE GUIDELINES

Because of the risk of explosion and the presence of bacteria and pathogenic viruses or of dangerous chemical substances, it is recommended that water workers:

- never smoke or eat on the work site;
- carefully wash their hands and brush under their nails as often as possible, especially before smoking or eating;
- clean and disinfect any wound and protect it during work hours;

– never touch the face, especially the nose, eyes, and mouth, with the hands;
– take a shower as often as possible.

From: Syndicat professionnel des distributeurs d'eau et Exploitants de réseaux d'assainissement, *Hygiène et Sécurité, Assainissement* (Paris, 1979), p. 101.

Table 10.2 Gases and Vapours Endangering Health and Safety of Sewer Workers

Gas	Explosive volume concentration (%)	Characteristics	Maximum volume concentration safe for humans (%)	
			60 min exposure	8 h exposure
NH_3	16-25	colourless, lighter than air, caustic (pungent) odour	0,03	0,01
CO_2	non-flammable	colourless, odourless	4-6	0,5
CO	12,5-74,2	colourless, odourless, tasteless	4,0	0,005
Cl_2	non-flammable	yellowish green, irritant, caustic (pungent) odour, support combustion	0,0004	0,0001
C_2H_6	3,1-5	colourless, odourless, flammable, tasteless, explosive provided sufficient O_2 (min 12%)	no limit	no limit
Gasoline fumes	1,3-7	odour perceived at 0,03%; flammable, explosive	0,4-0,7	variable
H_2S	4,3-46	smells of rotten eggs at low concentration, flammable, colourless, explosive	0,02-0,03	0,001
CH_4	5-15	colourless, odourless, flammable, explosive, tasteless	no limit	no limit
N_2	non-flammable	colourless, odourless, tasteless	no limit	no limit
O_2	non-flammable	colourless, odourless, tasteless, supports combustion	no limit	no limit

Adapted from: Water Pollution Control Federation, *Safety and Health in Wastewater Systems, Manual of Practice I* (Washington, 1983), pp. 8-5–8-6.

10.5 PROTECTING AGAINST DISEASES CAUSED BY VIRUSES, BACTERIA, PROTOZOA, AND WORMS

The general population, and especially water workers, have several ways to protect against bacterial diseases. Below is a brief description of these means.

10.5.1 Vaccination

Vaccination aims to immunize or protect humans against attacks by pathogenic micro-organisms by inoculation. The vaccine is made from micro-organisms that are either dead or weakened, or from other soluble products. The human body produces antibodies that effectively fight subsequent infection by pathogenic micro-organisms.

Vaccines are not available for all diseases, and vary in their effectiveness. Sewer workers and laboratory workers exposed to wastewater must also check with health authorities that vaccinations previously received are still effective (the effectiveness of some vaccines is not permanent), obtain booster shots when necessary, and receive all necessary vaccinations.

In Quebec, vaccines for polio and tetanus and for non-waterborne diseases such as diphtheria, measles, and whooping cough are mandatory. Vaccines for smallpox are no longer given, as this disease was eradicated in 1971.

10.5.2 Heating Water to Boiling Point

Boiling water transforms it from its liquid state to its vaporous state by heating and maintaining it at an adequate temperature. At atmospheric pressure, pure water evaporates at 100°C. This high temperature contributes to the destruction of some micro-organisms.

10.5.3 Avoidance of Skin Contact with Untreated Fresh Water in Warm Climates

Some agents of waterborne diseases, found in fresh water in warm-climate countries, can penetrate a swimmer's skin, reach the bloodstream, and affect vital organs (bilharziasis is an example). This is why tourists to these countries are discouraged from swimming in streams and lakes.

CONCLUSION

In addition to the usual risks encountered on construction sites, workers in water distribution, wastewater collection, water treatment, and wastewater treatment are exposed to specific risks of viruses and pathogenic bacteria and of toxic and explosive chemical substances. For their health and safety, water workers must:

- follow strict hygiene rules, at all times (frequent hand washing, showers, etc.);
- wear appropriate clothing;
- never undertake hazardous work alone.

PROBLEMS

10.1 Explain the dangers related to the presence of hydrogen sulphide (H_2S) and when these dangers occur.

10.2 Explain why a water worker working in a confined atmosphere must always be physically attached to a co-worker on the outside.

10.3 Explain, using the ideal gas law, why too much carbon dioxide (CO_2), a gas of low toxicity, causes asphyxia.

Chapter 11

LOADS ON PIPES AND SELECTION OF PIPES ABLE TO SUSTAIN SUCH LOADS

INTRODUCTION

Because sewer pipes are installed underground, they are subjected to loads that vary according to the burying depth, type of backfill, and various dead loads (fixed loads due to a constant mass, such as asphalt cover of a road) acting on the pipes. To these loads, we must add live loads (occasional moving loads) due to moving vehicles, because pipes are often installed under roads. This chapter demonstrates how to evaluate foreseeable total loads on the pipes and how to choose pipes that can sustain these loads.

11.1 CLASSIFICATION OF LOADING CONDITIONS

To compute the forces due to various loads exerted on a pipe, methods of construction and installation of the pipe must be known. This section describes these methods and defines their most important variables (section 11.2).

Trench conditions. The pipe is laid in a relatively narrow trench dug in undisturbed ground, then covered with backfill up to the initial ground level. As shown in figure 11.1, friction (shearing) forces between the backfill and the trench walls must be taken into account.

Embankment conditions. The pipe is covered with backfill well above ground level. To compute the force exerted on the pipe, we must take into account the various possible embankment conditions. a) The pipe is laid on natural soil then covered with backfill or laid in a trench with a width equal to or greater than the so-called transition width (see below). The vertical friction (shearing) forces exerted along the trench walls or between the backfill over the pipe and the backfill on each side of the pipe must then be computed differently. b) The pipe is laid in a shallow trench and backfilled much higher than ground level.

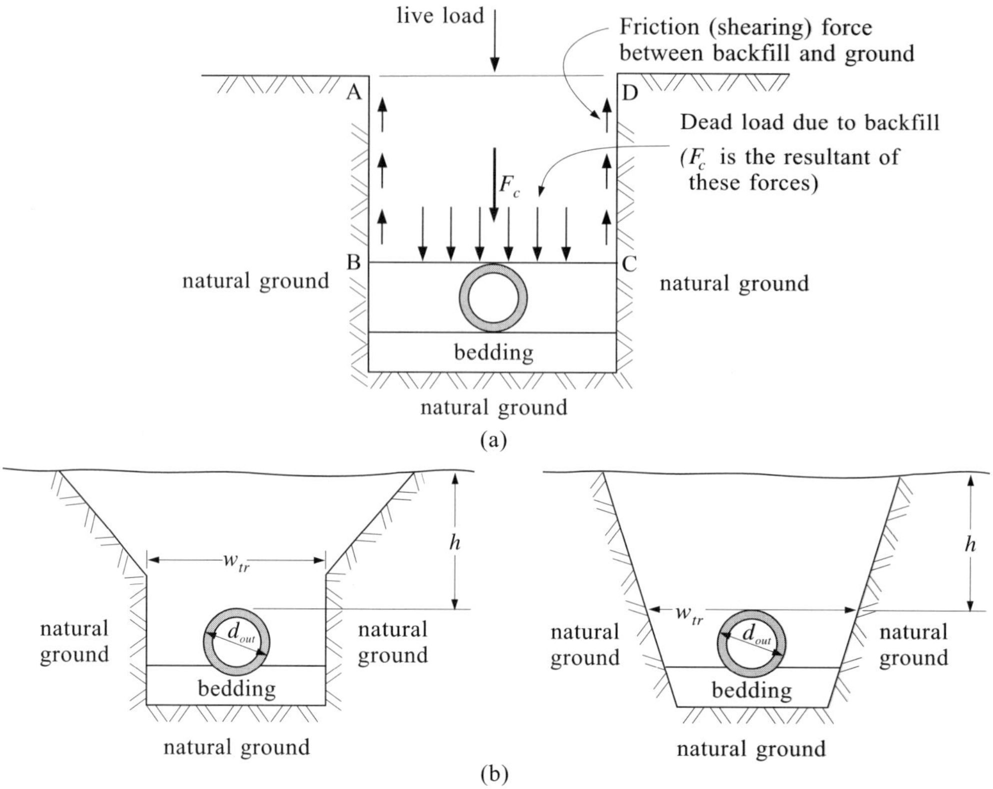

Figure 11.1 a) Trench conditions. Volume ABCD is the volume of the backfill with load, F_c, acting on the pipe. The friction (shearing) forces exerted between the backfill and the trench walls can tend upward or downward depending on whether the backfill material is more or less compacted than the natural soil; b) some of the symbols for variables used to compute forces due to dead loads

When the pipe crown (top) is above the original ground level, the pipe is said to be installed in *positive projecting conditions*. When the pipe crown (top) is below the original ground level, the pipe is said to be laid in *negative projecting conditions*. When the crown (top) is at original ground level, the pipe is said to be laid in *zero projecting conditions* (fig. 11.2).

Construction in tunnels. The pipe is laid in a tunnel dug into rock under the ground surface. For large flows, the pipe might be the tunnel itself, after the walls are made watertight and covered with concrete. The pipe may also be pushed into soft soil.

Transition width of a trench. This is the minimum trench width at which the friction (shearing) forces between the backfill and the trench walls are negligible for computation of the load due to backfill. This width can be estimated using the data in figure 11.3 (see also example 11.1).

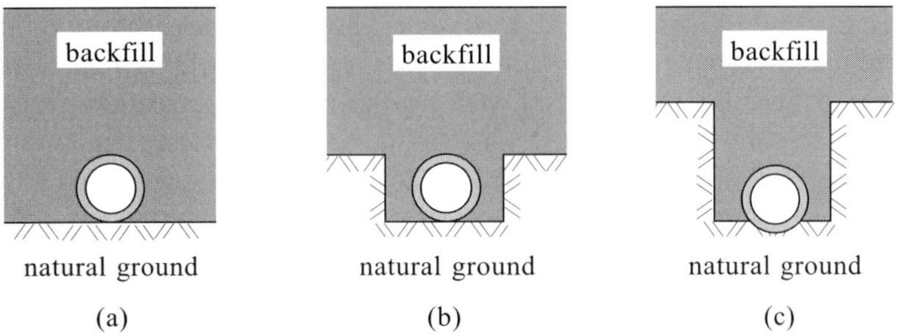

Figure 11.2 Embankment conditions: a) positive projecting conditions: the pipe crown (top) is above the natural ground level; b) zero projecting conditions: the pipe crown (top) is at the original ground level; c) negative projecting conditions: the pipe crown (top) is below the original ground level

11.2 VARIABLES AND SYMBOLS USED IN THIS CHAPTER

To understand the mathematical expressions used to compute the loads exerted on pipes, several variables must first be defined:

w_{tr}: width of the trench at the top of the pipe crown (top), regardless of the slope of the trench walls. After construction, when the supports (e.g. supporting walls) of the trench walls are left in place, the width w_{tr} is equal to the width of the trench minus the total thickness of these supports.

h: vertical distance separating the top of the pipe crown (top) and the ground surface, once the construction is finished (i.e., after backfill is added).

d_{out}: outside diameter of the pipe.

x: vertical distance separating the top of the pipe crown (top) and the original ground surface. The variable x is used to compute dead loads on positive and negative projecting pipes.

p: when the top of the crown of the buried pipe is above the original ground level (in the case of positive projecting condition) $p = x/d_{out}$. The value of p cannot exceed 1,0.

p': when the top of the buried pipe crown is below the original ground level (in the case of negative projecting condition), $p' = x/w_{tr}$.

r_{sd}: the settlement ratio, r_{sd}, expresses the relative settlement (compaction) of the backfill (in the trench) and of the original soil on each side of the trench. If r_{sd} is negative, the backfill will settle more than the natural soil; the forces exerted by the weight of the backfill on the pipe are therefore attenuated by the friction (shearing) forces between this backfill and the trench walls. For embankment conditions, r_{sd} expresses the ratio between the settlement of the backfill located over the pipe and the settlement of the backfill located on each side of the pipe. Table 11.1 suggests different values attributed to r_{sd}.

μ: coefficient of internal friction of the backfill material = tg ϕ (table 4.10).

μ': coefficient of friction between the backfill material and the trench walls = tg ϕ'. $\mu' \leq \mu$.

k: Rankine's coefficient, ratio of the lateral and vertical pressures exerted in the backfill material.

$$k = \frac{\sqrt{\mu^2 + 1} - \mu}{\sqrt{\mu^2 + 1} + \mu} = \frac{1 - \sin\phi}{1 + \sin\phi}$$

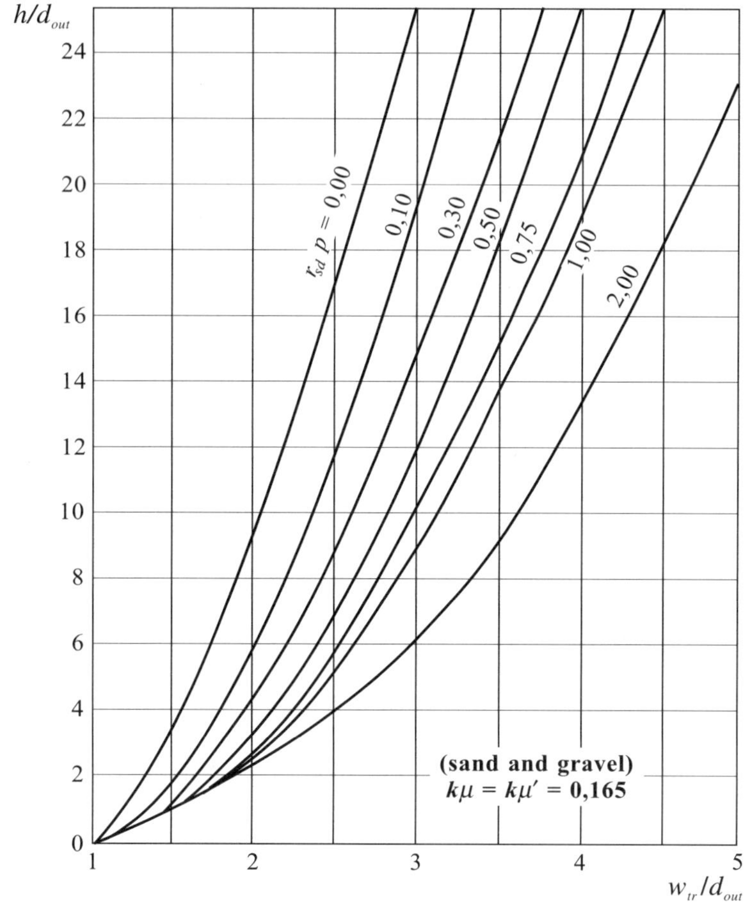

Figure 11.3 For a pipe of outside diameter d_{out} installed at a depth h, this graph enables the computation of the width of the trench for which the load due to the backfill would be the same as if the pipe were installed under embankment conditions. This trench width is the **transition width** of the trench. To use this graph, the pipe is considered to have been installed in positive projecting conditions – that is, laid on natural ground. "It may be difficult to determine an appropriate value of $r_{sd}p$ – $r_{sd}p$ = 0,5 is a reasonable good working value" (Spangler and Handy, 1973, p. 690). **This graph may be used for other granular material**

Adapted from: American Society of Civil Engineers and Water Pollution Control Federation, *Gravity Sanitary Sewer Design and Construction* (New York: ASCE and WPCF, 1982), p. 172.

Table 11.1 Design values for the settlement ratio, r_{sd}, as a function of the installation and foundation (bedding) conditions, and the value of p'

Installation and Foundation Conditions	Settlement ratio, r_{sd}
Positive projecting conditions:	
rock or unyielding soil	+1,0
sandy or ordinary soil	+0,7*
clay or yielding soil	+0,3
Zero projecting conditions	0,0
Negative projecting conditions	
$p' = 0{,}5$	-0,1
$p' = 1{,}0$	-0,3
$p' = 1{,}5$	-0,5
$p' = 2{,}0$	-1,0

* With good construction methods resulting in proper compaction of bedding and backfill materials, a settlement ratio design value of +0,5 is recommended.

Adapted from: Ontario Concrete Pipe Association Inc., *Concrete Pipe Design Manual* (1986), table 4.28, p. 4.96.

11.3 LOADS ON PIPES DUE TO BACKFILL MATERIAL AND NATURAL SOIL-TRENCH CONDITIONS

This section shows how to compute forces due to dead loads exerted on a pipe laid in a trench that is narrow enough to take into account the friction (shearing) forces between the backfill and the trench walls; for such a case, we say that trench conditions exist. **There are two main types of pipes: rigid (concrete and reinforced-concrete pipes) and flexible (pipes made of plastic, corrugated steel, copper, and ductile cast iron)**.

11.3.1 Rigid Pipes

The forces due to backfill material acting on pipes are calculated using the Marston equation:

$$F_c = C_d m g w_{tr}^2 \tag{11.1}$$

where F_c = force exerted on the pipe (N/m)

m = unit mass of the backfill material located over the top of the pipe (kg/m^3)

w_{tr} = width of the trench at the top level of the pipe crown (m)

C_d = load coefficient (Fig. 11.4), for which the value is a function of:
- the coefficient of internal friction of the backfill, μ
- the coefficient of friction between the backfill material and the trench walls, μ'
- the Rankine's coefficient, k
- the ratio h/w_{tr}

g = gravitational constant (9,81 m/s^2)

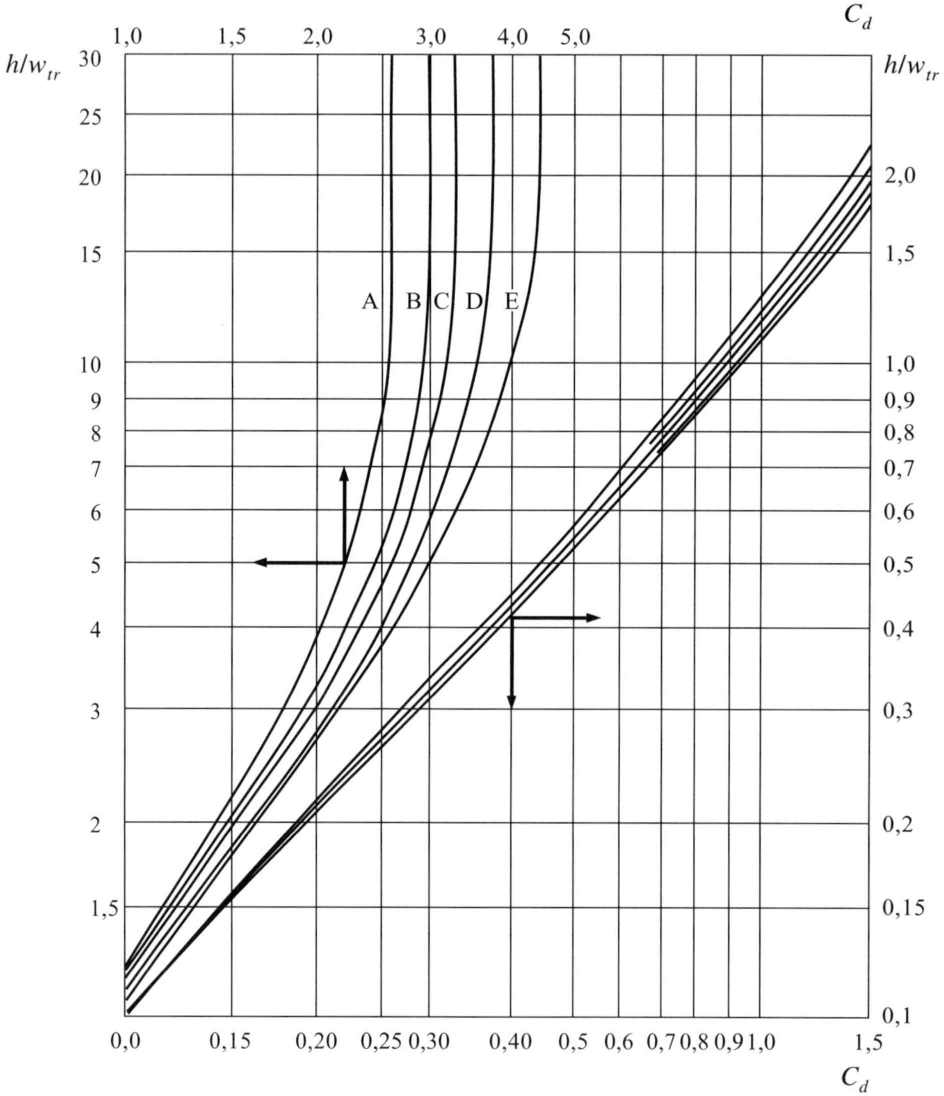

Figure 11.4 Values of the coefficient C_d used to compute the force due to backfill loads exerted on pipes laid in trenches (trench conditions).

Adapted from: American Society of Civil Engineers and Water Pollution Control Federation, *Gravity Sanitary Sewer Design and Construction* (New York: ASCE and WPCF, 1982), p. 170.

A: $k\mu' = 0{,}1924$ (backfill: granular material without cohesion)
B: $k\mu' \leq 0{,}165$ (backfill: sand and gravel)
C: $k\mu' \leq 0{,}150$ (backfill: saturated topsoil)
D: $k\mu' \leq 0{,}130$ (backfill: ordinary clay)
E: $k\mu' \leq 0{,}110$ (backfill: saturated clay)

EXAMPLE 11.1*

A circular concrete pipe 1200 mm in nominal diameter, d_{nom} (1220 mm real diameter), is to be installed in a trench with a width at the pipe crown level of 2,60 m. Over the pipe top (crown), the backfill will be sand and gravel; it will be 11 m high and weigh 1900 kg/m^3. The trench is excavated in sandy soil (which forms the bedding). Calculate the dead load exerted on the pipe (fig. 11.5).

Solution

Computation of the outside diameter of the pipe, d_{out}

A 1200 mm (48 in or 4 ft) pipe has a wall 5 in thick (4 + 1; chapter 8, section 8.4.1). As a result:

$$d_{out} = 1220 + 2\,(5 \text{ in} \times 25{,}4 \text{ mm/in}) = 1474 \text{ mm } (1{,}474 \text{ m})$$

Computation of the transition width of the trench, $(w_{tr})_{trans}$

For this computation, we must assume that the pipe is laid on the natural ground (embankment conditions case [a]) so that $p = 1{,}0$ ($p = x/d_{out} = d_{out}/d_{out} = 1{,}0$). Moreover, as it is in sandy soil (which forms the bedding), $r_{sd} = +0{,}7$. Finally, the ratio $h/d_{out} = 7{,}46$ (11,0 m/1,474 m).

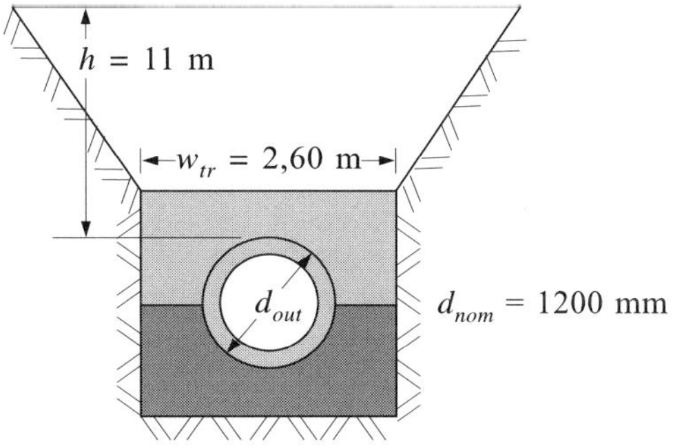

Figure 11.5 (example 11.1) Rigid pipe installed in a trench

---

* Adapted from Ontario Concrete Pipe Association Inc., *Concrete Pipe Design Manual* (1986), example 4.1, p. 4.16.

From the values of h/d_{out} (7,46) and $r_{sd}p$ (0,7), we find (fig. 11.3) that w_{tr}/d_{out} = 2,8, where:

$$\begin{aligned} w_{tr} &= 2{,}8d_{out} \\ &= 2{,}8 \times 1{,}454 \\ &= 4{,}13 \text{ m} \end{aligned}$$

This width is the transition width of the trench.

Because the width of the trench is smaller than 4,13 m, the pipe is laid in trench conditions.

Computation of the dead load exerted on the pipe

$$\begin{aligned} h/w_{tr} &= 11/2{,}60 \\ &= 4{,}23 \end{aligned}$$

For h/w_{tr} = 4,23, and for a backfill made of sand and gravel, C_d = 2,3 (fig. 11.4, curve B); the projected dead load is therefore:

$$\begin{aligned} F_c &= C_d mg w_{tr}^2 \\ &= 2{,}3 \times 1900 \times 9{,}81 \times (2{,}60)^2 \\ &= 289\,800 \text{ N/m} \end{aligned}$$

11.3.2 Flexible Pipes

The forces due to backfill acting on flexible pipes are computed using the following version of the Marston equation:

$$F_c = C_d mg w_{tr} d_{out} \tag{11.2}$$

This equation was developed taking into account that the lowering of a flexible pipe causes the backfill on top of the pipe to drop, which increases the friction forces between the backfill and the trench walls. This phenomenon causes a further reduction of the load due to backfill exerted on the pipe.

However, because of the lowering of the flexible pipe, it is observed that the weight of the prism of backfill located on top of the pipe sometimes exerts a load greater than the load calculated in equation 11.2 (see also on this subject section 11.2 on thermoplastic pipes). It is therefore recommended that the weight of the backfill prism be computed using a width corresponding to the outside diameter of the pipe and using the height of the backfill. If the calculated weight is greater than that obtained in equation 11.2, it will be used to determine the class of pipe.

11.4 LOADS ON PIPES DUE TO BACKFILL: EMBANKMENT CONDITIONS

When pipe is laid on the ground and buried, the trench in which it lies is very wide ($w_{tr} > (w_{tr})_{trans}$), or it is covered with backfill well over the natural ground, the vertical friction (shearing) forces exerted along the trench walls or between the backfill over the pipe and the backfill on each side of the pipe must be computed differently. These pipe installations are called embankment conditions. This section shows how to compute forces due to backfill loads for the following three conditions: pipes installed in positive projecting conditions, pipes installed in negative projecting conditions, and pipes installed in zero projecting conditions (fig. 11.2).

11.4.1 Positive Projecting Embankment Conditions

In this case, the pipe crown (top) is located above the natural ground level. The forces due to backfill are computed with the following equation:

$$F_c = C_c mgd_{out}^2 \tag{11.3}$$

where C_c is a coefficient for which the value is a function of h/d_{out} and of $r_{sd}p$ (fig. 11.6).

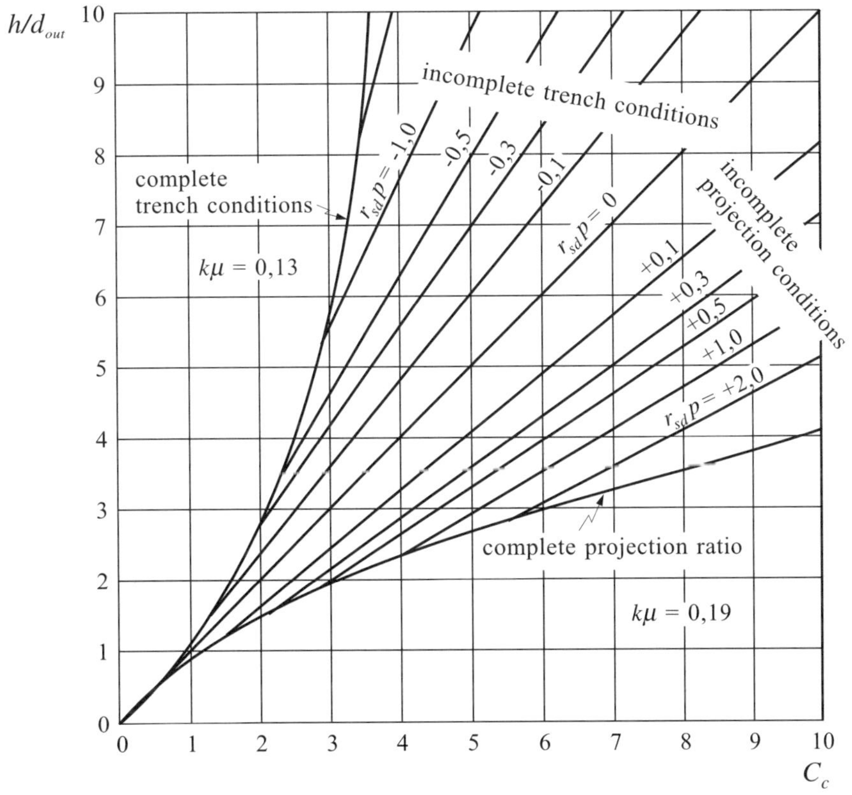

Figure 11.6 Value of the coefficient C_c used to compute of the force due to backfill exerted on **positive projecting pipes**. The values of $k\mu$ are recommended design values.

Adapted from: American Society of Civil Engineers and Water Pollution Control Federation, *Gravity Sanitary Sewer Design and Construction* (New York: ASCE and WPCF, 1982), p. 176.

EXAMPLE 11.2*

A concrete pipe with a nominal diameter of 1200 mm (1220 mm real diameter) is installed in positive projecting conditions where the natural soil is sandy, and is covered with 9 m of backfill weighing 1760 kg/m^3. Compute the dead load exerted on this pipe if the pipe crown (top) is located at $0{,}7d_{out}$ over the natural ground level (fig. 11.7).

Solution

Computation of the outside diameter of the pipe, d_{out}

The value of d_{out} has been computed to 1474 mm (example 11.1).

Computation of the dead load exerted on the pipe

For sandy soil (which forms the bedding), $r_{sd} = 0{,}7$; as $p = x/d_{out} = 0{,}7$, $r_{sd}p = 0{,}7 \times 0{,}7 = 0{,}49 \cong 0{,}5$. Moreover, $h/d_{out} = 9/1{,}474 = 6{,}1$. For $r_{sd}p = 0{,}5$ and $h/d_{out} = 6{,}1$; we find that $C_c = 9{,}2$ (fig. 11.6); the dead load is therefore:

$$\begin{aligned} F_c &= C_c mgd_{out}^2 \\ &= 9{,}2 \times 1760 \times 9{,}81 \times (1474)^2 \\ &= 333\,969 \text{ N/m} \end{aligned}$$

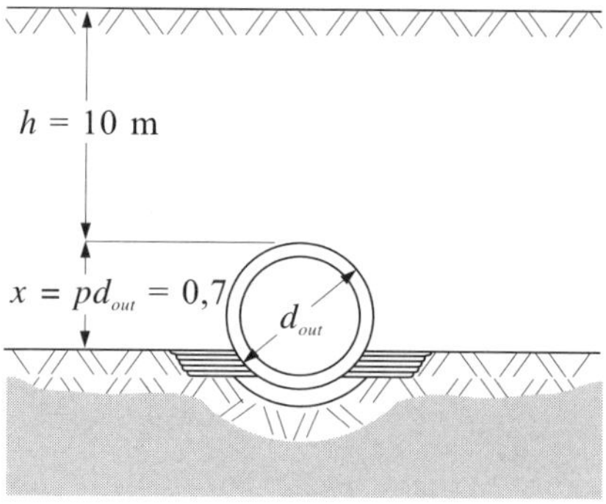

Figure 11.7 (example 11.2) Pipe installed in positive projecting conditions

11.4.2 Negative Projecting Embankment Conditions

Here, the pipe is laid in a shallow trench, and the pipe crown (top) is located below the original ground level. Then, it is covered with an embankment higher than the original ground level. The load exerted on this pipe is computed using the following equation:

$$F_c = C_n mgw_{tr}^2 \tag{11.4}$$

where C_n is a coefficient for which the value is a function of h/w_{tr}, r_{sd}, and p' (fig. 11.8).

* Adapted from Ontario Concrete Pipe Association Inc., *Concrete Pipe Design Manual* (1986), example 4.2, p. 4.18.

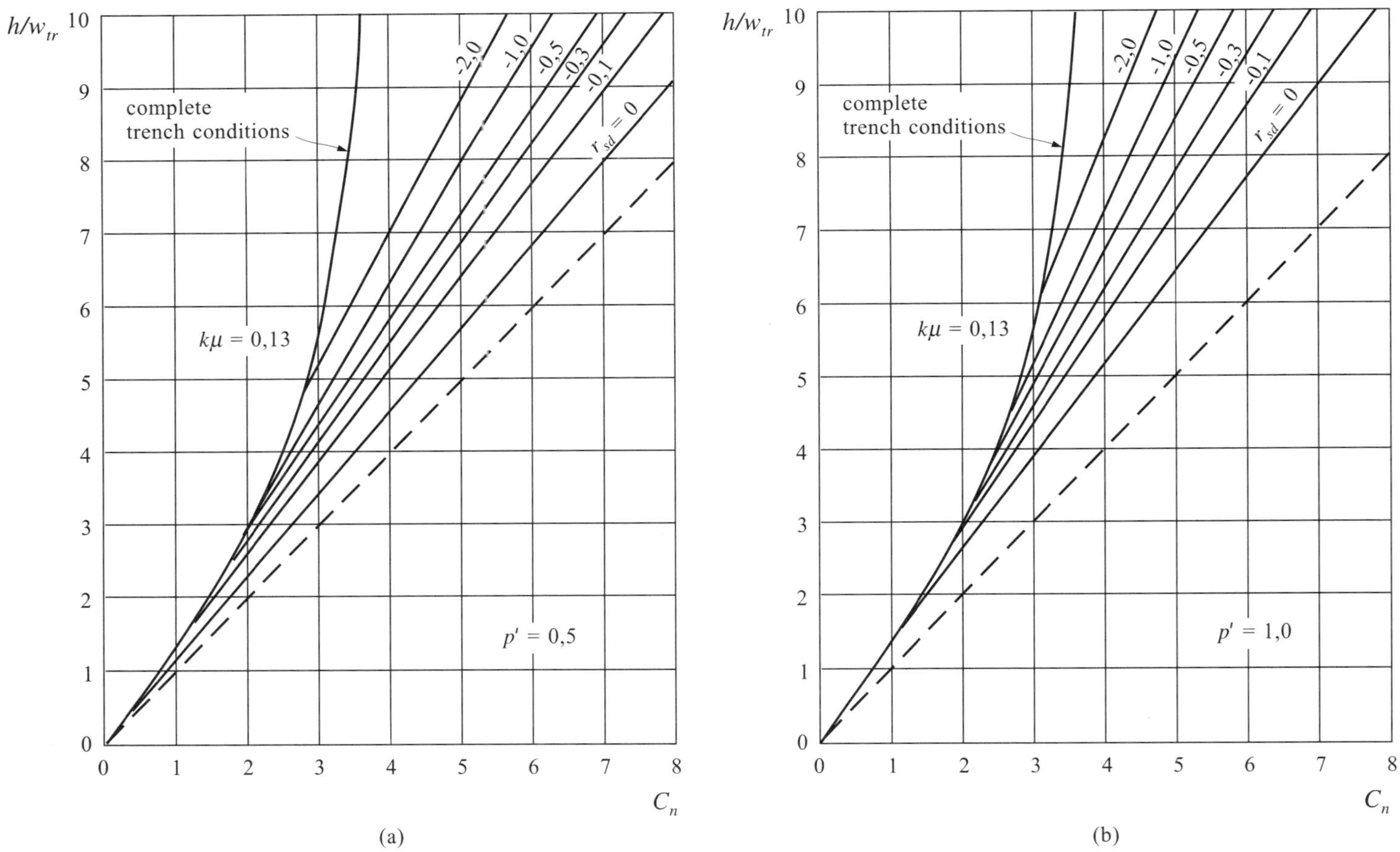

Figure 11.8 Value of the coefficient C_n used in computation of the force due to backfill exerted on negative projecting pipes: a) $p' = 0{,}5$; b) $p' = 1{,}0$; c) $p' = 1{,}5$; d) $p' = 2{,}0$.

Adapted from: American Society of Civil Engineers and Water Pollution Control Federation, *Gravity Sanitary Sewer Design and Construction* (New York: ASCE and WPCF, 1982), p. 176.

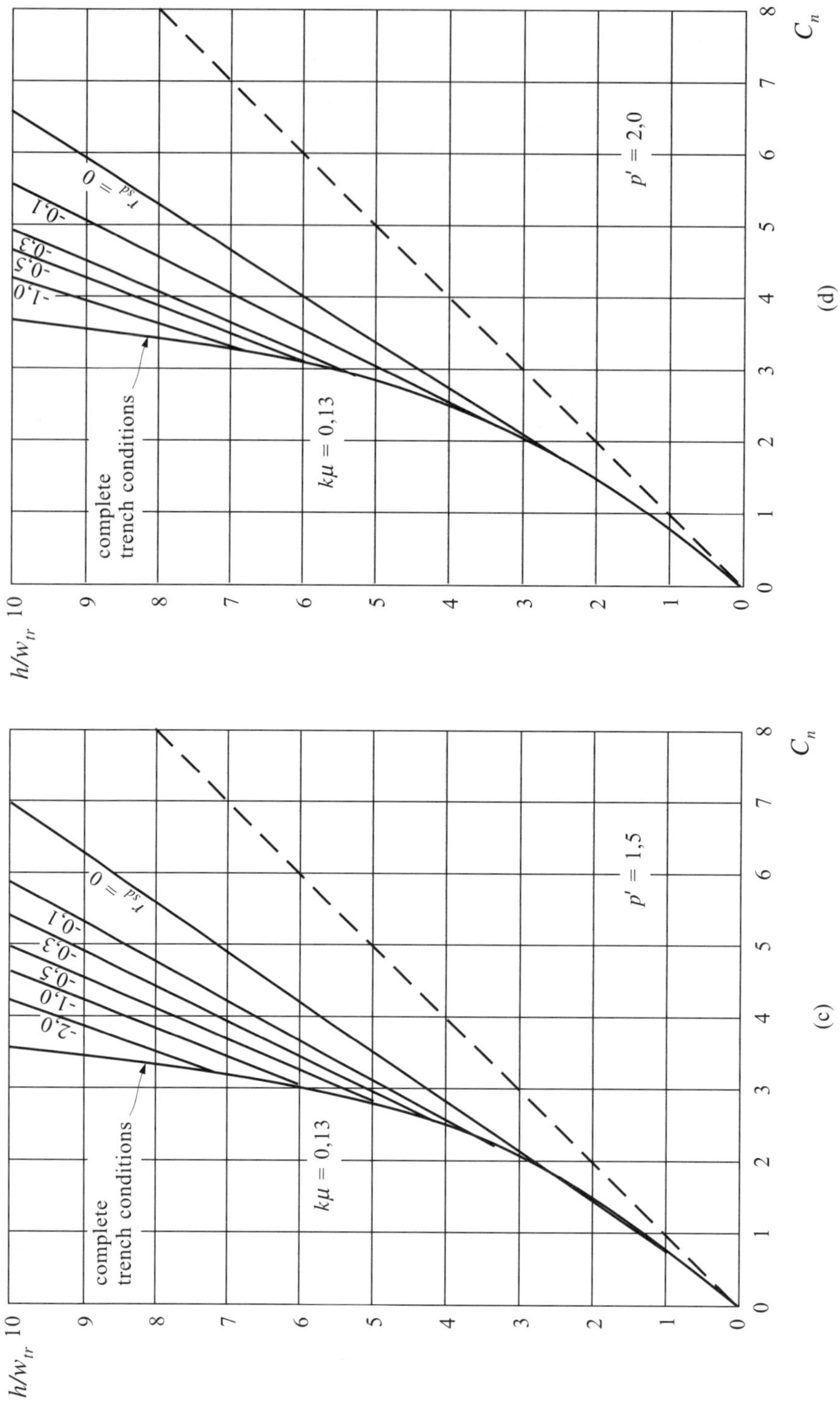

Figure 11.8 (continued)

11.4.3 Zero Projecting Embankment Conditions

Forces due to backfill are computed the same way as for a pipe installed in positive projecting conditions. However, in this case, as $p = 0$, $r_{sd}\,p = 0$ also.

EXAMPLE 11.3*

A concrete pipe with a nominal diameter of 1200 mm (1220 mm real diameter) is installed in negative projecting conditions where the natural soil is sandy. It is covered by 11 m of backfill that weighs 1900 kg/m^3. Compute the dead load exerted on this pipe if the width of the trench is 2 m and the vertical distance, x, separating the natural ground level and the horizontal plan over the pipe crown (top) is 2 m (fig 11.9).

Solution

The value of p' (x/w_{tr}) is 1,0 (2 m/2 m); for this value of p', a value of r_{sd} of -0,3 (table 11.1) is suggested; as a result, $r_{sd}\,p' = (-0{,}3) \times 1 = -0{,}3$. Moreover, $h/w_{tr} = 5{,}5$ (11 m/2 m). For $h/w_{tr} = 5{,}5$, $r_{sd} = -0{,}3$ and $p' = 1{,}0$, we find that $C_n = 3{,}6$ (fig. 11.8). The dead load is therefore:

$$
\begin{aligned}
F_c &= C_n m g w_{tr}^2 \\
&= 3{,}6 \times 1900 \times 9{,}81 \times (2{,}0)^2 \\
&= 268\,402 \;\; \text{N/m}
\end{aligned}
$$

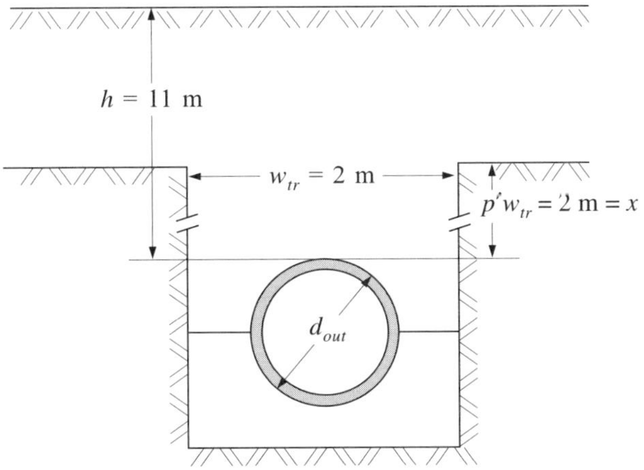

Figure 11.9 (example 11.3) Pipe installed in negative projecting embankment conditions

---

* Adapted from Ontario Concrete Pipe Association Inc., *Concrete Pipe Design Manual* (1986), example 4.3, p. 4.19.

11.5 FORCES DUE TO CONCENTRATED LIVE LOADS ON PIPES

Pipes must also sustain the effects of moving loads (cars, trucks, trains, etc.; fig. 11.10), called live loads. Figure 11.11 gives an example of the distribution in the ground of a force due to the live load of a truck; this is the load of two wheels installed side by side (load HS-20). The most critical loading configuration and the surface on which the forces apply vary as a function of the height of the backfill. Table 11.2 shows three critical configurations for live loads due to trucks. The pressure due to a live load exerted on a pipe is computed with the following equation:

$$p_L = \frac{PgI}{A} \qquad (11.5)$$

where p_L = average pressure in a horizontal plan under load P, (e.g. at the top of the pipe, P_a)

P = concentrated live load (kg)

I = impact factor

A = surface area on which the load is distributed at a given level (m^2)

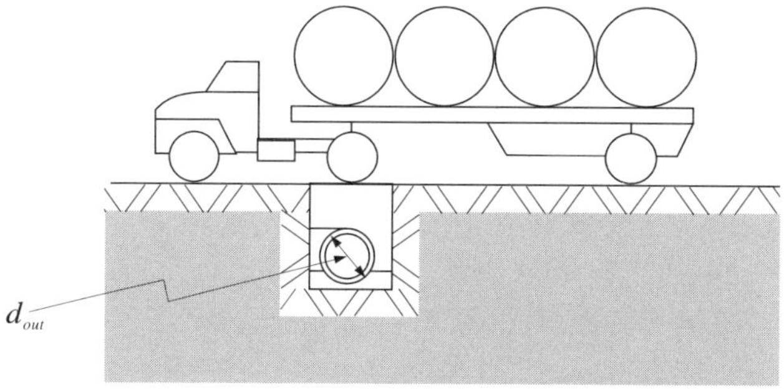

Figure 11.10 Live loads exerted on a pipe
Adapted from: *Concrete Pipe Design Manual* (op. cit), p. 4-26.

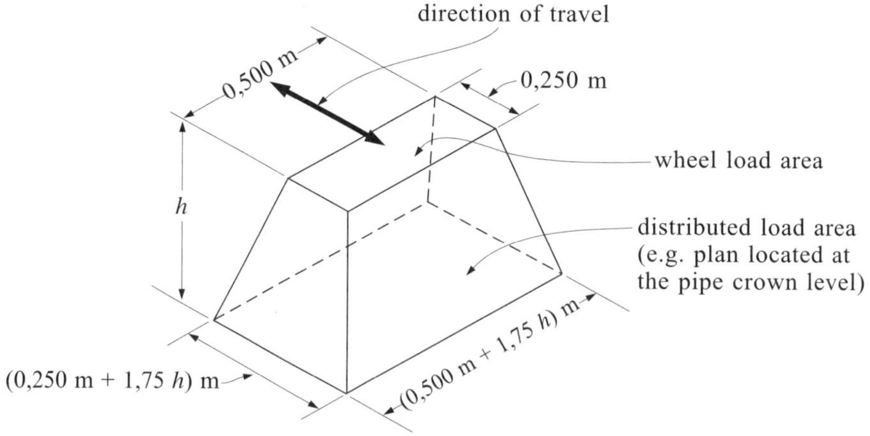

Figure 11.11 Distributed load area, single dual wheel (load HS-20)
Adapted from: *Concrete Pipe Design Manual* (op. cit), p. 4-8.

As the distance from the ground surface where the live load is exerted increases, the surface area over which the load is distributed increases. In fact, for a theoretical concentrated live load, the area is approximately equal to the base of the cone with its top located where the live load is applied, and its edge forms an angle, θ, with its rotation axis (fig. 11.12). For backfill normally used in roads, made of sand and gravel, an angle θ of 26,6° is suggested, for a slope of the edge of the cone of 2:1 (2 vertical to 1 horizontal). The base of the cone in question is on the horizontal plan for which we must find the pressure and resulting forces.

Table 11.2 Three critical live-load configurations of a truck

Height of backfill (m)	Live load of truck (kg)	Distributed load area (m²)
$h < 0,40$	7 250	$(0,25 + 1,75\,h)\,(0,50 + 1,75\,h)$
$0,40 \leq h \leq 1,25$	14 500	$(0,25 + 1,75\,h)\,(1,70 + 1,75\,h)$
$1,25 < h$	21 750	$(1,45 + 1,75\,h)\,(1,70 + 1,75\,h)$

Adapted from: *Concrete Pipe Design Manual* (op. cit), pp. 4–9.

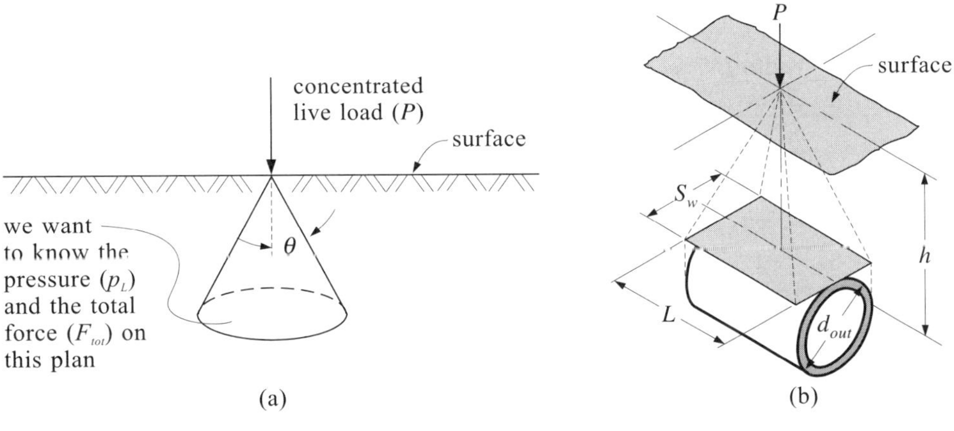

Figure 11.12 Influence of a surface live load on a pipe: a) distribution in the ground of the effect due to this load; b) area used in the computation of the force exerted on the pipe

Adapted from (fig. 11.12b): American Society of Civil Engineers and Water Pollution Control Federation, *Gravity Sanitary Sewer Design and Construction* (New York: ASCE and WPCF, 1982), p. 189.

Using the impact factor, I, the value of the load is multiplied to take into account the impact with which the live load is applied. Table 11.3 lists the values of design impact factors as a function of the depth at which the pipe is laid.

Table 11.3 Values of the impact factor multiplying live loads as a function of the installation depth (cover) of the pipe

Cover (height of the backfill [h] over the pipe crown [top]) **(m)**	**Impact Factor, *I***
0–0,30	1,3
0,31–0,60	1,2
0,61–0,90	1,1
0,91–over	1,0

The total force exerted on the pipe due to a live load can therefore be computed using the following equation:

$$F_{tot} = p_L \; L \, w \tag{11.6}$$

where F_{tot} = total force due to the surface live load exerted over the pipe crown (N)

p_L = pressure exerted in the horizontal plan immediately over the crown (top) of the pipe (Pa)

L = length of the pipe on which force F_{tot} is exerted immediately over the crown (top) of pipe (m)
This length is calculated as a function of the cone-shaped force distribution in the ground or any other justifiable distribution (fig. 11.12).

w = width of the pipe on which force F_{tot} is exerted immediately over the crown (top) of pipe; for a circular pipe, it is the outside diameter, d_{out} (m)

Then, we solve for the force per unit length that is due to a live load, W_L, using the following equation:

$$W_L = \frac{F_{tot}}{L} \tag{11.7}$$

where W_L = force due to the live load exerted on the pipe immediately over the crown (top) of the pipe (N/m)

L = length of the pipe on which force F_{tot} is exerted (m)

More simply, the force W_L can be computed using the following equation immediately over the crown (top) of the pipe:

$$W_L = \frac{C_s \, P \, I}{L} \tag{11.8}$$

where C_s = coefficient that is a function of the outside diameter of the pipe (in metres) and of the distance between the load and the top of the pipe, in metres (table 11.4)

P = concentrated live load (N)

I = impact factor (table 11.3)

L = pipe length on which the load is distributed immediately over the crown (top) of the pipe

Table 11.4 Values of coefficient C_s used to compute forces W_L (due to live loads: equation 11.8) and F_c (due to uniformly distributed loads: equation 11.9)

$\frac{w}{2h}^{*}$ or $\frac{d_{out}}{2h}^{**}$	$\frac{L}{2h}$ **(uniformly distributed load)*** **or (concentrated live load)****													
	0,1	**0,2**	**0,3**	**0,4**	**0,5**	**0,6**	**0,7**	**0,8**	**0,9**	**1,0**	**1,2**	**1,5**	**2,0**	**5,0**
0,1	0,019	0,037	0,053	0,067	0,079	0,089	0,097	0,103	0,108	0,112	0,117	0,121	0,124	0,128
0,2	0,037	0,072	0,103	0,131	0,155	0,174	0,189	0,202	0,211	0,219	0,229	0,238	0,244	0,248
0,3	0,053	0,103	0,149	0,190	0,224	0,252	0,274	0,292	0,306	0,318	0,333	0,345	0,355	0,360
0,4	0,067	0,131	0,190	0,241	0,284	0,320	0,349	0,373	0,391	0,405	0,425	0,440	0,454	0,460
0,5	0,079	0,155	0,224	0,284	0,336	0,379	0,414	0,441	0,463	0,481	0,405	0,525	0,540	0,548
0,6	0,089	0,174	0,252	0,320	0,379	0,428	0,467	0,499	0,524	0,544	0,572	0,596	0,613	0,624
0,7	0,097	0,189	0,274	0,349	0,414	0,467	0,511	0,546	0,584	0,597	0,528	0,650	0,674	0,688
0,8	0,103	0,202	0,292	0,373	0,441	0,499	0,546	0,584	0,615	0,639	0,674	0,703	0,725	0,740
0,9	0,108	0,211	0,306	0,391	0,463	0,524	0,574	0,615	0,647	0,673	0,611	0,742	0,766	0,784
1,0	0,112	0,219	0,318	0,405	0,481	0,544	0,597	0,639	0,673	0,701	0,740	0,774	0,800	0,816
1,2	0,117	0,229	0,333	0,425	0,505	0,572	0,628	0,674	0,711	0,740	0,783	0,820	0,849	0,868
1,5	0,121	0,238	0,345	0,440	0,525	0,596	0,650	0,703	0,742	0,774	0,720	0,861	0,894	0,916
2,0	0,124	0,244	0,355	0,454	0,540	0,613	0,674	0,725	0,766	0,800	0,800	0,894	0,930	0,956

* See uniformly distributed load, figure 11.13.

** See concentrated live load, figure 11.12.

From: American Society of Civil Engineers and Water Pollution Control Federation, *Gravity Sanitary Sewer Design and Construction* (New York: ASCE and WPCF, 1982), p. 190.

Table 11.5 Highway loads (kg/linear metre) exerted on a circular pipe – load HS-20

d_{nom} (of the pipe)	d_{out}	Height of backfill, *h*, above the pipe (m)												
(mm)	(m)	0,15	0,30	0,45	0,61	0,76	0,91	1,06	1,22	1,52	1,83	2,13	2,44	2,74*
300	0,40	5620	3090	2190	1610	1130	820	670	570	430	340	280	240	190
375	0,49	6310	3510	2590	1900	1340	980	800	670	520	420	340	280	240
450	0,58	6120	3880	2930	2170	1530	1120	920	770	590	480	390	330	280
525	0,67	5830	4200	3260	2410	1710	1250	1020	860	670	540	450	370	310
600	0,76	6100	4480	3570	2650	1890	1380	1130	950	740	590	490	420	360
675	0,85	5770	4370	3850	2870	2050	1500	1230	1040	830	650	540	450	390
750	0,94	5390	4210	4120	3080	2200	1600	1320	1120	880	710	580	490	420
900	1,12	4750	4180	4360	3470	2480	1830	1520	1280	1000	820	670	560	490
975	1,21	4480	3970	4240	3630	2620	1920	1600	1350	1060	860	710	610	520
1050	1,29	4260	3790	4120	3810	2740	2020	1800	1410	1120	910	760	640	550
1200	1,47	3850	3470	3900	3690	2960	2190	1830	1550	1220	1000	830	700	610
1350	1,65	3510	3200	3700	3510	3050	2350	1960	1670	1320	1090	910	770	650
1500	1,83	3230	2960	3650	3350	2920	2500	2080	1770	1410	1160	970	830	710
1650	2,01	2990	2750	3750	3210	2800	2440	2200	1870	1500	1240	1040	880	760
1800	2,19	2780	2570	3840	3260	2690	2340	2250	1980	1580	1310	1100	940	800
2100	2,54	2450	2290	4060	3410	2690	2170	2100	2020	1730	1430	1210	1030	890
2400	2,89	2190	2050	3590	3410	2800	2230	1980	1890	1860	1550	1310	1120	970
2550	3,07	2070	1960	3420	3260	2840	2280	2010	1850	1920	1590	1350	1160	1010
3000	3,61	1800	1710	3010	2870	2530	2200	2110	1900	2080	1740	1470	1280	1120
3300	3,96	1650	1580	2780	2680	2350	2050	1980	1920	2170	1810	1550	1340	1180
3600	4,32	1520	1460	2590	2480	2200	1930	1860	1800	2190	1900	1620	1400	1240

* When the height of the backfill, *h*, over the pipe exceeds 2,74 m, highway live load is negligible.

The values of the loads shown in this table include multiplication by an impact factor when the backfill height is less than 0,9 m.

From: TUBECON, *Canalisations et éléments préfabriqués et béton* (Association québécoise des fabricants de tuyaux de béton, 1985), p. 4-22.

Table 11.6 Railway live loads (kg/linear metre) exerted on a circular pipe

d_{nom} (of the pipe) (mm)	Height of backfill, h, above the pipe (m)																
	0,3	**0,6**	**0,9**	**1,2**	**1,5**	**1,8**	**2,1**	**2,4**	**2,7**	**3,1**	**3,7**	**4,3**	**4,9**	**5,5**	**6,1**	**7,6**	**9,1***
300	5 400	5 060	4 550	4 020	3 480	2 900	2 560	2 200	1 870	1 620	1 310	1 070	880	730	620	430	310
375	6 590	6 160	5 530	4 910	4 260	3 650	3 120	2 680	2 290	1 980	1 510	1 310	1 070	890	760	520	370
450	7 770	7 280	5 530	5 790	5 010	4 300	3 690	3 170	2 710	2 340	1 870	1 530	1 250	1 060	910	610	450
525	8 960	8 380	7 530	6 650	5 770	4 950	4 240	3 650	3 120	2 690	2 160	1 770	1 440	1 200	1 040	710	510
600	10 130	9 480	8 530	7 540	6 530	5 610	4 810	4 120	3 530	3 050	2 440	2 010	1 640	1 370	1 180	800	580
675	11 320	10 590	9 520	8 440	7 290	6 260	5 370	4 610	3 940	3 410	2 720	2 230	1 830	1 530	1 320	890	640
750	12 500	11 700	10 520	9 300	8 060	6 900	5 920	5 090	4 360	3 750	3 010	2 470	2 020	1 700	1 460	1 000	710
900	14 880	13 910	12 500	11 060	9 600	8 210	7 050	6 060	5 180	4 460	3 590	2 950	2 400	2 010	1 730	1 180	850
1 050	17 260	16 070	14 490	12 810	11 110	9 520	8 180	7 010	6 000	5 180	4 150	3 410	2 780	2 340	2 010	1 370	980
1 200	19 640	18 300	16 520	14 570	12 650	10 830	9 300	7 980	6 830	5 890	4 720	3 870	3 170	2 650	2 280	1 550	1 120
1 350	22 020	20 530	18 450	16 368	14 170	12 140	10 420	8 930	7 650	6 610	5 300	4 340	3 540	2 980	2 560	1 730	1 250
1 500	24 400	22 770	20 530	18 150	15 620	13 450	11 530	9 890	8 470	7 310	5 860	4 810	3 930	3 290	2 830	1 920	1 380
1 650	26 630	25 000	22 470	19 790	17 260	14 760	12 660	10 860	9 300	8 020	6 430	5 280	4 310	3 620	3 110	2 110	1 520
1 800	29 020	27 230	24 400	21 580	18 750	16 070	13 780	11 830	10 120	8 740	7 010	5 740	4 690	3 930	3 380	2 290	1 650
2 100	33 780	31 550	28 420	25 150	21 720	18 750	16 070	13 750	11 750	10 150	8 140	6 680	5 460	4 570	3 930	2 680	1 920
2 400	38 540	36 010	32 440	28 720	24 850	21 280	18 300	15 620	13 410	11 580	9 280	7 620	6 220	5 210	4 480	3 050	2 190
2 550	40 920	38 240	34 370	30 360	26 340	22 620	19 340	16 670	14 220	12 280	9 850	8 080	6 610	5 530	4 760	3 240	2 320
3 000	47 320	44 340	39 880	35 270	30 500	26 190	22 470	19 190	16 520	14 220	11 400	9 360	7 630	6 400	5 510	3 750	2 680
3 300	52 080	48 660	43 750	36 690	33 480	28 720	24 700	21 130	18 150	15 620	12 530	10 280	8 410	7 040	6 040	4 120	2 950
3 600	56 840	53 120	47 760	42 260	36 600	31 400	26 930	23 060	19 790	17 110	13 670	11 220	9 170	7 680	6 590	4 490	3 210

* When the height of the backfill, h, over the pipe exceeds 9,1 m, the live load for the railway is negligible.

The impact factor used to compute the values of the loads shown in this table varies from 1,4 (for a backfill height equal to zero) to 1,0 (for a backfill height equal to or greater than 3,0 m).

From: TUBECON, *Canalisations et éléments préfabriqués et béton* (Association québécoise des fabricants de tuyaux de béton, 1985), p. 4-23.

Table 11.4 shows the values of the coefficient C_s, used for computation of a force W_L (equation 11.8) exerted on a pipe supporting a concentrated live load. It is also used for computation of force F_c due to a dead load uniformly distributed on the ground surface (section 11.6) and symmetrical with respect to the pipe ($d_{out}/2h$, concentrated live load; $w/2h$, uniformly distributed load).

In practice, however, these computations are not done to evaluate forces due to the live loads of cars and trucks. Instead, we use tables that list pre-computed values: table 11.5 for live loads exerted on unpaved roads (the most unfavourable conditions existing during the construction of roads, before they are paved), and table 11.6 for live loads due to trains. These tables give the values of forces acting on pipes of various diameters installed under a road or a railway track, and for various heights of backfill over the top of the pipe crown (top). These values take the impact factor into account when the height of the backfill is less than 0,9 m (roads) or 3,0 m (railways) .

11.6 FORCES ON PIPES DUE TO HORIZONTAL DEAD LOADS UNIFORMLY DISTRIBUTED AND SYMMETRICAL WITH RESPECT TO THE PIPES

To compute the forces due to a load that is uniformly distributed on the surface (fig. 11.13), above the pipe, the following equation can be used:

$$F_c = C_s mgd_{out} \tag{11.9}$$

where F_c = force exerted on a unit length of the pipe (N/m)
C_s = coefficient of variable value according to dimensions (L and w) of load and height of backfill (h) (table 11.4)

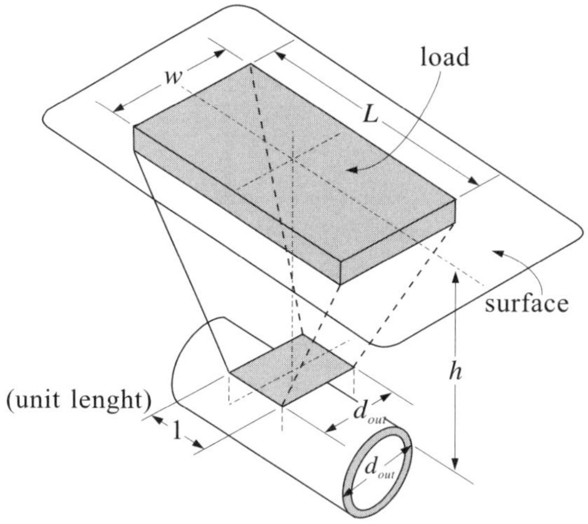

Figure 11.13 Dead load uniformly distributed over the pipe and symmetrical with respect to the pipe

From: American Society of Civil Engineers and Water Pollution Control Federation, *Gravity Sanitary Sewer Design and Construction* (New York: ASCE and WPCF, 1982), p. 189.

L = length of load uniformly distributed in pipe direction (m)
w = width of load perpendicular to pipe direction
m = uniformly distributed horizontal load (kg/m^2)
g = gravitational constant (9,81 m/s^2)
d_{out} = outside diameter of pipe

11.7 CUMULATIVE EFFECT OF DEAD LOADS AND LIVE LOADS ON BURIED PIPES

Pipes must sustain the combined effect of dead loads and live loads. To compute the forces resulting from these combined loads, the following expression is suggested:

$$F_{tot} = \frac{\Sigma F_c + \Sigma W_L}{L_f d_{out}} \times \text{FS} \tag{11.10}$$

where F_{tot} = total force exerted on pipe (N/m·mm outside diameter)
ΣF_c = sum of forces due to dead loads and backfill (N/m)
ΣW_L = sum of forces due to live loads (N/m)
d_{out} = outside diameter of pipe (mm)
FS = safety factor (1,0 is suggested)
L_f = bedding factor (factor that takes into account the type of foundation – that is, bedding at the bottom of the trench; section 11.10)

If forces F_c and W_L are expressed in kg instead of N, the result must be multiplied by g (9,81 N/kg) to obtain the total force, F_{tot}, in N. The nominal diameter can also be used in equation 11.10 to solve the total force per mm of the nominal diameter.

11.8 BEARING CAPACITY OF REINFORCED CONCRETE PIPES

When all forces exerted on a pipe are known, we must then choose a pipe with an appropriate bearing capacity. Concrete pipes are divided into 5 classes (I to V), each with a different bearing capacity. Table 11.7 lists the bearing capacity of each class of reinforced-concrete pipe. Manufacturers have computed these capacities as a function of the nominal diameter.

Table 11.7 Bearing capacity of the various classes of reinforced-concrete pipe

Bearing capacity	**Classe**				
	I	**II**	**III**	**IV**	**V**
Bearing capacity D_{03}* (N/m·mm d_{nom})**	40	50	65	100	140
Bearing capacity prior to crushing (N/m·mm d_{nom})	60	75	100	150	175

* D_{03} = force leading to the formation of cracks of 0,3 mm wide over a length of 30 cm.
** Newton per meter of length per mm of diameter.

The difference between the various classes of concrete pipes is due in part to the amount of steel reinforcement added to the pipe walls by the manufacturer (fig. 11.14). When the reinforcement is not circular or is not uniformly distributed throughout the section, the pipe must be installed according to the manufacturer's specifications because the strength is not the same in all directions.

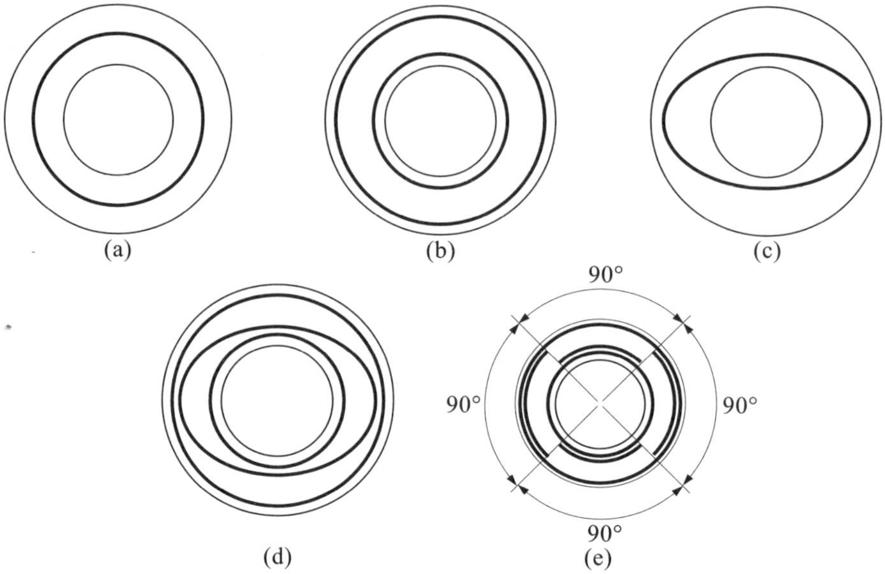

Figure 11.14 Various cage configurations of reinforced concrete pipe: a) single circular cage; b) double circular cage; c) single elliptical cage; d) combination of elliptical and circular cages; e) quadrant-shaped cage.

From: TUBECON, *Canalisations et éléments préfabriqués en béton* (Association québécoise des fabricants de tuyaux de béton, 1995), chapter 3.

11.9 MINIMUM DIMENSIONS OF TRENCHES' VERTICAL CROSS SECTIONS

Proper installation of pipes requires that the trench sections be of appropriate dimensions, taking into account the following parameters: natural ground into which the trench is excavated; slope of the trench walls; space required by the equipment available to compact the backfill; size of the pipes.

11.9.1 Slopes of Trench Walls

In Quebec, the slopes of trench walls must meet the requirements of the Health and Safety Commission (Commission de la santé et de la sécurité du travail: CSST). The slope, s, can be 10/1 (gradient of 10 per unit length) if the trench is excavated in rock and 2/1 if excavated in another material. In the latter case, the Health and Safety Commission also allows for the walls to be vertical up to 1,2 m from the bottom of the trench, then the slope must be 1/1 up to the surface. The use of any other slope must be approved by a specialist in soil mechanics.

11.9.2 Width of the Bottom of the Trench and Depth of Sewer-Pipe Bedding

The width, w_{tr}, at the bottom of the trench and the thickness, d, of the bedding under the pipe invert are a function of the diameter of the pipe to be installed and the slope of the trench walls (table 11.8). Figure 11.15 shows a typical trench cross section.

Table 11.8 Thickness of the bedding under the pipe and width at the bottom of the trench* for various pipe diameters

Nominal diameter, d_{nom}	Excavation made entirely in a material other than rock (wall slope 2/1)		Excavation in rock (wall slope 10/1)	
	Thickness of bedding under pipe	Width of bottom trench	Thickness of bedding under pipe	Width of bottom trench
150 and less	150	900	150	950
200	150	900	150	1000
225	150	900	150	1000
250	150	900	150	1050
300	150	950	150	1100
350	150	1000	150	1150
375	150	1000	150	1200
400	150	1000	150	1200
450	150	1050	150	1250
500	150	1100	150	1300
525	150	1100	150	1350
600	150	1150	150	1450
675	200	1350	150	1650
750	200	1400	150	1750
900	200	1500	150	1900
1050	200	1600	150	2050
1200	200	1750	150	2200
1350	250	1800	150	2400
1500	250	1900	150	2550
1650	250	2000	150	2700
1800	300	2050	200	2850
1950	300	2200	200	3000
2100	300	2300	200	3200
2250	300	2400	200	3350
2400	300	2500	200	3500
2700	300	2700	200	3800
3000	300	2950	200	4150
3600	300	3400	200	4800

*For soft soils, these values may have to be increased.

For large diameters, the fact that the bottom of the trench is smaller than the diameter of the pipe is not an error. The geometry of the trench leaves a minimum distance of 600 mm between the pipe and the trench wall as required for backfill compaction purposes (section 8.10).

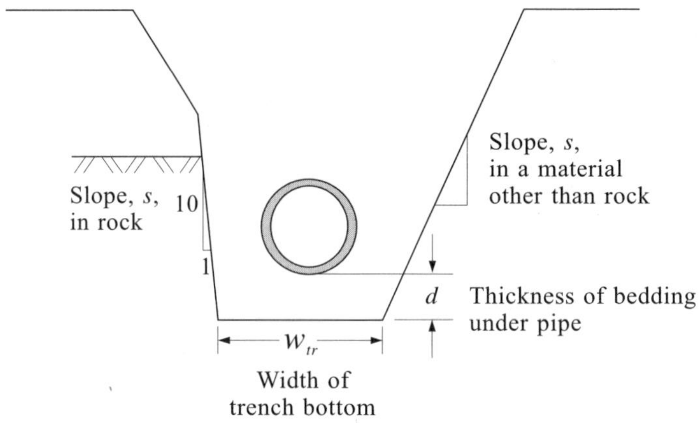

Figure 11.15 Typical trench cross section

11.10 SEWER-PIPE BEDDING (LAYING OUT OF TRENCH BOTTOM)

A pipe can better sustain forces exerted on it if the bottom of the trench in which it is laid is prepared so that the lower part (invert) of the pipe is in full contact with the material of the trench bottom. The choice of foundation material is also very important. Moreover, when the bedding rises on each side of the pipe, it should reinforce the sides of the pipe. Figure 11.16 shows the four major classes of foundations or beddings (A, B, C, and D), of which three (A, B and C) can be built in two different ways. In each case, a foundation factor, L_f, is included in the computation of the total force, F_{tot}, exerted on the pipes (equation 11.10). Table 11.9 gives the values of L_f for the four foundation classes shown in figure 11.16.

Table 11.9 Various values of L_f for the foundation (bedding) classes of reinforced-concrete pipes

Class of bedding*	**Bedding factor, L_f**
A	4,8, 3,4 or 2,8, depending on quantity of steel reinforcement
B	1,9
C	1,5
D	1,1

* In Quebec, bedding classes B and C are the most common.

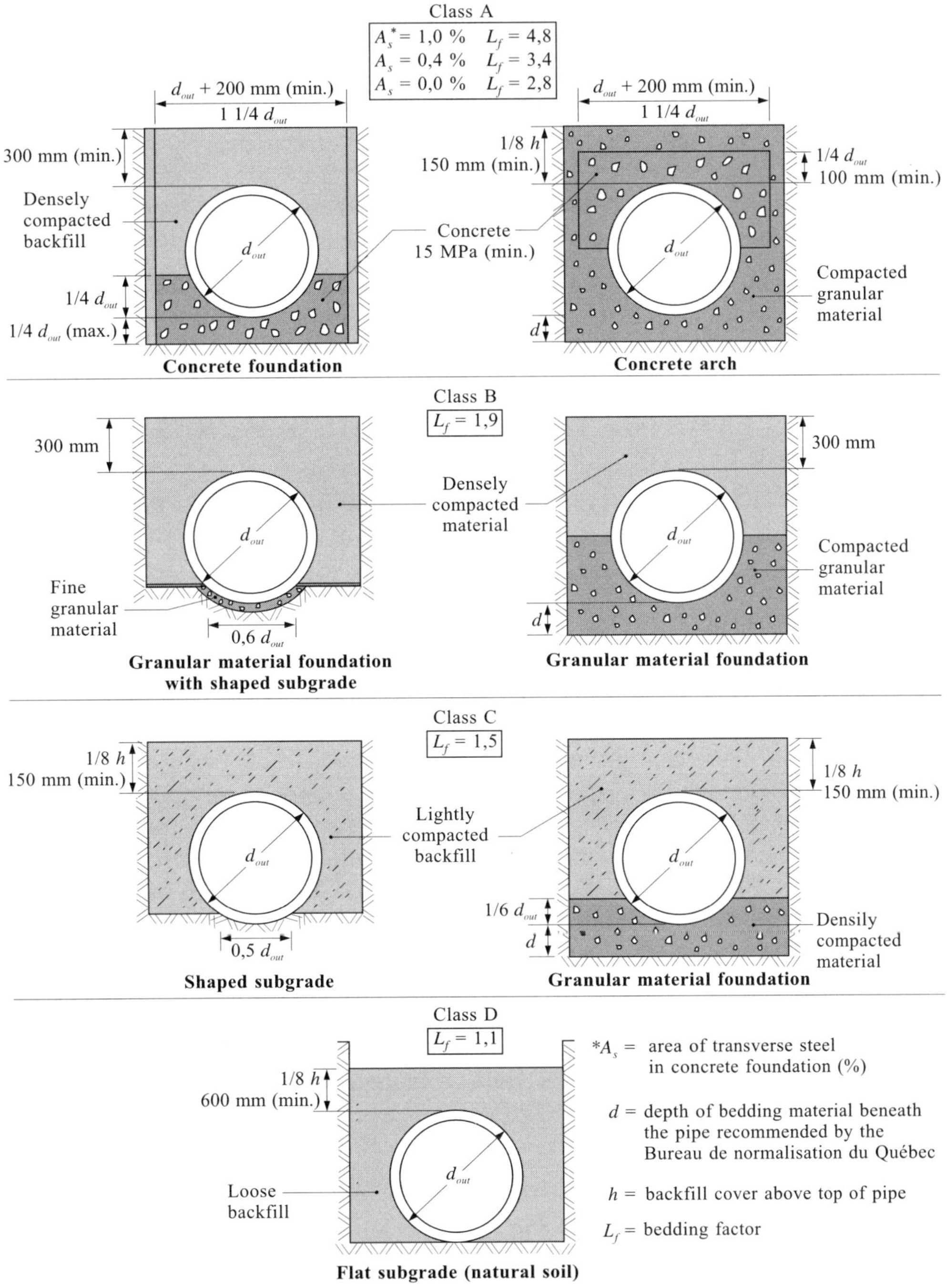

Figure 11.16 Various bedding classes for installation of reinforced-concrete pipes and related bedding factors (L_f) From: TUBECON, *Canalisations et éléments préfabriqués en béton* (Association québécoise des fabricants de tuyaux de béton, 1985), p. 4-90.

11.11 BACKFILL MATERIAL

Backfill is laid in layers of 300 mm, and each layer is compacted according to standards, such as modified Proctor (for pipes that are not laid under roads, some layers are not compacted). For pipes with diameters of less than 900 mm, backfill composed of granular material is laid up to 300 mm over the pipe crown (top). Above 300 mm, reclaimed compressible material is used, with the exception of organic soil. For pipes with diameters greater than 900 mm, granular backfill is laid for up to half the diameter of the pipe, then, as in the previous case, reclaimed material is used. The dimension of the granular material varies from 0 to 10 mm or from 0 to 20 mm (these numbers refer to the openings of the sieves used to classify materials).*

11.12 DEFLECTION OF THERMOPLASTIC SEWER PIPES UNDER LOADS

Thermoplastic (also called plastic) pipes have been on the market for many decades. Thermoplastics soften enough under heat to be shaped and reshaped without loosing their characterictics. Plastic pipes are used in many cases, such as electricity and plumbing. Their use for water has increased considerably as their mechanical characteristics have unproved (mainly resistance to impacts) and their cost has fallen. Many types of plastics are used for pipes, including ABS (Acrilonitrile-Butadiene-Styrene), PE (PolyEthylene) and PVC (PolyVinyl Chloride). In the case of PVC pipes, various materials are added to polyvinyl chloride like dyes, stabilisers, antioxydants and lubricants. Sewer pipes and drinking water pipes are produced in 4,0 and 6,1 m long lengths respectively. They thus require fewer joints than pipes made of other materials and that are consequently shorter. As a result, infiltration and exfiltration due to impaired joints are less significent. Moreover, because of their light weight and flexibility, plastic pipes are easier and less costly to install. Finally, when new, their Manning's roughness coefficient is only 0,009 and their Hazen-Williams coefficient is 150: consequently their hydraulic capacity is greater than pipes of similar diameter but other material. Municipalities largely use PVC pipes and – in a much smaller proportion – PE pipes. Therefore, due to the wide use of PVC pipes, this section is restricted to this type of pipe. The following material is mainly taken from Unibell PVC Pipe Association, *Handbook of PVC Pipe*, 1991 and also from *Tuyaux et raccords rigides en polychlorure de vinyle pour adduction et distribution de l'eau sous pression*, Bureau de normalisation du Québec, 1993.

* From: Bureau de normalisation du Québec, *Devis normalisé* (NQ 1809-300, 1987), section 8.2.2.1, p. 26.

11.12.1 Pipe Classes

Manufacturers identify their plastic pipes with 2 or 3 letters followed by a number, for example, DR 18 or SDR 21 (DR and SDR stand for Dimension Ratio and Standard Dimension Ratio) and have both the same meaning; the number is the ratio of the average outside diameter of the pipe, d_{out}, to the thickness, t, of the pipe wall. Consequently, a DR 21 pipe has a d_{out}/t of 21.

11.12.2 Resistance to Inside Hydraulic Pressures

Drinking water pipes must be able to support internal hydraulic pressure, as high as 1000 kPa according to Fire Underwriters Survey of Canada, even without outside loads such as those due to backfill. **The reference temperature is 73,4°F or 23°C.**

Manufacturers and control bodies have identified many internal hydraulic pressures which drinking water pipes must be able to resist.

- Nominal pressure (pressure rating or rated pressure), p_N , is the pressure during normal operation of the network including the brief shock pressures of a water hammer.
- Water hammer pressures (shock pressures, shock working pressures), p_S, are short lasting pressures – 60 to 70 s – due to rapid variations of water velocity in pipes, for example when valves are opened or closed or when pumps start or stop.
- Maximum working pressure (maximum operating pressure, maximum service pressure, pressure class), p_{MS}, is the normal maximum operating pressure without the short lasting shock pressure due to water hammer:

$$p_{MS} = p_N - p_S \tag{11.11}$$

- Hoop stress, T, is the transverse stress induced in pipe walls by internal hydrostatic pressure.
- Hydrostatic Design Basis, *HDB*. Manufacturers test their pipes under internal hydrostatic pressures during very long periods during which the plastic walls are subjected to deformation by creeping. *HDB* is the constant transversal stress that causes bursting of the pipes after 100 000 h (11,4 years). Usually manufacturers supply pipes with a *HDB* of 4000 Lb/in^2 (27,6 Mpa).
- Hydrostatic Design Stress, *HDS*, is Hydrostatic Design Basis, *HDB*, divided by a security factor, *FS*:

$$HDS = \frac{HDB}{FS} \tag{11.12}$$

With an appropriate value of *HDS*, the pipe has an infinite life.

The relation between hoop stress, T, in Mpa and any long term internal hydraulic pressure, p, in Mpa is the following:

$$T = \frac{p}{2}(DR - 1)$$

In 1852, Lame showed (in Moser, Buried Pipe Design, McGraw-Hill, p. 107) the following equation:

$$\frac{2HDS}{p_N} = \frac{d_{out}}{t} - 1 = SDR\,(\text{or } DR) - 1$$

from which:

$$p_N = \frac{2HDS}{SDR - 1} \tag{11.13}$$

Exemple 11.4

A thermoplastic PVC pipe has the following characteristics:

d_{out} = 8,625 in
t = 0,411 in
HDB = 4000 Lb/in^2
FS = 2,0

Find the long-term internal hydraulic pressure this PVC pipe can sustain at 73°F (23°C).

Solution

$SDR = d_{out}/t$ = 8,625/0,411 in = 21 in
$HDS = HDB/FS$ = 4000/2,0 = 2000 Lb/in^2

From equation 11.13:

$$p_N = \frac{2HDS}{SDR - 1} = \frac{2 \times 2000}{21 - 1}$$
$$= 200 \text{ Lb/in}^2 \ (1380 \text{ kPa})$$

For design purposes, this pipe can sustain an internal hydraulic pressure of 200 Lb/in^2 including water hammer pressures, p_s.

11.12.3 Pressures Due to Water Hammer

Manufacturers consider a water velocity variation of 2 ft/s (≅ 0,61 m/s). The table 11.10 presents surge pressures created by each 1 ft/s (≅ 0,3 m/s) water velocity variation in various PVC pipe classes.

Table 11.10 Surge pressures in plastic pipes due to 1 ft/s (≅ 0,3 m/s) flow velocity variation

DR (d_{out}/t)	13,5	14,0	17,0	18,0	21,0	25,0	26,0	32,5	41,0
Pressure surge:									
Lb/in^2	20,2	19,8	17,9	17,4	16,0	14,7	14,4	12,8	11,4
kPa	139	137	123	120	110	101	99	88	79

From equation 11.13, one finds:

$$p_N = \frac{2HDS}{SDR - 1} = \frac{2(HDS/FS)}{SDR - 1} = \frac{2(HDS/FS)}{(d_{out}/t) - 1}$$

and

$$FS \times p_N = \frac{2t}{d_{out} - 1} \times HDS$$

Since equation 11.11:

$$P_{MS} = p_N - p_S$$

then:

$$FS \times (p_{MS} + p_S) = \frac{2t}{d_{out} - 1} HDB$$

11.12.4 Temperature's Effects

When the pipe temperature exceeds 23°C (73,4°F), the internal hydraulic pressure, p_N or p_{MS}, pipe can sustain is the pressure at 23°C multiplied by the appropriate factors in the table 11.11.

Table 11.11

Temperature:							
°F	80	90	100	110	120	130	140
°C	27	32	38	43	49	54	60
Multiplying factor	0,88	0,75	0,62	0,50	0,40	0,30	0,22

Example 11.5

Compute the internal hydraulic pressure a plastic PVC pipe can resist based on the following information:

d_{out} = 9,050 in
t = 0,503 in
HDB = 4000 Lb/in^2 (27 590 kPa)
FS = 2,5
Maximum flow velocity variation: 2 ft/s

Solution

DR or $SDR = d_{out}/t$ = 9,050/0,503 ≅ 18 in

From equation 11.12:

$$HDS = \frac{HDB}{FS} = \frac{4000}{2,5} = 1600 \text{ Lb/in}^2 \text{ (psi)}$$

Surge pressure due to flow velocity variation (water hammer):

2 × 17,4 (table 11.10) = 34,8 (≅ 35 Lb/in^2)

From equations 11.11 and 11.13:

$$p_N = (p_{MS} + p_S) = \frac{2HDS}{SDR - 1}$$

$$\begin{aligned} p_{MS} &= \frac{2HDS}{SDR - 1} - p_S \\ &= \frac{2 \times 1600}{18 - 1} - 35 \\ &= 153 \text{ Lb/in}^2 \left(\cong 150 \text{ Lb/in}^2\right) \end{aligned}$$

The pipe under study can support an internal hydraulic pressure of 150 Lb/in^2 plus a surge pressure of 35 Lb/in^2.

11.12.5 Deflection of Buried Plastic Sewer Pipes

A plastic sewer pipe is expected to deflect under backfill load (and in absence of internal hydraulic pressure) without any sign of structural distress such as injurious cracking. Deflection is expressed as a percentage of vertical diameter. This percentage must not exceed 2,5% according to the ministry of Transport of Quebec (Canada). Other bodies accept 5% and sometimes even 7,5%.

Pipe Stiffness (PS). Pipe stifness is the vertical force, *F*, in Lb/linear inch per inch of vertical deflection, *DY*, of pipe. In the case of a flexible pipe not buried in soil, pipe stiffness is defined as:

$$PS = \frac{F}{\Delta Y} = \frac{EI}{0{,}149r^3} = \frac{6{,}71EI}{r^3} \tag{11.14}$$

When the pipe wall has a uniform thickness, equation 11.14 can be rewritten:

$$PS = \frac{F}{\Delta Y} = \frac{6{,}71Et^3}{12r^3} = 0{,}559\,\frac{t^3}{r^3} \tag{11.15}$$

where PS = Pipe stifness [Lb/in-in (psi)]
F = Force (Lb/linear inch)
ΔY = Vertical deflection (in)
E = Modulus of elasticity of the pipe material (psi)
I = Moment of inertia of the wall cross section per unit length of pipe (in^4/linear inch = in^3)
R = mean radius of pipe (in)
t = wall thickness (in)

Equation 11.15 can be rewritten as follows to take into account the variable $DR(D_o/t)$:

$$PS = 4{,}47\,\frac{E}{(DR-1)^3} \tag{11.16}$$

Initial buried PVC pipe deflection under dead and live loads. For dead loads on PVC pipes, manufacturers suggest the exclusive use of the prism of soil load over the pipes and ignore friction forces along the sides of the trenches. This approach generally overestimates the dead loads as well as the initial pipe deflection, which is then considered the long term deflection. This is why manufacturers suggest 1,0 for the value of the deflection lag factor D_L in equations 11.17, 11.18 and 11.19.

For pipes with uniform thickness:

$$\frac{\Delta Y}{d} \times (100) = \frac{D_L KP(100)}{0{,}149\left(\dfrac{F}{\Delta Y}\right) + 0{,}061E'} \tag{11.17}$$

Equation 11.17 can be rewritten as a function of the variable *DR*:

$$\frac{\Delta Y}{d} \times (100) = \frac{D_L KP(100)}{2E/3(DR-1)^3 + 0{,}061E'} \tag{11.18}$$

In those equations:

D_L = Deflection lag factor (suggested value: 1,0)
K = Bedding constant (in general, a value of $K = 0{,}1$ is assumed)
P = Prism of soil load over the pipe (Lb/in^2 or psi)
E = Modulus of elasticity of the pipe material (Lb/in^2 or psi)
E' = Modulus of soil reaction (Lb/in^2 or psi)

If the liquid limit of the soil type-pipe bedding is larger than 50%, one should consult a specialist in soil engineering or, as a security measure, use 0 for E'. Moreover if the liquid limit is lower than 50%, E' varies as followed:

- from 50 to 100 for dumped bedding;
- from 200 to 400 for slightly compacted bedding;
- from 400 to 1000 moderately compacted bedding;
- from 1000 to 2000 for very compacted bedding.

Moreover, for coarse-grained soils with little or no fines, the values of E' in the four foregoing cases are respectively 200, 1000, 2000, and 3000 Lb/in^2 (psi). For crushed rock, the four values are 1000, 3000, 3000, and 3000 Lb/in^2.

N.B. In the cases of pipes installed under roads, live loads must be added to dead loads when computing pipe deflection.

Example 11.6

Compute the vertical deflection percentage of a PVC pipe *DR* 35 with a uniform wall thickness, based on the following information:

- Backfill material: fine-grained soil weighing 120 Lb/ft^3;
- Liquid limit of bedding material: 50%;
- Height of backfill: 10 ft;
- Modulus of elasticity of the pipe material (E): 400 000 Lb/in^2 (psi);
- Modulus of soil reaction (E'): 200 Lb/in^2 (psi);
- Pressure due to backfill material (load due to prism of soil over the pipe) at the top pipe level: 8,33 Lb/in^2(psi);
- Deflection lag factor (D_L): 1,0.

Solution

From table 11.12, one finds the pipe stiffness ($PS = F/DY$) of a *DR* 35 PVC pipe is 46 Lb/in^2 (psi). Then, with equation 11.17, one computes the pipe deflection, $\Delta Y/d \times (100)$, to be 4,40%. With equation 11.18, one also finds a pipe deflection of 4,4%.

Table 11.12 PVC Pipe Stifness (*PS*) (Lb/in^2 or psi)

DR* or *SDR	**Minimum Modulus of elasticity 400 000 psi**	**Minimum Modulus of elasticity 500 000 psi**
64	7	9
51	14	18
42	26	32
41	28	35
35	46	57
33,5	52	65
32,5	57	71
28	91	114
26	115	144
25	129	161
21	224	279
18	364	455
17	437	546
14	815	1019
13,5	916	1145

Tables 11.13, 11.14, and 11.15 give PVC pipe deflections under roads and railways. Computations were made with equation 11.17, where the numerator has been increased by KW', the pressure at the top of the pipe level due to live load (K = the bedding constant = 0,1; W' = live load in Lb/in^2 or psi at the crown level). It is assumed that the backfill weighs 120 Lb/ft^3, that the modulus of elasticity of the pipe material (E) is 400 000 Lb/in^2 and that the value of the bedding constant, K, is 0,1; equation 11.17 then becomes equation 11.19:

$$\frac{\Delta Y}{d} \times (100) = \frac{\left(D_L KP + KW'\right) 100}{2E/3(DR - 1) + 0{,}061E'} \qquad (11.19)$$

Table 11.13 Calculated deflections ($\Delta Y/d$) of PVC (AWWA C900) pipes buried under roads and railways – DR 14

Height of cover (ft)	**2**	**4**	**6**	**8**	**10**	**12**	**14**	**16**	**18**	**20**
Live load	*H20* ; *E80*	*H20*; *E80*	*H20*; *E80*	*H20*; *E80*	*H20*; *E80*	*H20*; *E80*	*H20*; *E80*	*H20*; *E80*	*H20*; *E80*	*H20*; *E80*
E' **(lb/in^2)**										
50	0,58; 2,26	0,49; 1,75	0,51; 1,66	0,59; 1,43	0,67; 1,28	0,80; 1,25	0,94; 1,27	1,07; 1,35	1,21; 1,49	1,34; 1,51
200	0,54; 2,10	0,45; 1,63	0,48; 1,55	0,55; 1,34	0,62; 1,19	0,75; 1,16	0,87; 1,19	1,00; 1,26	1,12; 1,33	1,25; 1,40
400	0,50; 1,93	0,42; 1,49	0,44; 1,42	0,50; 1,22	0,57; 1,09	0,69; 1,07	0,80; 1,09	0,91; 1,15	1,03; 1,22	1,14; 1,29
1000	0,40; 1,54	0,34; 1,19	0,35; 1,13	0,40; 0,98	0,46; 0,87	0,55; 0,85	0,64; 0,87	0,73; 0,92	0,82; 0,97	0,91; 1,03
2000	0,30; 1,15	0,25; 0,89	0,26; 0,85	0,30; 0,73	0,34; 0,66	0,41; 0,64	0,48; 0,65	0,55; 0,69	0,62; 0,73	0,68; 0,77

Table 11.14 Calculated deflections ($\Delta Y/d$) of PVC (AWWA C900) pipes buried under roads and railways – DR 18

Height of cover (ft)	**2**	**4**	**6**	**8**	**10**	**12**	**14**	**16**	**18**	**20**
Live load	*H*20; *E*80	*H*20; *E*80	*H*20: *E*80	*H*20; *E*80	*H*20; *E*80	*H*20; *E*80	*H*20; *E*80	*H*20; *E*80	*H*20; *E*80	*H*20; *E*80
***E*′(lb/in²)**										
50	1,26; 4,89	1,07; 3,79	1,11; 3,60	1,28; 3,10	1,45; 2,79	2,09; 2,71	2,04; 2,76	2,33; 2,93	2,62; 3,10	2,91; 3,27
200	1,09; 4,22	0,92; 3,38	0,97: 3,11	1,12; 0,69	1,25; 2,40	1,51; 2,34	1,76; 2,38	2,01; 2,53	2,26; 2,67	2,51; 2,82
400	0,92; 3,58	0,78; 2,77	0,82: 2,63	0,93; 2,27	1,06; 2,03	1,28; 1,98	1,48; 2,01	1,69; 2,14	1,91; 2,26	2,12; 2,38
1000	0,63; 2,44	0,53; 1,89	0,56. 1,79	0,63; 1,55	0,72; 1,38	0,87; 1,35	1,01; 1,37	1,16; 1,46	1,30; 1,54	1,45; 1,63
2000	0,41; 1,59	0,35; 1,23	0,36. 1,17	0,42; 1,01	O,47; 0,91	0,57; 0,88	0,66; 0,90	0,76; 0,95	0,85; 1,01	0,95; 1,06

Table 11.15 Calculated deflections ($\Delta Y/d$) of PVC (AWWA C900) pipes buried under roads and railways – DR 25

Height of cover (ft)	**2**	**4**	**6**	**8**	**10**	**12**	**14**	**16**	**18**	**20**
Live load	*H*20; *E*80	*H*20; *E*80	*H*20; *E*80	*H*20; *E*80	*H*20; *E*80	*H*20; *E*80	*H*20; *E*80	*H*20; *E*80	*H*20; *E*80	*H*20; *E*80
E'(lb/in^2)										
50	3,24; 12,56	2,73; 9,73	2,86; 9,23	3,29; 7,96	3,73; 7,15	4,48; 6,97	5,22; 7,09	5,97; 7,52	6,71; 7,96	7,46; 8,39
200	2,30; 8,91	1,94; 6,96	2,05; 6,60	2,34; 5,71	2,66; 5,10	3,21; 4,97	3,71; 5,03	4,23; 5,34	4,76; 5,65	5,29; 5,95
400	1,65; 6,47	1,41; 4,97	1,47; 4,75	1,68; 4,10	1,91; 3,66	2,30; 3,57	2,67; 3,63	3,05; 3,85	3,43; 4,07	3,82; 4,29
1000	0,90; 3,51	0,76; 2,71	0,80; 2,58	0,91; 2,23	1,04; 1,99	1,25; 1,94	1,45; 1,97	1,66; 2,09	1,87; 2,21	2,08; 2,34
2000	0,51; 1,99	0,43; 1,54	0,45; 1,46	0,52; 1,26	0,59; 1,13	0,71; 1,10	0,83; 1,12	0,94; 1,19	1,06; 1,26	1,18; 1,33

CONCLUSION

Pipes transporting flows with specific velocities must be able to sustain forces due to dead loads (exerted mainly by backfill material) and live loads (exerted mainly by passing cars, trucks, or trains). As manufacturers make pipes that are guaranteed to sustain maximum loads, the engineer and contractor must choose among them to satisfy field conditions while taking into account large variations in load distribution in the soil.

This chapter has dealt mainly with the bearing capacity of reinforced concrete pipes. For pipes made of other materials (plastic, cast iron, etc.), the engineer must verify with the manufacturer whether the pipes produced are flexible or rigid, and what guarantees are offered in terms of the maximum force the pipes can sustain. For pressure pipes (backflow and water-distribution pipes), we must also take into account the internal pressure resistance of the pipe itself (even without backfill load) and of the connecting joints.

PROBLEMS

11.1 A reinforced-concrete pipe with a nominal diameter of 600 mm is installed in a trench. The width of the trench is equal to the outside diameter of the pipe plus 60 cm (fig. 11.17). The pipe is covered with 425 cm of backfill made of sand and gravel, and the mass of the backfill is 1900 kg/m^3. The bedding is also made of sand and gravel (r_{sd} = +0,7). Is this pipe laid in trench conditions or in embankment conditions? What is the force, expressed in N/m, exerted by the backfill on the pipe? Choose the appropriate bedding and class for this pipe.

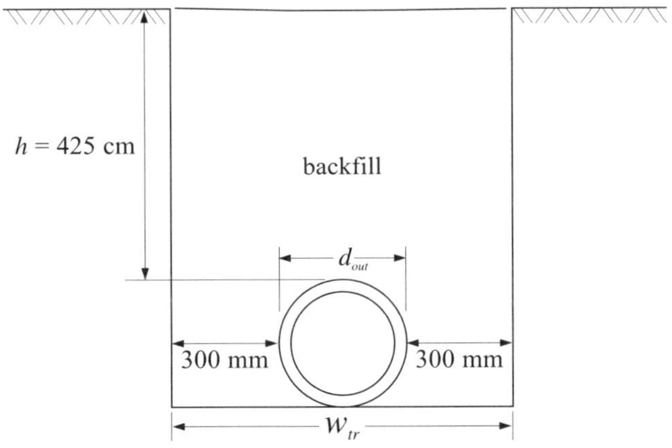

Figure 11.17 (problem 11.1)

11.2 A flexible pipe with an outside diameter of 910 mm is laid in a 1350 mm wide trench and covered with 365 cm of backfill composed of sand and gravel weighing 1900 kg/m^3 (fig. 11.18). The bedding is also made of sand and gravel (r_{sd} = +0,7). What is the force, expressed in N/m, exerted by the backfill on the pipe? What can you conclude about the width of this trench?

11.3 A reinforced-concrete pipe with a nominal diameter of 1200 mm is buried up to mid-diameter in ordinary soil. It is then covered with a 975 cm layer of backfill composed of equal parts of sand and clay weighing a total of 2000 kg/m^3 (fig. 11.19). The bedding is made of well compacted sandy soil (r_{sd} = +0,5, see bottom note of table 11.1). What is the force, expressed in N/m, exerted by the backfill on this pipe?

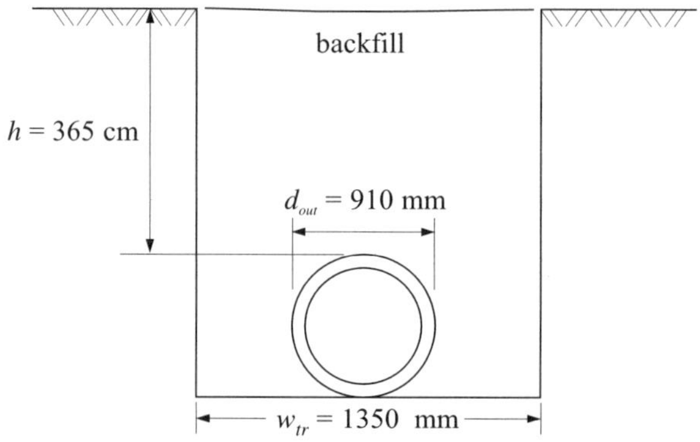

Figure 11.18 (problem 11.2)

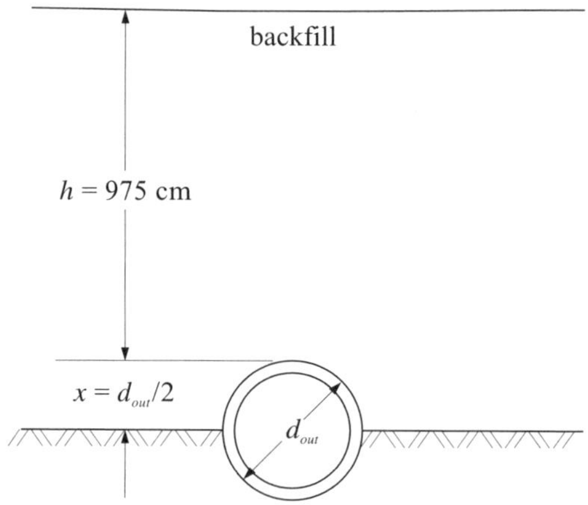

Figure 11.19 (problem 11.3)

11.4 A reinforced-concrete pipe with a nominal diameter of 1200 mm is laid in negative projecting conditions so that the pipe crown (top) is located 215 cm under the natural ground level in which the trench is excavated (fig. 11.20). The width of this trench is equal to the outside diameter of the pipe plus 60 cm. The value of r_{sd} is -0,3. What is the force, expressed in N/m, exerted by a backfill of 2000 kg/m^3 on this pipe crown (top)?

11.5 A reinforced-concrete pipe with a nominal diameter of 600 mm must be installed 91 cm under the surface of a road. What is the maximum force due to live loads that this pipe has to sustain?

11.6 A reinforced-concrete pipe with a nominal diameter of 600 mm must be installed 90 cm under a railway track. What is the maximum force due to live loads that this pipe has to sustain?

11.7 Solve the problem in example 11.2 with the following burying depths (h): a) h = 8 m, b) h = 6 m; c) h = 4 m; d) h = 2 m; e) h = 1 m.

11.8 Solve the problem in example 11.2 with the following values for p (i.e., embedding of the pipe in natural soil): a) p = 0; b) p = 0,1; c) p = 0,3; d) p = 0,5; e) p = 0,9; f) p =1,0.

11.9 Solve the problem in example 11.3 assuming that the pipe is covered with 9 m of backfill material.

11.10 Solve problem 11.9 assuming that the pipe is laid on natural soil – that is, the ground level is at the same height as the pipe bottom (invert), all other conditions being identical.

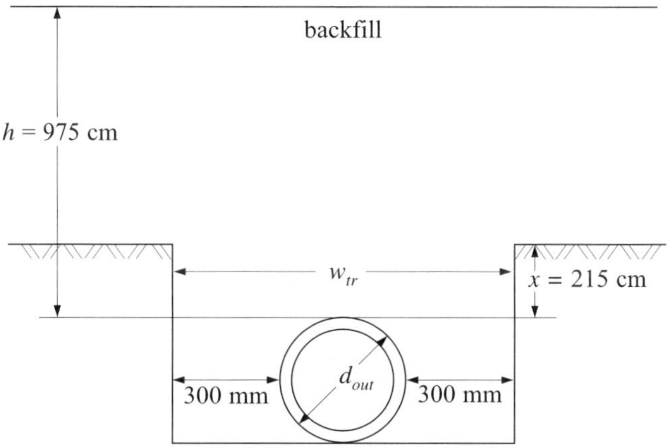

Figure 11.20 (problem 11.4)

ANSWERS TO PROBLEMS

Chapter 1

1.1 74 179

1.2 Geometric growth

1.3 $r = 0{,}0226 = 2{,}26\%$; $K_g = 0{,}0223$

1.4 Geometric growth; $r = 0{,}045 = 4{,}5\%$; population on January 1, 1999 = 29 180

1.5 $s = 312\ 745$; $a = 0{,}9743$; $b = -0{,}051$; $P_{2010} = 305\ 510$

Chapter 2

2.1 $s = 0{,}003$ m/m (0,0029)

2.2 $s = 0{,}0036$ m/m; $h_{min} = 0{,}141$ m; $v_{min} = 0{,}834$ m/s

2.3 $d_{min} = 45$ cm

2.4 $p_B = 382{,}7$ kPa

2.5 a) 0,1757 m^3/s; 0,601 m/s

b) 0,64 m^3/s; 2,19 m/s

c) 0,9075 m^3/s; 3,11 m/s

d) 0,0879 m^3/s

Chapter 3

3.1 15 000 000 L/d; 625 000 L/h and 174 L/s

3.2 a) 11 920 m^3/d c) 8 920 m^3/d

b) 6 720 m^3/d d) 9 720 m^3/d

3.3 812 121,2 L/d × 365 d = 296 424 242,4 L/year, thus $74 106,06/year

Chapter 4

4.1 3 000 L/min

4.2 a) 225 L/s b) $d_{min} = 276$ mm (without fire flow)

c) $d = 409$ mm (without fire flow); $v = 1{,}15$ m/s

4.3 $(Q_h)_{max} = 0{,}17$ L/s $Q_{(\text{water supply units})} = 4{,}7$ L/s

A pipe with a diameter of 51,7 mm (commercial diameter: 75 mm) should be sufficient, but head losses should be considered to ensure that this is so.

4.4 a) 22 000 L/min

b) Two fire hydrants equipped with three spouts

Chapter 5

5.1 L_{equ} (305 mm) = 908,79 m; Q_{ABD} = 29,24 L/s; Q_{ACD} = 45,76 L/s;

$(H_L)_{ABD} = (H_L)_{ACD}$ = 4,0 m

5.2 L_{equ} (305 mm) = 281,82 m; Q_{ABC} = 224,55 L/s; Q_{ADC} = 75,45 L/s;

$(H_L)_{ABC} = (H_L)_{ADC}$ = 12,73 m

5.3 L_{equ} (205 mm) = 1,20 m; Q_{AD} = 6,00 L/s; Q_{ABCD} = 69,00 L/s;

$(H_L)_{AD} = (H_L)_{ABCD}$ = 0,34 m

5.4 Answer, after 5 approximations:

Pipe	***Q* (L/s)**
A–B	+154,2
B–C	+110,2
C–G	+98,2
G–F	+50,2
F–E	+32,2
E–D	-55,8
D–A	-85,8

Pressure at node *F*: between 28,34 and 29,25 m

5.5 $Q_{1\text{-}2} = Q_{1\text{-}3} = Q_{2\text{-}4} = Q_{3\text{-}5} = Q_{4\text{-}6} = Q_{5\text{-}7}$ = 15 L/s; $Q_{4\text{-}5} = Q_{6\text{-}7} = 0$

5.6 All answers would be the same.

5.7

S_2 = 72,1 L/s	p_2 = 448,8 kPa	$Q_{N_1-N_2}$ – 110,1 L/s	$L_{N_1-N_3}$ – 288 m
S_4 = 24,1 L/s	p_3 = 518 kPa	$Q_{N_1-N_3}$ = 49,9 L/s	
S_6 = 4,4 L/s		$Q_{N_2-N_4}$ = 38 L/s	
S_7 = 4,4 L/s		$Q_{N_3-N_5}$ = 33,9 L/s	
S_8 = 9,5 L/s		$Q_{N_4-N_6}$ = 13,9 L/s	
S_9 = 9,5 L/s		$Q_{N_5-N_7}$ = 13,9 L/s	
		$Q_{N_6-N_8}$ = 9,5 L/s	
		$Q_{N_7-N_9}$ = 9,5 L/s	
		$Q_{N_4-N_5}$ = 0	
		$Q_{N_6-N_7}$ = 0	
		$Q_{N_8-N_9}$ = 0	

Chapter 6

6.1 $(Q_{dom})_{avg} = 384$ m^3/d; $(Q_{san})_{avg} = 516$ m^3/d; $(Q_{san})_{min} = 169{,}5$ m^3/d; $(Q_{san})_{max} = 1644$ m^3/d

6.2 3,24, source: table 6.1; 2,49, source: table 6.2

6.3 actual $(Q_{san})_{max} = 12\ 011$ m^3/d; ultimate $(Q_{san})_{max} = 14\ 572$ m^3/d

Chapter 7

7.1 $0{,}4 \leq R \leq 0{,}53$

7.2 $0{,}7 \leq R \leq 0{,}91$

7.4 a) 0,092 AR m^3/s b) 0,069 AR m^3/s

7.5 a) 0,069 AR m^3/s b) 0,069 AR m^3/s

7.6 0,088 m^3/s

7.7

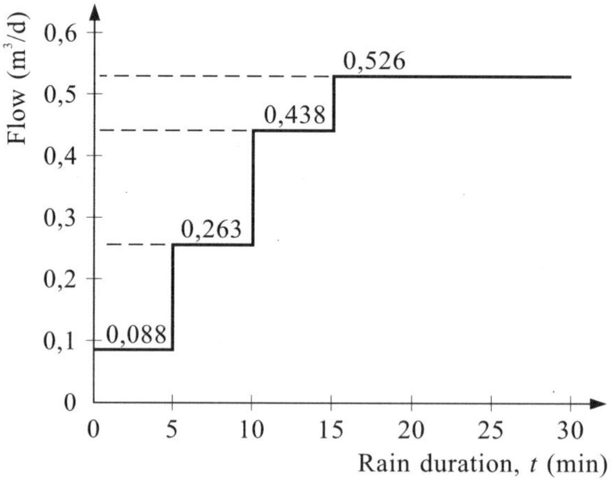

7.8 a) from A to C: 24,44 min; from B to C: 15,75 min

b) from A to C: 12,42 min; from B to C: 3,52 min

We choose a time of entry of 12,42 min; this time is calculated with the Kirpich equation, because the route travelled by the water is longer than 365 m.

7.9 Asphalt surface: $Q_{max} = 0{,}913$ m^3/s; grass surface: $Q_{max} = 0{,}265$ m^3/s. The shortest time of entry, the one on the asphalt surface, is critical because the runoff coefficient of each surface varies considerably and the runoff coefficients are therefore not similar everywhere. This fact does not respect one assumption of the rational equation.

7.10 Asphalt surface: $Q_{max} = 0{,}779$ m^3/s; grass surface: $Q_{max} = 0{,}2225$ m^3/s

7.11 $t_c = 10{,}27$ min a) 0,54 m^3/s b) 0,35 m^3/s

7.12 $Q_{max} = 2{,}54\ m^3/s$. The capacity of the main sewer is inadequate.

7.13 a) 0,0442 m^3/s d) 0 g) 0,0248 m^3/s j) 0,0262 m^3/s

b) 0,0201 m^3/s e) 0,0310 m^3/s h) 0,0289 m^3/s

c) 0,0281 m^3/s f) 0,012 m^3/s i) 0,0263 m^3/s

7.14 0,0866 m^3/s

7.15 $Q = 0{,}228\ m^3/s$

7.16 a) The maximum flow of 0,196 m^3/s is obtained for a rainfall of 5 min.

b) 0,180 m^3/s

7.17 $d = 2{,}06$ m; $v = 3$ m/s; $s = 0{,}0037$ m/m

7.18 a) 98,92 m b) $t_e = 5{,}86$ min c) $Q = 0{,}043\ m^3/s$

Chapter 8

8.1 125,195 m

8.2 29,95 m

8.3 $d_{min} = 545$ mm; $H = 3{,}25$ m

8.4 0,453 m^3/s

8.5 a) System collecting some rainwater; flooding of basements during periods when water runs high in streams.

b) Only one: pseudo-sanitary connection.

c) Domestic wastewater, water collected from flat roofs, foundation drains, basement drains, and in-ground driveway drains.

d) Pseudo-sanitary pipe: water collected from pseudo-sanitary connections; storm pipe: all other runoffs, including street runoff.

Chapter 9

9.1 a) Less important, as the intensity of the rainfall is lower.

b) $Q_{20\ \text{min}} = Q_{10\ \text{min}} \times (68{,}26/99{,}29)$

c) No, the difference decreases.

9.2 DTS = $Q_{dom} + Q_{inf}$

Inflow is evaluated at night, after a rainfall.
Infiltration is evaluated at night, following several days without rain.

9.3 Combined system:

- The main sewer is designed to collect the flow of a rainfall with a return period of 10 years.

– The intercepting sewer is designed to collect the following maximum flow:

$$Q_{max} = P \times Q_{avg}/\text{person} \times \text{FP}_{max} + Q_{inf}$$

Sanitary system:

– The main and intercepting sewers are designed to collect the following maximum flow:

$$Q_{max} = P \times Q_{avg}/\text{person} \times \text{FP}_{max} + Q_{inf} + Q_{capt}$$

9.4

$(R_p)_1 - (R_p)_2$	$d = 400$ mm	$s = 0{,}0029$ m/m
$(R_p)_2 - (R_p)_3$	$d = 400$ mm	$s = 0{,}01$ m/m
$(R_p)_3 - (R_p)_4$	$d = 750$ mm	$s = 0{,}001\ 25$ m/m
$(R_p)_4 - (R_p)_5$	$d = 750$ mm	$s = 0{,}0067$ m/m

Levels of inner inverts:

$(R_p)_1$: 197,395 m
$(R_p)_2$: 196,96 m
$(R_p)_3$: 195,45 m ($d = 400$ mm)
$(R_p)_3$: 195,105 m ($d = 750$ mm)
$(R_p)_4$: 194,9175 m
$(R_p)_5$: 193,9125 m

Chapter 11

11.1 The pipe is installed in trench conditions.

Force exerted by the backfill over the pipe: 67 424 N/m.

Bedding class B: $L_f = 1{,}9$,

$$\text{thus } \frac{67\ 424\ \text{N/m}}{1{,}9 \times 600\ \text{mm}\ (d_{nom})} = 59\ \text{N/m} - d_{nom}$$

A class III pipe is chosen.

11.2 Force exerted by the backfill over the pipe:

– with the Marston equation: 41 216 N/m;
– with the weight of the backfill directly over the pipe: 61 909 N/m.

The trench is too narrow for the backfill installation: transition width of the trench = 2047,5 mm.

11.3 391 751 N/m

11.4 252 941 N/m

11.5 From table 11.5: 1380 kg/m × 9,81 N/kg = 13 538 N/m

11.6 From table 11.6: 8530 kg/m × 9,81 N/kg = 83 679,3 N/m

11.7

Burying Depth, *h* (m)	**Loads, F_c (N/m)**
1	26 259
2	71 274
4	150 050
6	232 578
8	315 106
9 (see example 11.2	348 867

11.8

P	**Loads, F_c (N/m)**
0	228 827
0,1	270 090
0,3	303 852
0,5	326 359
0,9	352 618
1,0	367 623

11.8 F_c = 216 212 N/m

11.9 F_c = 384 717 N/m

Appendix A

FLOWING CONDITIONS GRAPHS* FOR VARIOUS SEWER PIPES

The following graphs show flows and velocities for various pipe sections not flowing full. Each of these graphs was experimentally plotted for a specific pipe. Although they are theoretically valid only for the pipes used in the experiment, the graphs may be used for pipes of different dimensions and characteristics, as the error induced is minimal.

* From (figs. A.1 and A.2): Swan and Horton, *Hydraulic Diagrams for the Discharge of Conduits and Canals* (McGraw-Hill, 1922), 53 p.

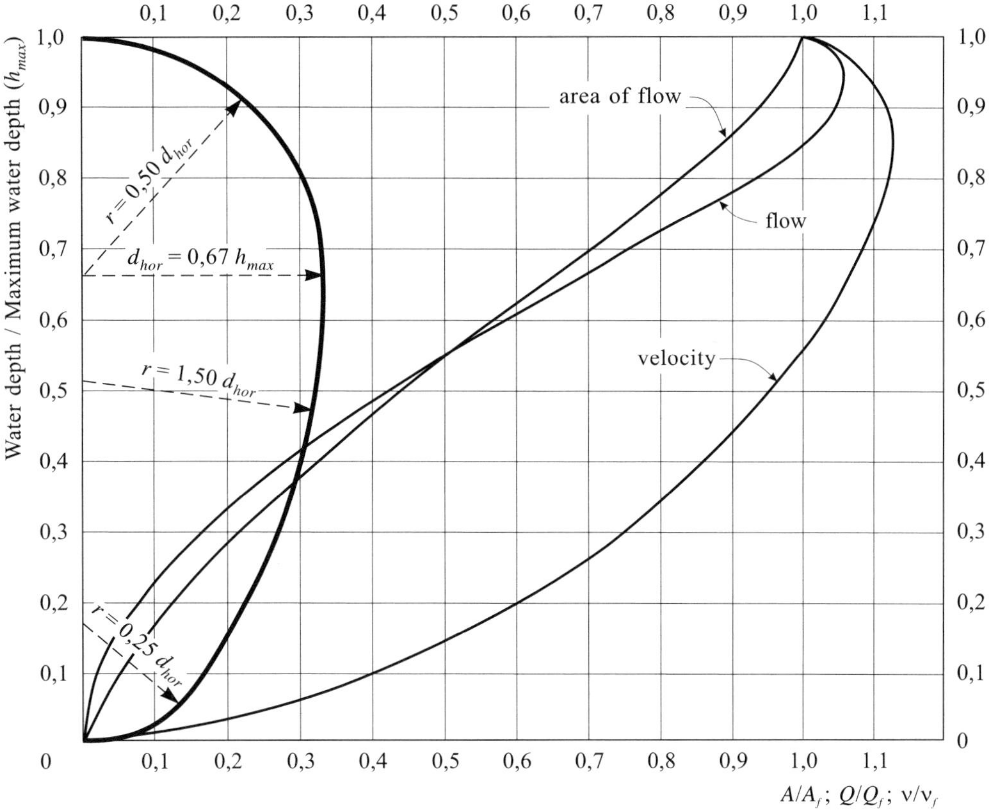

Figure A.1 Flow conditions (uniform) in oval sewer pipes not flowing full (the index f indicates "flowing full"); n (Kutter) = 0,015; maximum depth of water (h_{max}) = 1,254 × horizontal diameter, d_{hor}, of the circular pipe of equivalent capacity; d_{hor} = widest flow section = 0,836 × diameter of the equivalent pipe

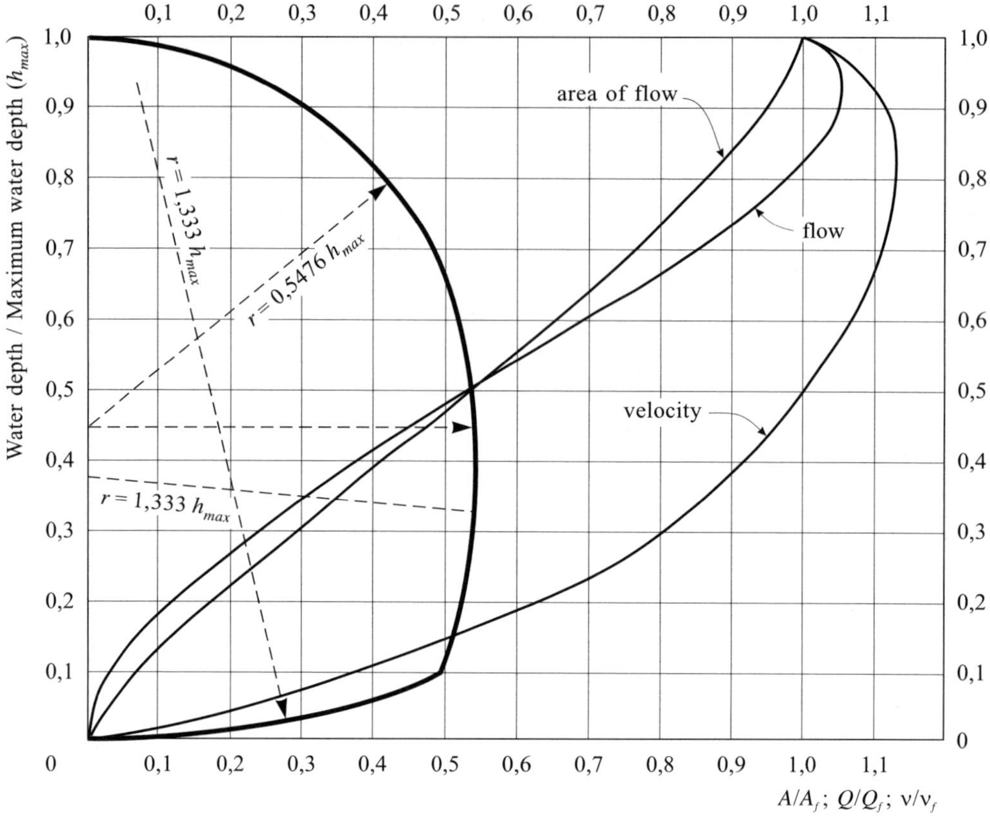

Figure A.2 Flow conditions (uniform) in horse-shoe-shaped sewer pipes not flowing full (the index *f* indicates "flowing full"); *n* (Kutter) = 0,015; h_{max} = 0,939 × diameter of the circular pipe of equivalent capacity

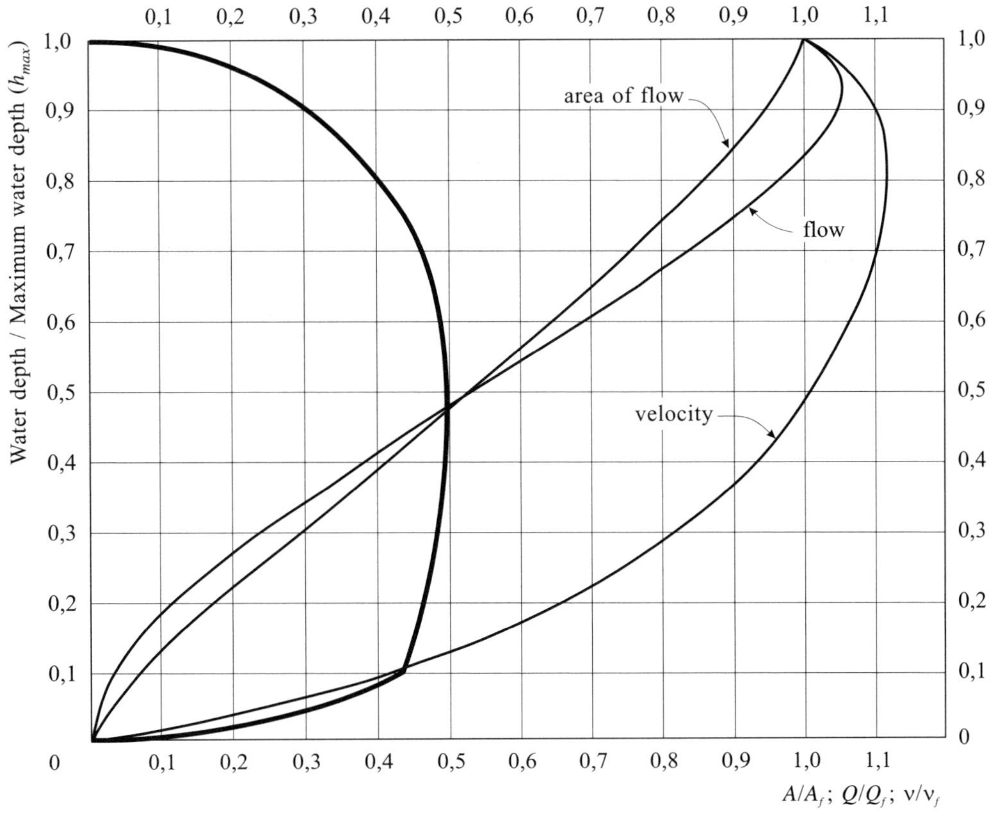

Figure A.3 Flow conditions (uniform) in horse-shoe-shaped sewer pipes in the Montreal region not flowing full (the index f indicates "flowing full"); n (Kutter) = 0,015; h_{max} = 0,97 × diameter of the circular pipe of equivalent capacity

Appendix

B

GRAPHS* OF THE HYDRAULIC CAPACITY OF CONCRETE CULVERTS OF VARIOUS DIAMETERS

* From: TUBECON, *Canalisations et éléments préfabriqués en béton*, Association québécoise des fabricants de tuyaux de béton (1995), chapter 4.

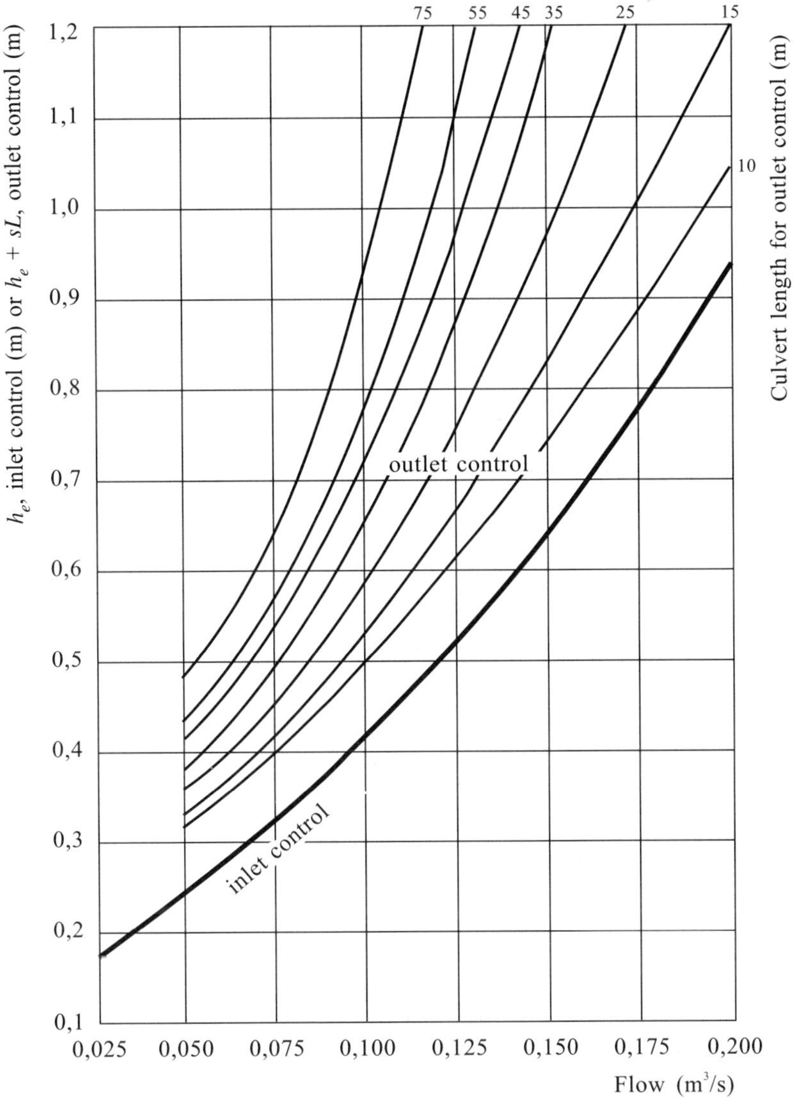

Figure B.1 Hydraulic capacity of concrete culverts with a diameter of 300 mm and of various lengths; roughness coefficient, n, is 0,012 and projecting inlet.

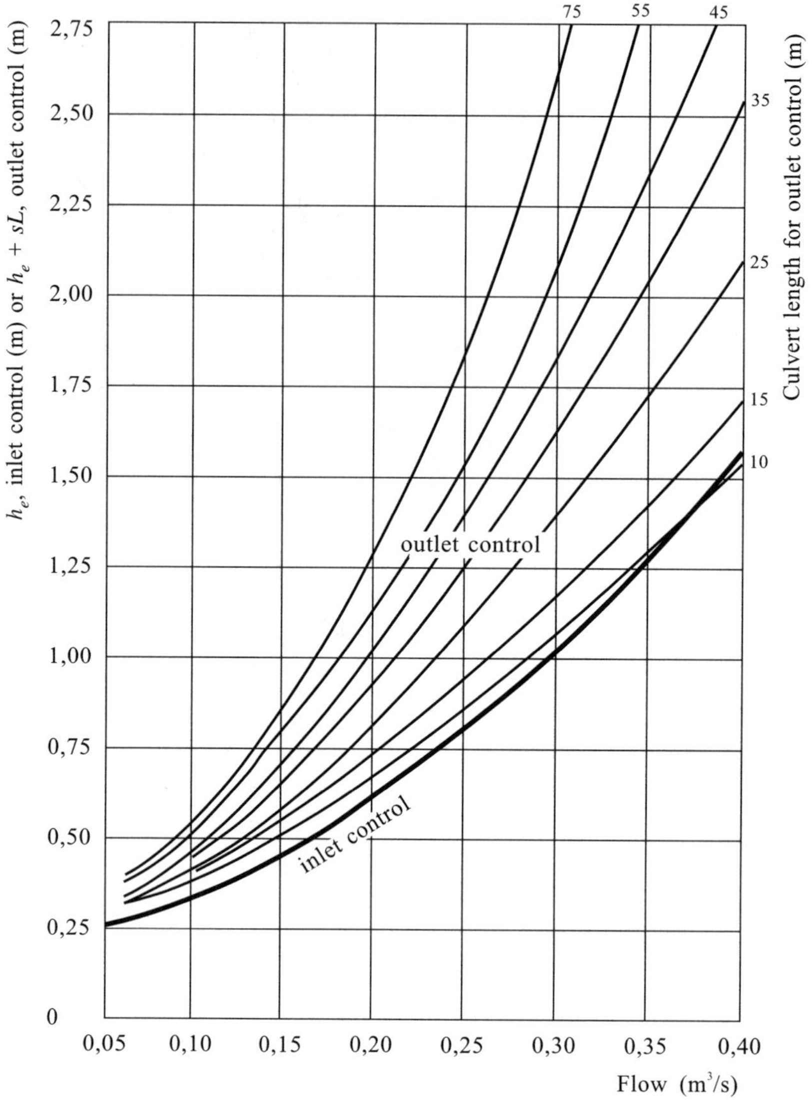

Figure B.2 Hydraulic capacity of concrete culverts with a diameter of 375 mm and of various lengths; roughness coefficient, n, is 0,012 and projecting inlet.

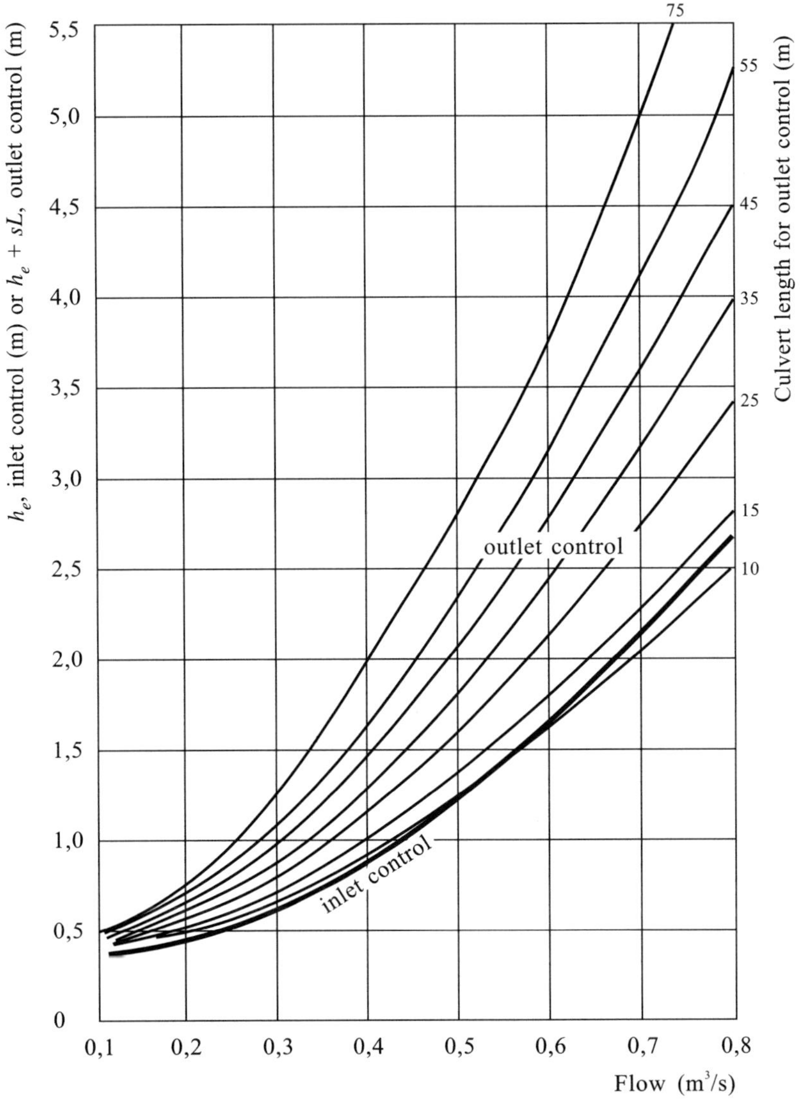

Figure B.3 Hydraulic capacity of concrete culverts with a diameter of 450 mm and of various lengths; roughness coefficient, *n*, is 0,012 and projecting inlet.

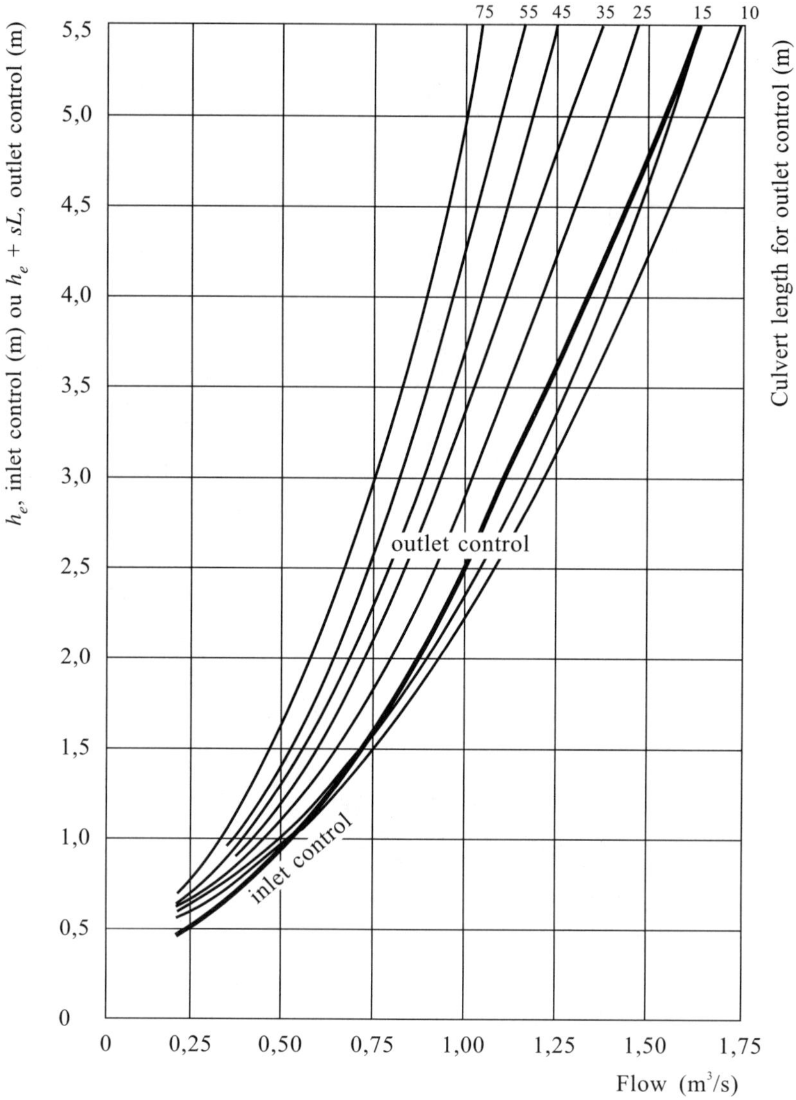

Figure B.4 Hydraulic capacity of concrete culverts with a diameter of 525 mm and of various lengths; roughness coefficient, n, is 0,012 and projecting inlet.

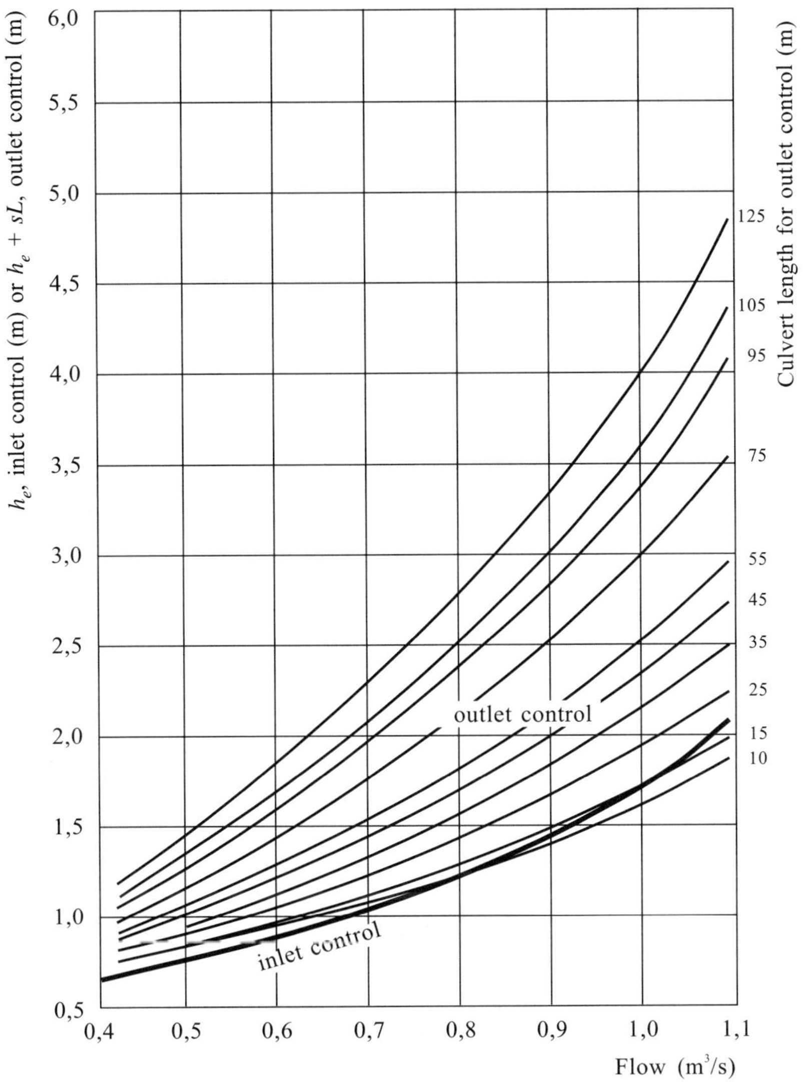

Figure B.5 Hydraulic capacity of concrete culverts with a diameter of 600 mm and of various lengths; roughness coefficient, *n*, is 0,012 and projecting inlet.

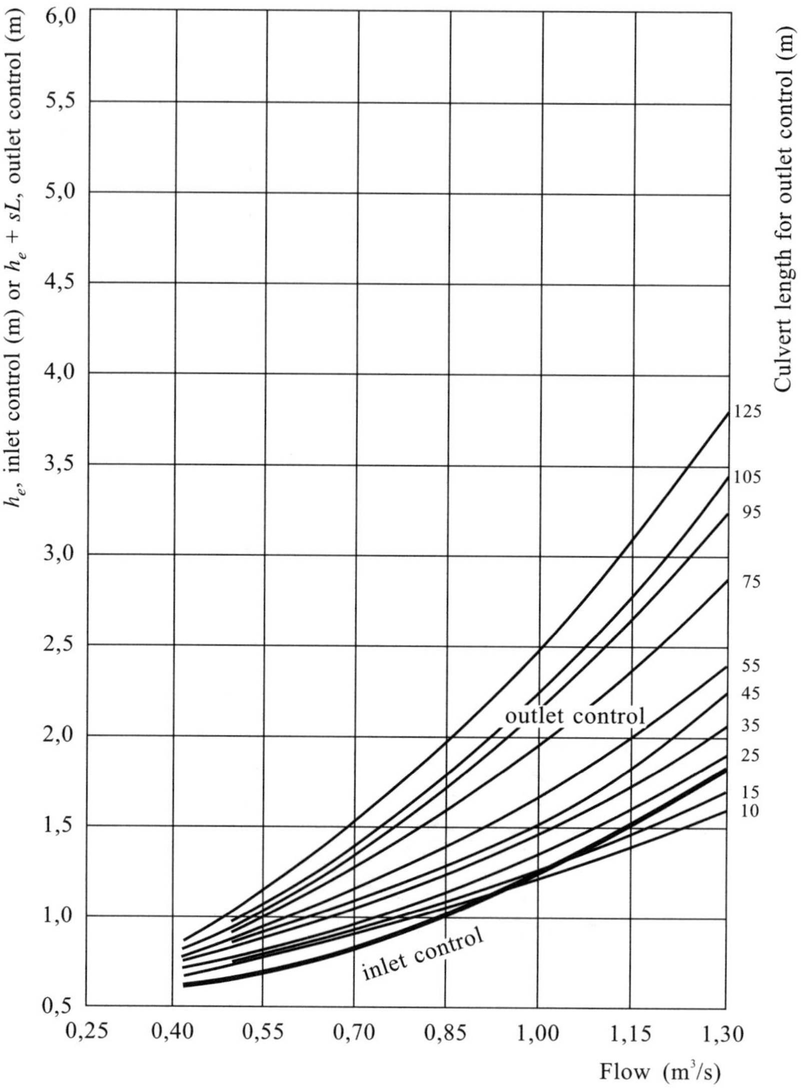

Figure B.6 Hydraulic capacity of concrete culverts with a diameter of 675 mm and of various lengths; roughness coefficient, n, is 0,012 and projecting inlet.

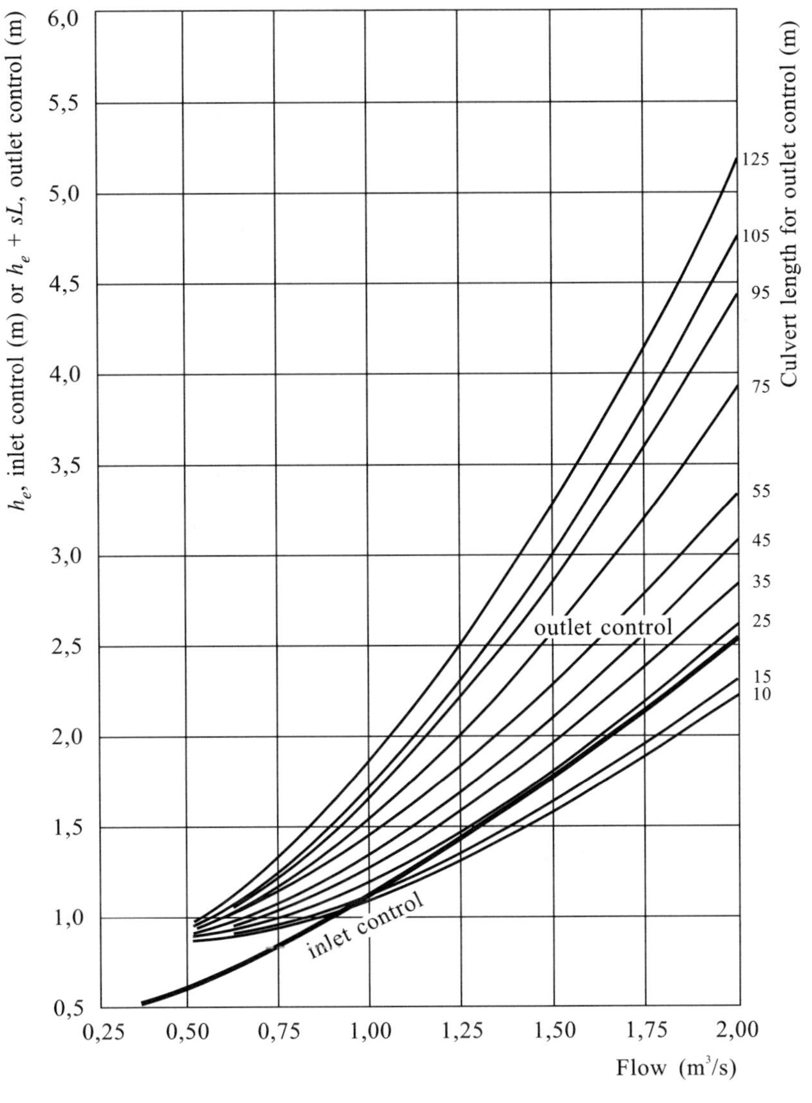

Figure B.7 Hydraulic capacity of concrete culverts with a diameter of 750 mm and of various lengths; roughness coefficient, *n*, is 0,012 and projecting inlet.

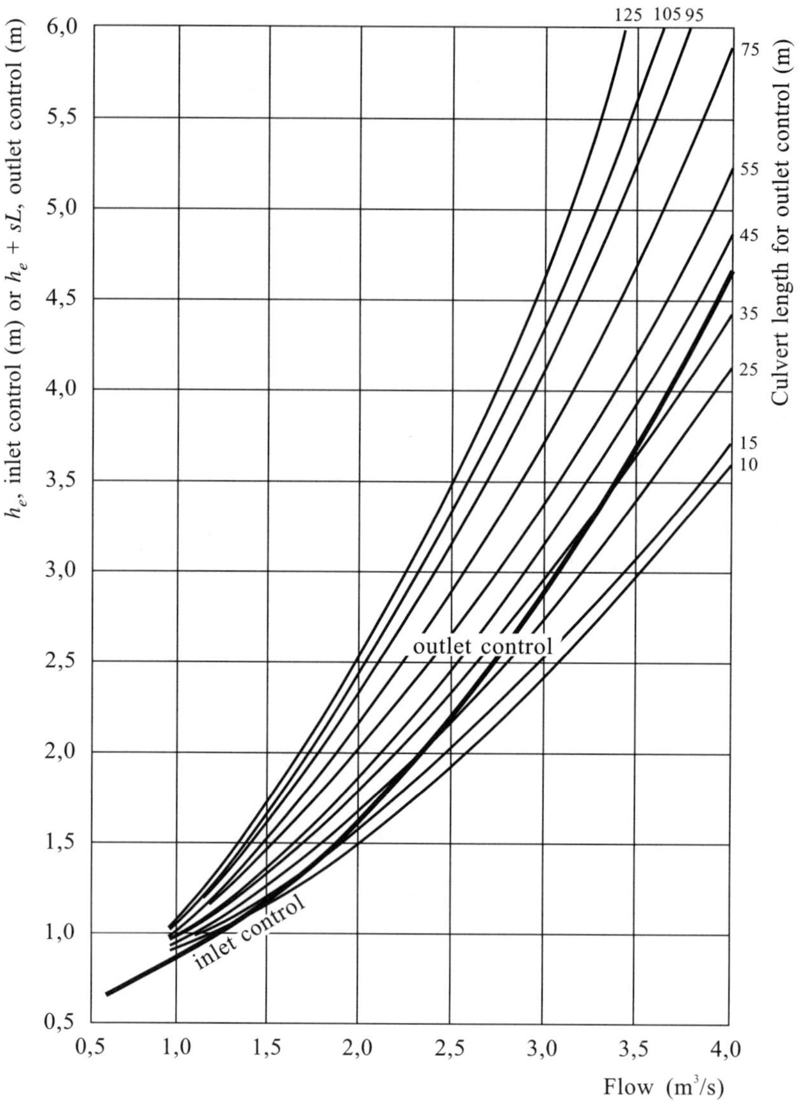

Figure B.8 Hydraulic capacity of concrete culverts with a diameter of 900 mm and of various lengths; roughness coefficient, *n*, is 0,012 and projecting inlet.

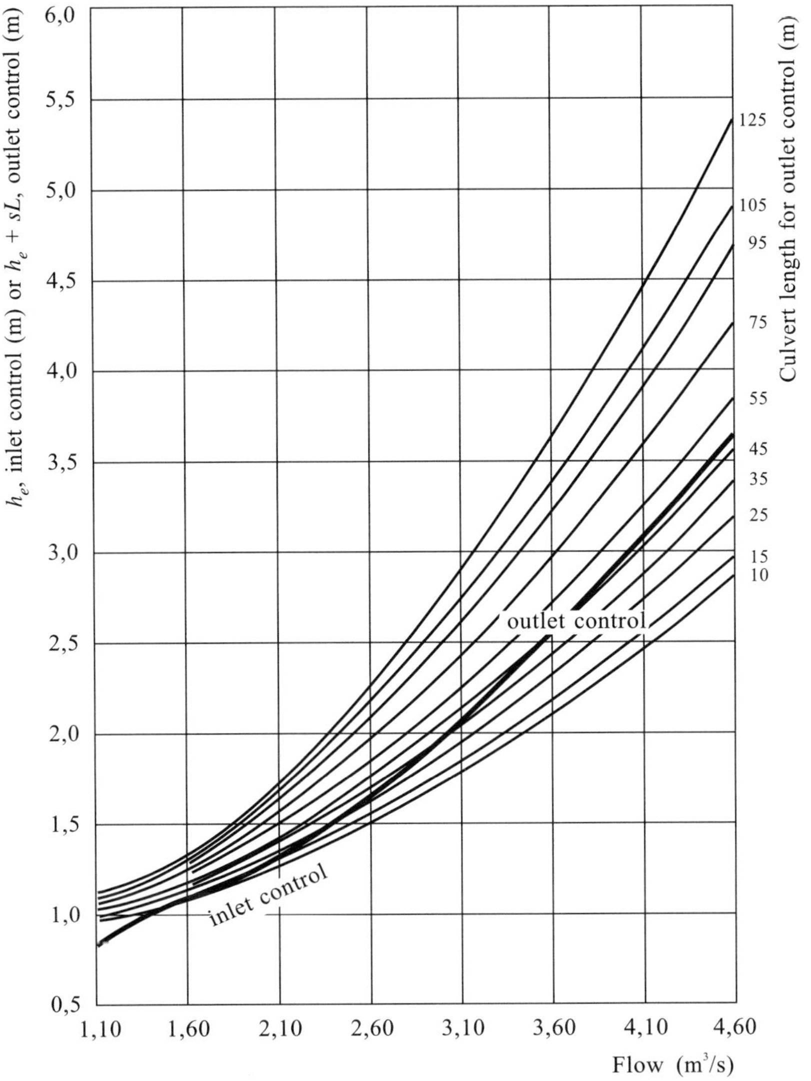

Figure B.9 Hydraulic capacity of concrete culverts with a diameter of 1050 mm and of various lengths; roughness coefficient, *n*, is 0,012 and projecting inlet.

Appendix C

UNITS

Length

1 m = 3,28 ft
1 km = 0,62 mile
1 km = 1 000 m

1 in = 2,54 cm
1 ft = 12 in
1 mile = 1,61 km

Surface

1 ha = 10 000 m^2
1 ha = 2,47 acres
1 km^2 = 247,1 acres
1 km^2 = 0,386 $mile^2$

1 acre = 43 560 ft^2
1 acre = 0,41 ha
1 $mile^2$ = 2,59 km^2

Volume

1 mL = 1 cm^3
1 L = 1000 mL
1 m^3 = 1000 L

1 ft^3 = 6,23 CA gal
1 ft^3 = 7,48 US gal

Pressure

1 Pa = 1 N/m^2
1 N = 1 $kg \cdot m/s^2$
1 kPa = 1000 N/m^2
1 kPa = 0,145 lb/in^2
1 m of water (5°C) = 9,81 kPa

1 atm = 101,33 kPa
1 lb/in^2 = 2,31 ft of water

1 ft of water = 0,43 lb/in^2
1 bar = 14,5 lbf/in^2
1 bar = 100 kPa

Flow

1 m^3/s = 22,8 × 10^6 US gal/d
1 m^3/d = 264,2 US gal/d

10^6 US gal/d = 3785 m^3/d

Volume mass of water (at 5°C)

1000 kg/m^3

Weight of water (5°C)

9810 N/m^3
10 lb/CA gal
8,33 lb/US gal

Weight of different materials

Clay: 16,0
Sand and gravel: 18,5
Cobble: 22,0

Concrete: 23,0
Water: 9,81

Appendix D

GRIDS FOR THE COMPUTATIONS OF STORM AND SANITARY SEWER SYSTEMS

Table D.1 Computations for a storm sewer system

Computation of flow										**Computation of sewer characteristics**						**Ground elevations**		**Inner invert elevations (minimum cover of 2 m)**	
Sewer line (between manholes)	Time of concentration	Increment of tributary area (ΔA)	Runoff coefficient of (ΔA)	Equivalent impermeable area of ΔA	Cumulative area	Cumulative impermeable area	Runoff coefficient of tributary area A	Intensity of precipitation	Maximum flow	Slope of sewer	Nominal diameter of sewer	Flow velocity in sewer flowing full	Length of sewer line	Flowing time in sewer line - pipe flowing full	Hydraulic capacity of sewer flowing full	Upstream	Downstream	Upstream	Downstream
	t_c (min)	ΔA (ha)	R_T	ΔAR_T (ha)	A (ha)	AR (ha)	R	I (mm/h)	Q_{max} (m^3/s)	s (m/m)	d_{nom} (mm)	v_f (m/s)	L (m)	t_f (min)	Q_p (m^3/s)	(m)	(m)	(m)	(m)

Table D.2 Computations for a sanitary sewer system

Average of tributary population										Computation of sewer characteristics					Ground elevations		Inner invert elevations (minimum cover of 2 m)	
Sewer line (between manholes)	Increment of tributary area	Increment of tributary population	Total population tributary	Average domestic flow	Maximum peak factor	Maximum domestic flow	Increment of parasitic flow	Total parasitic flow	Maximum sanitary flow	Slope of sewer	Nominal diameter of sewer	Flow velocity in sewer flowing full	Length of sewer line	Hydraulic capacity of sewer flowing full	Upstream	Downstream	Upstream	Downstream
	ΔA (ha)	ΔP	P	$(Q_{dom})_{avg}$ (L/s)	FP_{max}	$(Q_{dom})_{max}$ (L/s)	ΔQ_{par} (L/s)	Q_{par} (L/s)	$(Q_{san})_{max}$ (L/s)	s (m/m)	d_{nom} (mm)	v_f (m/s)	L (m)	Q_f (m^3/s)	(m)	(m)	(m)	(m)

Appendix E

GUIDELINES FOR THE CREATION OF WATER SYSTEMS FOR FIRE PROTECTION

WATER SUPPLY FOR PUBLIC FIRE PROTECTION

A GUIDE TO RECOMMENDED PRACTICE

PREFACE

This guide summarizes the more significant recommendations of Fire Underwriters Survey with respect to fire protection requirements in municipal water works system design. It reflects the manner in which the FUS surveyor assesses the water supply aspect of a municipality's fire risk potential during surveys on behalf of the Canadian property insurance industry and represents the accumulated experience of many years study of actual fires. Water supply is one of a number of components evaluated by FUS in the community fire protection system. Recommendations applying to fire departments and code enforcement are covered in other publications of Fire Underwriters Survey. FUS local offices are prepared to assist municipal officials or their consultants with advice on specific problems, as time limits permit, in accordance with the intent of this guide. The minimum size water supply credited by FUS must be capable of delivering not less than 1000 L/min for two hours or 2000 L/min for one hour in addition to any domestic consumption at the maximum daily rate. Static suction supplies to fire department pumpers are recognized as a supplement to the piped system.

In the FUS assessment of a water supply system the major emphasis is placed upon its ability to deliver **adequate** water to control major fires throughout the community on a **reliable** basis via sufficient and suitable **hydrants**. What is ultimately available to the fire department is the critical test in this fire protection evaluation.

Rates of flow for firefighting purposes are expressed in litres per minute as this is the adopted unit for the firefighting field.

With this edition of the Guide, quantities are specified in S.I. units only.

PART I

GENERAL

ADEQUACY AND RELIABILITY. An adequate and reliable water supply for firefighting is an essential part of the fire protection system of a municipality. This is normally a piped system in common with domestic potable water service for the community.

A water supply system is considered to be fully adequate if it can deliver the necessary fire flow at any point in the distribution gridiron for the applicable time period specified in the table "Required Duration of Fire Flow'" with the consumption at the maximum daily rate (average rate on maximum day of a normal year). When this delivery is also possible under certain emergency or unusual conditions as herein specified, the system is considered to be reliable. In cities of population in excess of 250,000 (or smaller places with high fire incidence and severe hazard conditions) it is usually necessary to consider the possibility of two simultaneous major fires in the area served by the system.

Fire flows are amounts of water necessary to control fires. These are determined as shown in Part II. System design should contemplate meeting the required fire flows existing or probable with the possible exception of gross anomalies where there is no fire threat to the remainder of the community. In these cases, the properties should preferably be modified in hazard to reduce the required flow as part of a coordinated community fire protection system.

The protection of buildings by automatic sprinkler systems is a significant contribution to the fire protection of the community and should be encouraged, not penalized by onerous service charges or metering requirements.

In order to provide reliability, duplication of some or all parts of the system will be necessary, the need for duplication being dependent upon the extent to which the various parts may reasonably be expected to be out of service as a result of maintenance and repair work, an emergency or some unusual condition. The introduction of storage, either as part of the supply works or on the distribution system , may partially or completely offset the need for duplicating various parts of the system, the value of the storage depending upon its amount, location and availability.

STORAGE. In general, storage reduces the requirements of those parts of the system through which supply has already passed. Since storage usually fluctuates, the normal daily minimum maintained is the amount that should be considered as available for fires. Because of the decrease in pressure when water is drawn down in standpipes, only the portion of this normal daily minimum storage that can be delivered at a residual pressure of 150 kPa at the point of use is considered as available. As well as the quantity available, the rate of delivery of water to the system from storage for the fire flow period is critical to this consideration.

PRESSURE. The principal requirement to be considered is the ability to deliver water in sufficient quantity to permit fire department pumpers to obtain an adequate supply from hydrants. To overcome friction loss in the hydrant branch, hydrant and suction hose, a minimum residual water pressure of 150 kPa on the street main is required during flow. Under conditions of exceptionally low suction losses a lower residual may be possible. This includes the use of 100 mm and larger outlets for fire department pumper use and hydrants with large waterways.

Higher sustained pressure is of importance in permitting direct continous supply to automatic sprinkler systems, to building standpipe and hose systems, and in maintaining a water plan such that no portion of the protected area is without water, such as during a fire at another location. Residual pressures that exceed 500 kPa during large flows are of value as they permit short hose-lines to be operated directly from hydrants without supplementary pumping.

SUPPLY WORKS

NORMAL ADEQUACY OF SUPPLY WORKS. The source of supply, including impounding reservoirs, and each part of the supply works should normally be able to maintain the maximum daily consumption rate plus the maximum required fire flow. Each distribution service within the system should similarly support its own requirements. In large cities where fire frequency may result in simultaneous fires, additional flow must be considered in accordance with the potential. Filters may be considered as capable of operating at a reasonable overload capacity based upon records and experience. In general, overload capacity will not exceed 25 per cent, but may be higher in well designed plants operating under favourable conditions.

The absolute minimum supply available under extreme dry weather conditions should not be taken as the measure of the normal ability of the source of supply such as supply from wells. The normal or average capacity of wells during the most favourable nine month period should be considered, or the normal sustained flow of surface supplies to the source.

RELIABILITY OF SOURCE OF SUPPLY. The effect on adequacy must be considered for such factors as frequency, severity and duration of droughts, physical condition of dams and intakes; danger from earthquakes, floods, forest fires, and ice dams or other ice formations; silting-up or shifting of channels; possibility of accidental contamination of the watershed or source; absence of watchmen or electronic supervision where needed; and injury by physical means. Where there is a risk of disruption, special precautions or alternate supplies should be arranged.

Where supply is from wells, some consideration should be given to the absolute minimum capacity of the wells under the most unfavorable conditions; also to the length of time that the supply from the wells would be below the maximum daily consumption rate, and the likelihood of this condition recurring every year or only at infrequent intervals. It should be recognized that some water is generally available from wells and that the most extreme conditions are not as serious as a total interruption of the supply, as would be the case in the breaking of a dam or shifting of a channel. The possibility of clogging, salinity, and the need for periodic cleaning and overhauling must be considered. Dependence upon a single well, even where records are favorable, may be considered a feature of unreliability.

Frequent cleaning of reservoirs and storage tanks may be considered as affecting reliability.

Continuity of and delay in implementing water supplies obtained from systems or sources not under the control of the municipality or utility should be considered also from these aspects.

GRAVITY SYSTEMS. A gravity system delivering supply from the source to distribution directly without the use of pumps is advantageous from a fire protection point of view because of its inherent reliability, but a pumping system can also be developed to a high degree of reliability.

PUMPING

RELIABILITY OF PUMPING CAPACITY. Pumping capacity, where the system or service is supplied by pumps, should be sufficient, in conjunction with storage when the two most important pumps are out of service, to maintain the maximum daily consumption rate plus the maximum required fire flow at required pressure for the required duration. For smaller municipalities (usually up to about 25,000 population) the relative infrequency of fires is assumed as largely offsetting the probability of a serious fire occurring at times when two pumps are out of service. (The most important pump is normally, but not always, the one of largest capacity, depending upon how vital is its contribution to maintaining flow to the distribution system.)

To be adequate, remaining pumps in conjunction with storage should be able to provide required fire flows for the specified durations at any time during a period of five days with consumption at the maximum daily rate. Effect of normal minimum capacity of elevated storage located on the distribution system and storage of treated water above lowlift pumps should be considered. The rate of flow from such storage must be considered in terms of any limitation of water main capacity. The availability of spare pumps or prime movers that can quickly be installed may be credited, as may pumps of compatible characteristics which may be valved from another service.

POWER SUPPLY FOR PUMPS. Electric power supply to pumps should be so arranged that a failure in any power line or the repair or replacement of a transformer, switch, control unit or other device will not prevent the delivery, in conjunction with elevated storage, of required fire flows for the required durations at any time during a period of two days with consumption at the maximum daily rate.

Power lines should be underground from the station or substation of the power utility to water plants and pumping stations and have no other consumers enroute. The use of the same transmission lines by other consumers introduces unreliability because of the possibility of interruption of power or deterioration of power characteristics.

Overhead power lines are more susceptible to damage and interruption than underground lines and introduce a degree of unreliability that depends upon their location and construction. In connections with overhead lines, consideration should be given to the number and duration of lightning, wind, sleet, and snow storms in the area; the type of poles or towers and wires; the nature of the country traversed; the effect of earthquakes, forest fires, and floods; the lightning and surge protection provided; the extent to which the system is dependent upon overhead lines; and the ease of and facilities for repairs.

The possibility of power system or network failures affecting large areas should be considered. In-plant auxiliary power or internal combustion driver standby pumping are appropriate solutions to these problems in many cases, particularly in small plants where high pumping capacity is required for fire protection service. When using automatic starting, prime movers for auxiliary power supply and pumping should have controllers labelled or listed by Underwriters' Laboratories of Canada to establish their reliability.

FUEL SUPPLY. At least a five day supply of fuel for internal combustion engines or boilers used for regular domestic supply should be provided. Where long hauls, condition of roads, climatic conditions, or other circumstances could cause interruptions of delivery longer than five days, a greater storage should be provided. Gas supply should be from two independent sources or from duplicate gas-producer plants with gas storage sufficient for 24 hours. Unreliability of regular fuel supply may be offset in whole or in part by suitable provisions for the use of an alternate fuel or power supply.

BUILDINGS AND PLANT

BUILDINGS AND STRUCTURES. Pumping stations, treatment plants, control centers and other important structures should be located, constructed, arranged, and protected so that damage by fire, flooding, or other causes will be held to a minimum. They should contain no combustible material in their construction, and, if hazards are created by equipment or materials located within the same structure, the hazardous section should be suitably separated by fire resistive partitions or fire walls.

Buildings and structures should have no fire exposures. If exposures exist, suitable protection should be provided. Electrical wiring and equipment should be installed in accordance with the Canadian Electrical Code. All internal hazards should be properly safeguarded in accordance with good practice. Private in-plant fire protection should be provided as needed.

MISCELLANEOUS SYSTEM COMPONENTS, PIPING AND EQUIPMENT. Steam piping, boiler-feed lines, fuel-piping (gas or oil lines to boilers as well as gas, oil or gasoline lines to internal-combustion engines), and air lines to wells or control systems should be so arranged that a failure in any line or the repair or replacement of a valve, fuel pump, boiler-feed pump, injector, or other necessary device, will not prevent the delivery, in conjunction with storage, of the required fire flows for the specified duration at any time during a period of two days with consumption at the maximum daily rate.

Plants should be well arranged to provide for effective operation. Among the features to be considered are: ease of making repairs and facilities for this work, danger of flooding because of broken piping; susceptibility to damage by spray; reliability of priming and chlorination equipment; lack of semi-annual inspection of boilers or other pressure vessels; dependence upon common nonsectionalized electric bus bars; poor arrangement of piping; poor condition or lack of regular inspections of important valves; and factors affecting the operation of valves or other devices necessary for fire service such as the design, operation, and maintenance of pressure regulating valves, altitude valves, air valves, and other special valves or control devices, provision of power drives, location of controls, and susceptibility to damage.

Reliability of treatment works is likely to be influenced by the removal from service of at least one filter or other treatment unit; the reduction of filter capacity by turbidity, freezing, or other conditions of the water; the need for cleaning basins; and the dependability of power for operating valves, wash-water pumps, mixers, and other appurtenances.

OPERATIONS. Reliability in operation fo the supply system and adequate response to emergency or fire demands are essential. Instrumentation, controls and automatic features should be arranged with this in mind. Failure of an automatic system to maintain normal conditions or to meet unusual demands should result in the sounding of an alarm where remedial action will be taken.

The operating force should be competent, adequate, and continuously available as may be required to maintain domestic and fire service.

EMERGENCY SERVICES. Emergency crews, provided with suitable transportation, tools, and equipment, should be continuously on duty in the larger systems and be readily available upon call in small systems. Spare pipe and fittings and construction equipment should be readily available. Alarms for fires in buildings should be received by the utility at a suitable location where someone is always on duty who can take appropriate action as required, such as placing additional equipment in operation, operating emergency or special valves, or adjusting pressures. Receipt of alarms may be by fire alarm circuit, radio, outside alerting device, or telephone, but where special operations are required, the alarm service should be equivalent to that needed for a fire station.

Response of an emergency crew should be made to major fires to assist the fire department in making the most efficient use of the water system and to ensure the best possible service in the event of a water main break or other emergency. The increase of pressures by more than 25 per cent for fires is considered to increase the possibility of breaks.

PIPING

RELIABILITY OF SUPPLY MAINS. Supply mains cut off for repair should not drastically reduce the flow available to any district. This includes all pipe lines or conduits on which supply to the distribution system is dependent, including intakes, suction or gravity lines to pumping stations, flow lines from reservoirs, treatment plant piping , force mains, supply and arterial mains, etc. Consideration should be given to the greatest effect that a break, joint separation or other failure could have on the delivery of the maximum daily consumption rate plus required fire flow at required pressure over a three day period. Aqueducts, tunnels or conduits of substantial construction may be considered as less susceptible to failure and equivalent to good mains with a long history of reliability.

INSTALLATION OF PIPE. Mains should be in good condition and properly installed. Pipe should be suitable for the service intended. Asbestos-cement, poly-vinyl chloride (PVC), cast and ductile iron, reinforced concrete and steel pipe manufactured in accordance with appropriate Canadian Standards Association or ANSI/AWWA standards, or any pipes listed by Underwriters' Laboratories of Canada for fire service are considered satisfactory. Normally, pipe rated for a maximum working pressure of 1000 kPa is required. Service records, including the frequency and nature of leaks, breaks, joint separations, other failures and repairs, and general conditions should be considered as indicators of reliability. When mains are cleaned they should be lined.

Mains should be so laid as not to endanger one another, and special construction should be provided to prevent their failure at stream crossings, railroad crossings, bridges, and other points where required by physical conditions; supply mains should be valved at one and one half kilometer intervals and should be equipped with air valves at high points and blow offs at low points. Mains should not be buried extremely deep or be unusually difficult to repair, though depths to ten feet may be required because of frost conditions.

The general arrangement of important valves, of standard or special fittings, and of connections at cross-overs, intersections, and reservoirs, as well as at discharge and suctions headers, should be considered with respect to the time required to isolate breaks. The need for check valves on supply or force mains and for other arrangements to prevent flooding of stations or emptying of reservoirs at the time of a break in a main should also be considered, as well as the need for relief valves or surge chambers. Accessibility of suitable material and equipment and ease of making repairs should be considered.

Arterial feeder mains should provide looping throughout the system for mutual support and reliability, preferably not more than 1000 metres between mains. Dependence of a large area on a single main is a weakness. In general the gridiron of minor distributors supplying residential districts should consist of mains at least 150 mm in size and arranged so that the lengths on the long sides of blocks between intersecting mains do not exceed 200 metres. Where longer lengths of 150 mm pipe are necessary 200 mm or larger intersecting mains should be used. Where initial pressures are unusually high, a satisfactory gridiron may be obtained with longer lengths of 150 mm pipe between intersecting mains.

Where deadends and a poor gridiron are likely to exist for a considerable period or where the layout of the streets and the topography are not well adapted to the above arrangement, 200 mm pipe should be used. Both ability to meet the required fire flows and reliability of a reasonable supply by alternate routing must be taken into account in this consideration.

VALVES. A sufficient number of valves should be installed so that a break or other failure will not affect more than 400 metres of arterial mains, 150 metres of mains in commercial districts, or 250 metres of mains in residential districts. Valves should be maintained in good operating condition. The recommended inspection frequency is once a year, and more frequently for larger valves and valves for critical applications.

A valve repair that would result in reduction of supply is a liability, but because of the probable infrequency of occurrence, it might be considered as introducing only a moderate degree of unreliability even if it resulted in total interruption. The repair of a valve normally should be accomplished in two days. Valves opening opposite to the majority are undesirable and when they do occur they should be clearly identified.

HYDRANTS

SIZE, TYPE AND INSTALLATION. Hydrants should conform to American Water Works Association Standard for Dry Barrel Fire Hydrants or Underwriters Laboratories of Canada listing. Hydrants should have at least two 65 mm outlets. Where required fire flows exceed 5000 L/min or pressures are low there should also be a large pumper outlet. The lateral street connection should not be less than 150 mm in diameter. Hose threads, operating and cap nuts on outlets should conform to Provincial Standard dimensions. A valve should be provided on lateral connections between hydrants and street mains.

Hydrants that open in a direction opposite to that of the majority are considered unsatisfactory. Flush hydrants are considered undesirable because of delay in getting into operation; this delay is more serious in areas subject to heavy snow storms. Cisterns are considered unsatisfactory as an alternative to pressure hydrants. The number and spacing of hydrants should be as indicated in the table titled "Standard Hydrant Distribution".

INSPECTION AND CONDITION. Hydrants should be inspected at least semi-annually and after use; inspection should include operation at least once a year. Where freezing temperatures occur, the semi-annual inspections should be made in the spring and fall of each year. Because of the possibility of freezing they should be checked frequently during extended periods of severe cold. Hydrants should be kept in good condition and suitable records of inspections and repairs be maintained. Hydrants should be painted in highly visible colors so that they are conspicuous and be situated with outlets at least twelve inches above the grade. There should be no obstruction that could interfere with their operation. Snow should be cleared promptly after storms and ice and snow accumulations removed as necessary.

HYDRANT DISTRIBUTION. Hydrant locations and spacing should be convenient for fire department use. Hydrants should be located at intersections, in the middle of long blocks and at the end of long dead end streets. To allow for convenient utilization of water supplies, distribution density of hydrants should be in accordance with the required fire flows indicated in the table titled "Standard Hydrant Distribution" (page 11). The maximum recommended spacing of hydrants in commercial, industrial, institutional and multi-family residential areas is 100 metres; in single family residential areas 200 metres is recommended. In areas where fire apparatus have access (eg. large properties, private developements etc.), hydrants should be required by bylaw. The planning of hydrant locations should be a cooperative effort between the water utility and fire department.

RECORDS

PLANS AND RECORDS. Complete, up-to-date plans and records essential for the proper operation and maintenance of the system should be available in convenient form, suitably indexed and safely filed. These should include plans of the source as well as records of its yield and a reliable estimate of the safe yield; plans of the supply works including dams, intakes, wells, pipelines, treatment plants, pumping stations, storage reservoirs and tanks; and a map of the distribution system showing mains, valves, and hydrants. Plans and maps should be in duplicate and stored at different locations.

Detailed distribution system plans, in a form suitable for field use, should be available for maintenance crews. Records of consumption, pressures, storage levels, pipes, valves, hydrants, and of the operations of the supply works and distribution system, including valve and hydrant inspections and repairs should be maintained.

STANDARD HYDRANT DISTRIBUTION

Fire Flow Required litres per minute	Average Area per Hydrant - m²
2,000	16,000
4,000	15,000
6,000	14,000
8,000	13,000
10,000	12,000
12,000	11,000
14,000	10,000
16,000	9,500
18,000	9,000
20,000	8,500
22,000	8,000
24,000	7,500
26,000	7,000
28,000	6,500
30,000	6,000
32,000	5,500
34,000	5,250
36,000	5,000
38,000	4,750
40,000	4,500
42,000	4,250
44,000	4,000
46,000	3,750
48,000	3,500

REQUIRED DURATION OF FIRE FLOW

Fire Flow Required litres per minute	Duration hours
2,000 or less	1.0
3,000	1.25
4,000	1.50
5,000	1.75
6,000	2.0
8,000	2.0
10,000	2.0
12,000	2.5
14,000	3.0
16,000	3.5
18,000	4.0
20,000	4.5
22,000	5.0
24,000	5.5
26,000	6.0
28,000	6.5
30,000	7.0
32,000	7.5
34,000	8.0
36,000	8.5
38,000	9.0
40,000 and over	9.5

Interpolate for intermediate figures.

Area refers to surface area of blocks and bounding streets. For a street without adjacent streets, a depth of one-half block is used.

A water supply system is considered to be adequate for fire protection when it can supply water as indicated above with consumption at the maximum daily rate. Certain types of emergency supplies may be included where reasonable conditions for their immediate use exist. Storage on the system is credited on the basis of the normal daily minimum maintained insofar as pressure permits its delivery at the rate considered.

PART I I

GUIDE FOR DETERMINATION OF REQUIRED FIRE FLOW (INSURANCE SERVICES OFFICE 1974 VERSION)

N.B. It should be recognized that this is a "guide" in the true sense of the word, and requires a certain amount of knowledge and experience in fire protection engineering for its effective application. Its primary purpose is for the use of surveyors experienced in this field, but it is made available to municipal officials,consulting engineers and others interested as an aid in estimating fire flow requirements for municipal fire protection. Some examples of typical fire flow requirements are shown in the Appendix.

Required Fire Flow may be described as the amount and rate of water application required in fire fighting to confine and control the fires possible in a building or group of buildings which comprise essentially the same fire area by virtue of immediate exposure. This may include as much as a city block.

1. An estimate of the fire flow required for a given fire area may be determined by the formula:

$$F = 220\, C\sqrt{A}$$

where

F = the required fire flow in litres per minute.
C = coefficient related to the type of construction.
= 1.5 for wood frame construction (structure essentially all combustible).
= 1.0 for ordinary construction (brick or other masonry walls, combustible floor and interior).
= 0.8 for noncombustible construction (unprotected metal structural components, masonry or metal walls).
= 0.6 for fire-resistive construction (fully protected frame, floors, roof).

Note: For types of construction that do not fall within the categories given, coefficients shall not be greater than 1.5 nor less than 0.6 and may be determined by interpolation between consecutive construction types as listed above. Construction types are defined in the Appendix.

A = The total floor area in square metres (including all storeys, but excluding basements at least 50 per cent below grade) in the building being considered.

For fire resistive buildings, consider the two largest adjoining floors plus 50 per cent of each of any floors immediately above them up to eight, when the vertical openings are inadequately protected. If the vertical openings and exterior vertical communications are properly protected (one hour rating), consider only the area of the largest floor plus 25 per cent of each of the two immediately adjoining floors.

The fire flow as determined by the above shall not exceed
30,000 L/min for wood frame construction
30,000 L/min for ordinary construction
25,000 L/min for noncombustible construction
25,000 L/min fire-resistive construction
except that for a normal 1-storey building of any type of construction the fire flow shall not exceed 25,000 L/min. The fire flow determined shall not be less than 2,000 L/min.

For one family and two family dwellings not exceeding 2 stories in height, see Note J.

2. The value obtained in No. 1 may be reduced by as much as 25% for occupancies having a low contents fire hazard or may be increased by up to a 25% surcharge for occupancies having a high fire hazard. Those may be classified as to contents as follows:

Non-Combustible	-25%	Free Burning	+15%
Limited Combustible	-15%	Rapid Burning	+25%
Combustible	No charge		

As a guide for determining low or high fire hazard occupancies, see the lists in the Appendix.
The fire flow determined shall not be less than 2,000 L/min.

3. The value obtained in Number 2 above may be reduced by up to 50% for complete automatic sprinkler protection depending upon adequacy of the system. Where buildings are either fire resistive or non-combustible construction, and have a **very low** fire hazard, the reduction may be up to 75%. The percentage reduction made for an automatic sprinkler system will depend upon the extent to which the system is judged to reduce the possibility of fires spreading within and beyond the fire area. Normally this reduction will not be the maximum allowed without proper system supervision including water flow and control valve alarm service.

4. To the value obtained in No.2 above a percentage should be added for structures exposed within 45 metres by the fire area under consideration. This percentage shall depend upon the height, area, and construction of the building(s) being exposed, the separation, openings in the exposed building(s), the length and height of exposure, the provision of automatic sprinklers and/or outside sprinklers in the building(s) exposed, the occupancy of the exposed building(s), and the effect of hillside locations on the possible spread of fire.

   The charge for any one side generally should not exceed the following limits for the separations shown:

Separation	Charge	Separation	Charge
0 to 3 m	25%	20.1 to 30 m	10%
3.1 to 10 m	20%	30.1 to 45 m	5%
10.1 to 20 m	15%		

   The total percentage shall be the sum of the percentages for all sides, but shall not exceed 75%.

   The fire flow shall not exceed 45,000 L/min nor be less than 2,000 L/min.

Note A: The guide is not expected to necessarily provide an adequate value for lumber yards, petroleum storage, refineries, grain elevators, and large chemical plants but may indicate a minimum value for these hazards.

Note B: Judgement must be used for business, industrial, and other occupancies not specifically mentioned.

Note C: Consideration should be given to the configuration of the building(s) being considered and accessibility by the fire department.

Note D: Wood frame structures separated by less than 3 metres shall be considered as one fire area.

Note E: Fire Walls: - In determining floor areas, a fire wall that meets or exceeds the requirements of the current edition of the National Building Code of Canada (provided this necessitates a fire resistance rating of 2 or more hours) may be deemed to subdivide the building into more than one area or may, as a party wall, separate the building from an adjoining building.

Normally any unpierced party wall considered to form a boundary when determining floor areas may warrant up to a 10% exposure charge.

Note F: High one storey buildings: When a building is stated as 1 = 2, or more storeys, the number of storeys to be used in the formula depends upon the use being made of the building. For example, consider a 1 = 3 storey building. If the building is being used for high-piled stock, or for rack storage, the building would probably be considered as 3 stories and, in addition, an occupancy percentage increase may be warranted. However, if the building is being used for steel fabrication and the extra height is provided only to facilitate movement of objects by a crane, the building would probably be considered as a one storey building and an occupancy credit percentage may be warranted.

Note G: If a building is exposed within 45 metres, normally some surcharge for exposure will be made.

Note H: Where wood shingle or shake roofs could contribute to spreading fires, add 2,000 L/min to 4,000 L/min in accordance with extent and condition.

Note I: Any non-combustible building is considered to warrant an 0.8 coefficient.

Note J: Dwellings: For groupings of detached one family and small two family dwellings not exceeding 2 stories in height, the following short method may be used. (For other residential buildings, the regular method should be used.)

Exposure distances	Suggested required fire flow Wood Frame	Masonry or Brick
Less than 3 m	See Note "D"	6,000 L/min
3 to 10 m	4,000 L/min	4,000 L/min
10.1 to 30 m	3,000 L/min	3,000 L/min
Over 30 m	2,000 L/min	2,000 L/min

If the buildings are contiguous, use a minimum of 8,000 L/min. Also consider Note H.

OUTLINE OF PROCEDURE

A. Determine the type of construction.

B. Determine the ground floor area.

C. Determine the height in stories.

D. Using the fire flow formula, determine the required fire flow to the nearest 1,000 L/min.

E. Determine the increase or decrease for occupancy and apply to the value obtained in D above. Do not round off the answer.

F. Determine the decrease, if any, for automatic sprinkler protection. Do not round off the value.

G. Determine the total increase for exposures. Do not round off the value.

H. To the answer obtained in E, subtract the value obtained in F and add the value obtained in G.

The final figure is customarily rounded off to the nearest 1,000 L/min.

APPENDIX

TYPES OF CONSTRUCTION

For the specific purpose of using the Guide, the following definitions may be used:

Fire-Resistive Construction —Any structure that is considered fully protected, having at least 3-hour rated structural members and floors. For example, reinforced concrete or protected steel.

Non-combustible Construction —Any structures having all structural members including walls, columns, piers, beams, girders, trusses, floors, and roofs of non-combustible material and not qualifying as fire-resistive construction. For example, unprotected metal buildings.

Ordinary Construction —Any structure having exterior walls of masonry or such non-combustible material, in which the other structural members, including but not limited to columns, floor, roofs, beams, girders, and joists, are wholly or partly of wood or other combustible material.

Wood Frame Construction —Any structure in which the structural members are wholly or partly of wood or other combustible material and the construction does not qualify as ordinary construction.

OCCUPANCIES

Examples of Low Hazard Occupancies:

Apartments
Asylums
Churches
Clubs
Colleges & Universities
Dormitories
Dwellings
Hospitals
Hotels
Institutions
Libraries, except Large Stack Room Areas
Museums
Nursing, Convalescent and Care Homes
Office Buildings
Prisons
Public Buildings
Rooming Houses
Schools
Tenements

Generally, occupancies falling in National Building Code Groups A, B, C and D are of this class.

Examples of High Hazard Occupancies:

Aircraft Hangers
Cereal, Feed, Flour and Grist Mills
Chemical Works—High Hazard
Cotton Picker and Opening Operations
Explosives and Pyrotechnics Manufacturing
Shade Cloth Manufacturing
Foamed Plastics, Storage or use in Manufacturing
High Piled Combustibles Storage in Excess of 6.5 metres high
Linseed Oil Mills
Match Manufacturing
Oil Refineries
Paint Shops
Pyroxylin Plastic Manufacturing and Processing
Solvent Extracting
Varnish and Paint Works
Wood Working with Flammable Finishing
Linoleum and Oilcloth Manufacturing

Other occupancies involving processing, mixing, storage and dispensing flammable and/or combustible liquids. Generally, occupancies falling in National Building Code Group F, Divisions 1 and 2 would be in this class.

Experience has shown that the following credits should normally be applied for the occupancies listed:

Dwellings*, apartments, and dormitories	—	25%
Hospitals	—	20%
Elementary Schools	—	20%
Junior and Senior high schools	—	15%
Open parking garages	—	25%

*When applying the standard method.

For other occupancies, good judgment should be used, and the percentage increase will not necessarily be the same for all buildings that are in the same general category — for example "Colleges and Universities": this could range from a 25%

decrease for buildings used only as dormitories to an increase for a chemical laboratory. Even when considering high schools, the decrease should be less if they have extensive shops.

It is expected that in commercial buildings no percentage increase or decrease for occupancy will be applied in most of the fire flow determinations. In general, percentage increase or decrease will not be at the limits of plus or minus 25%.

EXPOSURES

When determining exposures it is necessary to understand that the exposure percentage increase for a fire in a building (x) exposing another building (y) does not necessarily equal the percentage increase when the fire is in building (y) exposing building (x). The Guide gives the maximum possible percentage for exposure at specified distances. However, these maximum possible percentages should not be used for all exposures at those distances. In each case the percentage applied should reflect the actual conditions but should not exceed the percentage listed.

The maximum percentage for the separations listed generally should be used if the exposed building meets all of the following conditions:

- **a.** Same type or a poorer type of construction than the fire building.
- **b.** Same or greater height than the fire building.
- **c.** Contains unprotected exposed openings.
- **d.** Unsprinklered.

FUS WATER SUPPLY FOR PUBLIC FIRE PROTECTION 1991

EXAMPLES OF REQUIRED FIRE FLOWS (REVISED)

For convenience in making general estimates some examples of required fire flows in typical buildings are provided below. In establishing fire flows for areas of a Municipality as yet undeveloped, but where a broad range of commercial, institutional, residential and industrial occupancies may be expected to be created under modern building code requirements, an outside design figure of 15,000 L/min appears likely to be suitable. When very large or high fire load buildings are probable, 25,000 L/min is more appropriate. It should be noted particularly that the tendency to install automatic sprinkler protection in large area and high hazard industrial and commercial buildings is a key factor in keeping required fire flows within ecomonically acceptable limits in many cases.

The following examples suppose no significant exposures to other buildings nor sprinkler protection unless specified. Where areas are given they are ground areas unless specified.

DETACHED DWELLINGS (TOTAL FLOOR AREAS)

— Under 100 m² = 2,000 L/min
  101 m² — 200 m² = 3,000 L/min
  Over 200 m² = 4,000 L/min

— Add for exposures to similar buildings on both sides:

Over 30 m	- nil
30 — 10 m	add 1,000 L/min
10 — 3 m	add 2,000 L/min
less than 3 m	see Note "D", if Frame. Brick, add 3,000 L/min.

— If wood shingle or shake roofs are prevalent,add 2,000 to 4,000 L/min.

— Modern residential subdivisions of 1 and 2 storey single family homes detached 3 to 6 m require usually 4,000 to 5,000 L/min.

— Old congested two and three family tenements detached less than 3 m and running the length of the block may require 15,000 to 25,000 L/min and should be calculated according to Note "D".

— Modern Row or Town House groups may require 6,000 to 10,000 L/min including adjoining exposures, providing required fire separations are adequate.

APARTMENT BUILDINGS

— 3 storeys, frame, 300 m² = 7,000 L/min and exposure coverage.

— 4 storeys, brick, 2,000 m² = 15,000 L/min and exposure coverage.

— 3 or more storeys, fire resistive, 5,600 m² with cut off shafts and stairs = 10,000 L/min and exposure coverage.

INSTITUTIONAL BUILDINGS

— 1 storey, fire resistive school of 2,300 m² = 5,000 L/min

— 3 storey, brick ordinary school of 2,300 m² = 15,000 L/min

— 3 or more storey, fire resistive hospital with adequate floor separations 1,000 m², no exposures = 4,000 L/min.

INDUSTRIAL BUILDINGS

— Typical industrial park, 1 storey ordinary, area 3,700 m² with average combustible contents fire load =14,000 L/min.

— Frame warehouse 1 storey, moderate contents fire load 3,700 m² = 20,000 L/min.

— Warehouse high fire load contents, brick non-combustible, 1 storey, 14,000 m² = 25,000 L/min. With full adequate automatic sprinkler protection (item 3, P.13) 13,000 L/min.

— Traditional 3 storey brick , ordinary factory with high fire load. 9,300 m² = 35,000 L/min.

COMMERCIAL BUILDINGS

- Old business district city blocks with largely contiguous brick ordinary construction, 2 to 4 storey with no exposures 14,000 to 25,000 L/min.
- Large mall shopping centre, no exposure, in one area of 16,000 m².

Non-combustible	—	22,000 L/min
Fire Resistive	—	17,000 L/min

- Moderately large shopping centre, no exposure, in one area of 9,300 m².

Non-combustible	—	17,000 L/min
Ordinary	—	22,000 L/min

- Small open shopping centre, or part of large one, 5,000 m².

Non-combustible	—	12,000 L/min
Ordinary	—	16,000 L/min
Frame	—	23,000 L/min

- Individual department store 4,600 m², 3 storeys.

Fire Resistive	—	16,000 L/min
Brick ordinary	—	26,000 L/min

- High-Rise, fire-resistive office building more than 9 storeys, 2,500 m².

Floors cut-off	—	8,000 L/min
Not cut-off	—	16,000 L/min

AUTOMATIC SPRINKLERS

The required flows shown above may be reduced by from 25 to 50% for buildings having full approved automatic sprinkler protection with satisfactory alarm supervisory service and to 75% where conditions of **very low** fire load and non-combustible or fire resistive construction are also present, as noted in Item 3 on page 13.

EXPOSURES

In addition, if not included in the description in the above, water for hosestreams covering any exposures is needed. There are many variables in this consideration. In general for significant exposures in the area of 3 to 20 metres distant, 15 to 25 percent of the required flow may be added per exposing side.

CONVERSION FACTORS

Multiply	By	To obtain
Centimetre	0.3937	Inches
Cubic Foot	0.0283	Cubic Metres
Cubic Metre	35.3145	Cubic Feet
Cubic Metre	219.97	Imperial Gallons
Cubic Metre	1000	Litres
Foot	0.3048	Metres
Horsepower	0.7457	Kilowatt
Imperial Gallon	4.546	Litres
Inch	2.54	Centimetres
Kilogram	2.2046	Pounds
Kilogram of Water	1	Litres
Kilopascal	0.1450	Pounds per sq. Inch
Kilowatt	1.341	Horsepower
Litre	0.21997	Imperial Gallons
Litre of Water	1	Kilograms
Metre	3.281	Feet
Metre of Water	10	Kilopascals
Pound	0.4536	Kilograms
Pound per sq. Inch	6.89476	Kilopascals
Gallon, U.S.	0.8327	Gallons, Imperial
Gallon, Imperial	1.201	Gallons, U.S.

BIBLIOGRAPHY

ANDERSLAND, O.B. and LADANYI, B. *An Introduction to Frozen Ground Engineering*. New York: International Thompson Publishing Company, 1994, 352 pages.

AMERICAN SOCIETY OF CIVIL ENGINEERS. *Design and Construction of Urban Stormwater Management.* Section 1. New York: American Society of Civil Engineers and Water Environment Federation, 1992.

AMERICAN SOCIETY OF CIVIL ENGINEERS and WATER POLLUTION CONTROL FEDERATION. *Gravity Sanitary Sewer Design and Construction.* New York: ASCE and WPCF, 1982.

ASSOCIATION AMÉRICAINE DE LA SANTÉ PUBLIQUE. *Prophylaxie des maladies transmissibles à l'homme*. French translation. Ottawa: Health and Welfare Canada, 1985.

BÉRON, P. and MARCHI, G. *Simulation d'un réseau de drainage urbain: le modèle SIRDU (ILLUDAS modifié).* GREMU-81/05. Montreal: École Polytechnique de Montréal, 1981.

CAUVIN, A. and DIDIER, G. *Distribution de l'eau dans les agglomérations.* Paris: Eyrolles, 1960.

COSTE, C. and LOUDET, M. *Guide de l'assainissement en milieu urbain et rural.* Paris: Éditions du Moniteur, 1980.

DUPONT, A. *Hydraulique urbaine*. Vol. 1: *Hydrologie, captage et traitement des eaux.* Vol. 2: *Ouvrages de transport, élévation et distribution des eaux.* Paris: Eyrolles, 1977.

FAIR, G.M., GEYER, J.C. and OKUN, D.A. *Water and Wastewater Engineering*, 2 volumes. New York: John Wiley & Sons, Inc., 1958.

GOUVERNEMENT DU QUÉBEC. *Devis normalisés: travaux de construction - clauses générales - conduites d'eau et égouts.* NQ 1809-300. Quebec City: Bureau de normalisation du Québec, 1987.

GRAY, D.M. *Manuel des principes d'hydrologie*. Section 1. Canadian National Committee, International Hydrology Decade, 1972.

HAMMER, M.J. *Water and Wastewater Technology*. 2d ed. New York: John Wiley & Sons, 1986.

HORTON, T. and SWAN, C.H. *Hydraulic Diagrams for the Discharge of Conduits and Canals*. New York: McGraw-Hill, 1922.

JEPPSON, R.W. *Analysis of Flow in Pipe Network.* Section 1. Ann Arbor Science, 1977.

KOCH, P. *L'Alimentation en eau des agglomérations*. Paris: Dunod, 1969.

LINSLEY, R.K. Jr, KOHLER, M.A. and PAULHUS, J.H. *Hydrology for Engineers*. 2d ed. New York: McGraw-Hill, 1975.

McGHEE, T.J. *Water Supply and Sewerage*. New York: McGraw-Hill, 1991.

METCALF and EDDY. *Wastewater Engineering: Collection and Pumping of Wastewater*. New York: McGraw-Hill, 1981.

MINISTÈRE DU TRANSPORT DU QUÉBEC. *Manuel de conception des ponceaux*. 1993.

OKUN, D.A. and PONGHIS, G. *Collecte et évacuation des eaux usées des collectivités*. Geneva: World Health Organization, 1976.

ONTARIO CONCRETE PIPE ASSOCIATION. *Concrete Pipe Design Manual*. Toronto, 1986.

PARCHER, M.J. *Wastewater Collection System Maintenance*. Lancaster, PA, USA: Technomic Publishing Co. Inc.

PATRY, G. and MARCHI, G. *Manuel de l'utilisateur du modèle ILLUDAS*. GREMU-80/07. Montreal: École Polytechnique de Montréal, 1979, 87 p.

PRAT, F. *Hydraulique urbaine*. Vol. 1: *Alimentation en eau potable*. Toulouse: Institut national Polytechnique de Toulouse, 1974.

RIVARD, G. *Gestion des eaux pluviales en milieu urbain: Concepts et applications*. Alias Communication Design Inc., 1998, 314 pages.

SIMON, A.L. *Practical Hydraulics*. 2d ed. New York: John Wiley & Sons, 1981.

SPANGLER, M.G. and HANDY, R.L. *Soil Engineering*, 3d ed. New York: Intext Educational Publishers, 1973.

STEPHENSON, D. *Pipeflow Analysis*, Section 1. Elsevier, 1984.

TERSTRIEP, M.L. and STALL, J.B. *The Illinois Urban Drainage Area Simulator*. Illinois State Water Survey, Bulletin 58, 1974.

TUBECON. *Canalisations et éléments préfabriqués en béton*. Association québécoise des fabricants de tuyaux de béton, 1995.

TWORT, A.C., LAW, F.M. and CROWLEY, F.W. *Water Supply*. 3d ed. Scotland: Edward Arnold, 1985.

VIESSMAN, W., JR. and HAMMER, M.J. *Water Supply and Pollution Control*. 5th ed. New York: Harper & Collins, 1993.

WHIPPLE, W. *et al*. *Stormwater Management in Urbanizing Areas*. Englewood Cliffs: Prentice-Hall, 1983.

WOOD, D.J. and CHARLES, C.O.A. "Hydraulic Network Analysis Using Linear Theory." *Journal of Hydraulics Division* [American Society of Civil Engineers], vol. 98, no. 7 (1972): 1157–70.

WORLD HEALTH ORGANIZATION. *Approvisionnement en eau des zones rurales et des petites agglomérations*. Geneva, 1961.

———. *Wastewater Engineering: Treatment, Disposal and Reuse*. 3d ed., rev. by G. Tchobanoglous and F.L. Burton. New York: McGraw-Hill, 1991.

INDEX

G

H

I

J

K

L

R

W

AGMV MARQUIS
Québec, Canada
1999